T0390291

The Future of Dynamic Structural Science

NATO Science for Peace and Security Series

This Series presents the results of scientific meetings supported under the NATO Programme: Science for Peace and Security (SPS).

The NATO SPS Programme supports meetings in the following Key Priority areas: (1) Defence Against Terrorism; (2) Countering other Threats to Security and (3) NATO, Partner and Mediterranean Dialogue Country Priorities. The types of meeting supported are generally "Advanced Study Institutes" and "Advanced Research Workshops". The NATO SPS Series collects together the results of these meetings. The meetings are co-organized by scientists from NATO countries and scientists from NATO's "Partner" or "Mediterranean Dialogue" countries. The observations and recommendations made at the meetings, as well as the contents of the volumes in the Series, reflect those of participants and contributors only; they should not necessarily be regarded as reflecting NATO views or policy.

Advanced Study Institutes (ASI) are high-level tutorial courses to convey the latest developments in a subject to an advanced-level audience

Advanced Research Workshops (ARW) are expert meetings where an intense but informal exchange of views at the frontiers of a subject aims at identifying directions for future action

Following a transformation of the programme in 2006 the Series has been re-named and re-organised. Recent volumes on topics not related to security, which result from meetings supported under the programme earlier, may be found in the NATO Science Series.

The Series is published by IOS Press, Amsterdam, and Springer, Dordrecht, in conjunction with the NATO Emerging Security Challenges Division.

Sub-Series

A.	Chemistry and Biology	Springer
B.	Physics and Biophysics	Springer
C.	Environmental Security	Springer
D.	Information and Communication Security	IOS Press
E.	Human and Societal Dynamics	IOS Press

http://www.nato.int/science
http://www.springer.com
http://www.iospress.nl

Series A: Chemistry and Biology

The Future of Dynamic Structural Science

edited by

Judith A.K. Howard

Department of Chemistry
Durham University
United Kingdom

Hazel A. Sparkes

School of Chemistry, University of Bristol
Bristol

Paul R. Raithby

Department of Chemistry
University of Bath
United Kingdom

and

Andrei V. Churakov

N.S. Kurnakov Institute of General and Inorganic Chemistry
Russian Academy of Sciences
Moscow, Russia

Published in Cooperation with NATO Emerging Security Challenges Division

Proceedings of the NATO Advanced Study Institute on
The Future of Dynamic Structural Science: Links to Energy and Environmental Security and Early Detection of CBRN Agents
Erice, Italy
30 May – 8 June 2013

Library of Congress Control Number: 2014931397

ISBN 978-94-017-8552-5 (PB)
ISBN 978-94-017-8549-5 (HB)
ISBN 978-94-017-8550-1 (e-book)
DOI 10.1007/978-94-017-8550-1

Published by Springer,
P.O. Box 17, 3300 AA Dordrecht, The Netherlands.

www.springer.com

Printed on acid-free paper

Sponsors

Without sponsorship, the course with which this book is associated would not have been able to go ahead, and we are very grateful for all financial support that was received.

We are very thankful to:

NATO Science for Peace and Security Programme

as well as

European Crystallographic Association
International Union of Crystallography
OPCW
PANalytical

Preface to ASI Volume

This volume contains papers presented at the 46th Course of the International School of Crystallography held at the Ettore Majorana Centre for Scientific Culture, Erice, Sicily from 30 May to 8 June 2013. The Course on 'The Future of Dynamic Structural Science – Links to Energy and Environmental Security and Early Detection of CBRN Agents' was designated a NATO Advanced Study Institute and the Directors of the Course acknowledge, with many thanks, the considerable financial support provided by the NATO Science for Peace and Security Programme.

The scientific program developed during the last years from our own researches and the idea to combine crystallographic experts with those from various spectroscopies was a bold decision, but it proved to be highly successful and the meeting was well received by participants and lecturers from both scientific areas. The cross-fertilisation of the fields has already spun off new international collaborations while this volume has been in preparation, and we expect to see more of these emerging in the future, as well as more meetings of this type in other countries and involving new participants, especially those who were unable to join us this year in Sicily.

We believe the course was timely for both the crystallographic and spectroscopic communities with the exciting new developments in sources, detectors, software and computational methodologies. Future time-resolved dynamic structural research needs a strong collaborative and interdisciplinary effort to achieve meaningful results and comprehensive analyses. These cannot be achieved by any one grouping alone, and we are delighted to have been a catalyst for new collaborations, and in particular to have brought the young scientists from different scientific areas together with the senior international team leaders, so that they too can appreciate the enormous value of the cross-disciplinary fertilisation of research for mutual societal benefit and to be able to contribute to its future.

Crystallography and spectroscopy are integrated across the sciences, and this course brought together chemists, physicists and biologists to discuss the synergy which exists in modern methods, and in particular the dynamic aspects that we are beginning to address within time resolved research. It is impossible to cover major diversity within one course, nor did we wish to become too focused in any one area, and we believe we have achieved a good scientific balance through the lectures presented, computer tutorials given by the lecturing staff to small groups of participants and wide ranging discussions in the lecture hall, the poster sessions and elsewhere throughout the course.

The crystallographic lectures covered both chemical and biological studies ranging from slow through to faster timescales as the week progressed. Topics covered included metastable species, linkage isomerism, dynamics under pressure in both chemical and biological species, single crystal to single crystal transformations, X-ray scattering from liquids, photo-excited species, magnetic properties, challenges for studying irreversible reactions, how to study biological processes during which radiation damage occurs, time resolved synchrotron experiments, use of pulsed X-ray beams and mechanical choppers, X-ray coherent diffraction imaging, timing tools and XFEL. Time-resolved spectroscopic techniques discussed included Raman, XAFS, fluorescence spectroscopy, absorption and emission spectroscopy. Approximately a third of the lecturers also gave demonstrations and/or workshops for the relevant software in order to introduce the participants to the currently available data analysis programs.

The course benefited greatly from additional financial support and sponsorship from a number of organisations and companies; these were The International Union of Crystallography (IUCr), The European Crystallographic Association (ECA), PANalytical and OPCW (Organisation for the Prohibition of Chemical Weapons, Nobel Peace Laureate 2013).

The planning and delivery of a complex course such as this depends on the sustained efforts of many people, and we would like to thank all of our lecturers for their time and dedication to the course in preparing lecture notes beforehand, teaching to the students during the course and now preparing manuscripts for this volume. We thank the participants who came from all over the world to join us in Erice and for their enthusiasm, oral presentations and poster contributions to the course, and not least we thank the local organisers and teams who worked so tirelessly on site to make the 46th course a great success. The Directors would like, in particular, to acknowledge the enormous support and leadership from the local organisers, Paola Spadon from the University of Padua and Annalisa Guerri from the University of Florence, and their indefatigable team of assistants, 'the orange scarves', who worked on site to help everyone throughout the meeting. We thank John Irwin and Erin Bolstad for incredible and invaluable IT support for the entire meeting, before and beyond. Also we are most grateful to the Ettore Majorana Centre and all the staff in Erice who run the centre and help in every practical way to support the participants throughout the course. Last, but by no means least, we thank Professor Sir Tom Blundell, Director of the International School of Crystallography, for this rewarding opportunity to organise and run the 46th Course in Erice this year.

We would like to dedicate this volume to the memory of our great friend and colleague for many years, Lodovico Riva di Sanseverino (1939–2010), who began these Erice schools almost 40 years ago in 1974 and who worked tirelessly and with a passion for crystallographic science throughout his long career.

Durham, UK Judith A.K. Howard
Bristol Hazel A. Sparkes
October 2013

Contents

Contributors

Director of the International School of Crystallography

Sir Thomas L. Blundell Department of Biochemistry, University of Cambridge, Cambridge, UK

Course Directors

Andrei V. Churakov N.S. Kurnakov Institute of General and Inorganic Chemistry, Russian Academy of Sciences, Moscow, Russia

Judith A.K. Howard Department of Chemistry, Durham University, Durham, UK

Paul R. Raithby Department of Chemistry, University of Bath, Bath, UK

Hazel A. Sparkes School of Chemistry, University of Bristol, Bristol

Organisers

Erin Bolstad Department of Biomedical/Pharmaceutical Science, University of Montana, Missoula, MT, USA

Annalisa Guerri Department of Chemistry, University of Florence, Sesto Fiorentino, Italy

John Irwin Leslie Dan Faculty of Pharmacy, University of Toronto, Toronto, ON Canada

Paola Spadon Department of Chemical Sciences, Padua University, Padua, Italy

Contributors to Chapters

P.A. Anfinrud NIDDK, Laboratory of Chemical Physics, NIH, Bethesda, MD, USA

Paul Barnes Department of Chemistry, University College London, London, UK

Biological Sciences, Birkbeck College, London, UK

Elena V. Boldyreva Institute of Solid State Chemistry and Mechanochemistry SB RAS, Novosibirsk State University, Novosibirsk, Russia

Savo Bratos Laboratoire de Physique Théorique de la Matière Condensée, Université Pierre & Marie Curie, Paris 05, France

Richard Briggs Center for Science at Extreme Conditions, School of Physics and Astronomy, University of Edinburgh, Edinburgh, UK

Emre Brookes Department of Biochemistry, University of Texas Health Science Center at San Antonio, San Antonio, TX, USA

Marco Cammarata Institut de Physique, Université de Rennes 1, Rennes, France

Paul Carey Department of Biochemistry, School of Medicine, Case Western Reserve University, Cleveland, OH, USA

Eric Collet Institut de Physique de Rennes, UMR 6251 UR1-CNRS, University Rennes 1, Rennes Cedex, France

Fabriza Foglia Department of Chemistry, University College London, London, UK

Kenneth P. Ghiggino School of Chemistry, University of Melbourne, Melbourne, VIC, Australia

Kristoffer Haldrup Department of Physics, NEXMAP Section, Centre for Molecular Movies, Technical University of Denmark, Lyngby, Denmark

Simon D.M. Jacques The School of Materials, The University of Manchester, Manchester, UK

Research Complex at Harwell (RCaH), Rutherford Appleton Laboratory, Harwell Oxford, Didcot, Oxon, UK

Menahem Kaftory Schulich Faculty of Chemistry, Technion – Israel Institute of Technology, Haifa, Israel

Yoshinori Kakitani Faculty of Science and Technology, Kwansei Gakuin University, Sanda, Japan

Surajit Kayal Department of Inorganic and Physical Chemistry, Indian Institute of Science, Bangalore, India

Dmitry Khakhulin European Synchrotron Radiation Facility, Experiments Division, Grenoble, France

Tae Kyu Kim Department of Chemistry, Pusan National University, Busan, Republic of Korea

Qingyu Kong Division Expériences, Synchrotron Soleil, Saint-Aubin, France

Yasushi Koyama Faculty of Science and Technology, Kwansei Gakuin University, Sanda, Japan

Jae Hyuk Lee Department of Chemistry, Kaist, Daejeon, Republic of Korea

Jean-Claude Leichnam Laboratoire de Physique Théorique de la Matière Condensée, Université Pierre & Marie Curie, Paris 05, France

Paul F. McMillan Department of Chemistry, University College London, London, UK

Filip Meersman Department of Chemistry, University College London, London, UK

Rousselot – Expertise Center, R&D Laboratory, Gent, Belgium

Martin Meedom Nielsen Department of Physics, NEXMAP Section, Centre for Molecular Movies, Technical University of Denmark, Lyngby, Denmark

Yuji Ohashi Ibaraki Quantum Beam Research Center (IQBRC), Tokai, Ibaraki, Japan

Arwen R. Pearson Astbury Centre for Structural Molecular Biology, The University of Leeds, Leeds, UK

Anna Polyakova Astbury Centre for Structural Molecular Biology, The University of Leeds, Leeds, UK

Nilesh Rai Department of Inorganic and Physical Chemistry, Indian Institute of Science, Bangalore, India

Paul R. Raithby Department of Chemistry, University of Bath, Bath, UK

Mattia Rocco Biopolymers and Proteomics Unit, IRCCS AOU San Martino-IST, Genova, Italy

Khokan Roy Department of Inorganic and Physical Chemistry, Indian Institute of Science, Bangalore, India

Friedrich Schotte NIDDK, Laboratory of Chemical Physics, NIH, Bethesda, MD, USA

Vukica Šrajer Center for Advanced Radiation Sources, The University of Chicago, Chicago, IL, USA

Simone Techert FS-Structural Dynamics of (Bio)Chemical Systems, Deutsches Elektronensynchrotron DESY, Hamburg, Germany

Structural Dynamics of (Bio)Chemical Systems, Institute for X-ray Physics, Georg August University, Göttingen, Germany

Structural Dynamics of biochemical Systems, Max Planck Institute for Biophysical Chemistry, Göttingen, Germany

Moniek Tromp Catalyst Characterisation, Chemistry, Catalysis Research Center, Technische Universität München, München, Germany

Siva Umapathy Department of Inorganic and Physical Chemistry, Indian Institute of Science, Bangalore, India

Ravi Kumar Venkatraman Department of Inorganic and Physical Chemistry, Indian Institute of Science, Bangalore, India

Michael Wulff European Synchrotron Radiation Facility, Experiments Division, Grenoble, France

Chapter 1
Single Crystal and Powder Methods for Structure Determination of Metastable Species

Paul R. Raithby

1.1 Introduction

1.1.1 Background

X-ray crystallography as we know it has been in existence for 100 years, since the publication of the paper by the *Braggs* in 1913, describing the *Law* that has been the cornerstone of the subject ever since. For single-crystal crystallography the experimental method relies on the measurement of a diffraction pattern, of discrete diffraction spots, and the interpretation of this pattern to provide a three-dimensional representation of the crystal structure, showing the positions of the atoms and the distances between them in the structure. Because, even now, the collection of a complete data set takes between minutes and hours, the structure obtained is time averaged over this period. Thus, the method is very good at providing the structure of the starting materials for a reaction, or for obtaining the structure of a crystalline reaction product, but it does not provide any information about how a solid-state reaction (or chemical process) occurs. The object of this book is to show how it is now becoming possible to introduce the *dimension of time* into the crystallographic experiment, so that we can *"watch chemistry happen"*.

1.1.2 Timescales

In order to achieve the objective of *watching chemistry happen* we have to think a little about the timescales on which various chemical processes happen and, therefore, what spectroscopic and diffraction probes would be suitable to investigate the

P.R. Raithby (✉)
Department of Chemistry, University of Bath, Claverton Down, BA2 7AY Bath, UK
e-mail: p.r.raithby@bath.ac.uk

J.A.K. Howard et al. (eds.), *The Future of Dynamic Structural Science*,
NATO Science for Peace and Security Series A: Chemistry and Biology,
DOI 10.1007/978-94-017-8550-1_1,

Table 1.1 Timescales associated with some common spectroscopic techniques

Technique	Energy of excited state (Hz)	Typical relaxation time (s)	Typical linewidth (Hz)
NMR (solution)	10^8	10	10^{-1}
ESR (solution)	10^{10}	10^{-5}	10^5
Rotational spectroscopy (gas phase)	10^{11}	10^{-4}	10^4
Vibrational spectroscopy (gas phase)	10^{14}	10^{-8}	10^8
Electronic spectroscopy (solution)	10^{16}	10^{-8}	10^8

structural changes. The timescales relating to some common spectroscopic techniques are presented in Table 1.1. When considering any experiment there are, in fact, four different times that we need to consider:

1. The time during which a single quantum of radiation or a particle interacts with a molecule.
2. The lifetime of the excited state molecule generated in the activation process.
3. The lifetime that the species being probed must have for it to be seen as a distinct species.
4. The total time of the experiment in which the species is observed.

The time taken for a quantum of radiation to interact with a molecule is effectively the time that it takes for a photon to pass by the molecule, which is usually slightly slower than the speed of light, and of the order of 10^{-16} to 10^{-19} s. This is a much shorter time than that required for a molecular vibration or a rotation to take place, so each photon "sees" a molecule as a snapshot with the molecule having a fixed electronic, vibrational and rotational state.

Most importantly in the context of this course, we must consider the total time during which we measure our data. In diffraction experiments, whether it be an X-ray, electron or neutron photon, it is diffracted in about 10^{-18} s, so it sees an effectively frozen structure, but, the experiment involves collecting data from many photons and each sees a molecule at a different stage of vibration or transformation. Thus, balancing the method of measurement against the timescale is very important.

1.2 Photocrystallography

1.2.1 The Principles

In this book, and particularly in this chapter, we are going to consider single-crystal and powder X-ray crystallography as our primary methods for *watching chemistry happen*. As you will probably be aware, there are certain restrictions if we want to use single-crystal crystallography, that is, we must use single-crystals that give a discrete diffraction pattern, and that must be retained throughout the experiment,

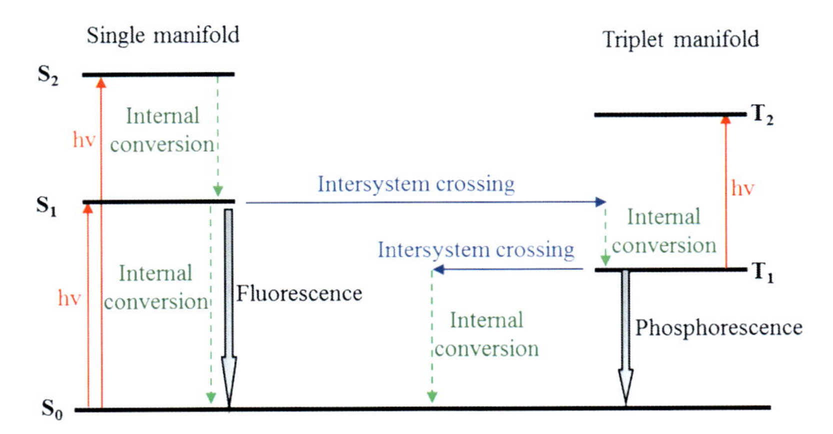

Fig. 1.1 Energy plot showing the photoactivation process

because if crystallinity is lost, the discrete pattern is lost and the results cannot be analysed. Thus, the reaction chemistry that can be carried out on a single crystal is limited to the introduction of external "**reagents**" that do not destroy the crystal. The introduction of reagents in solution is likely to degrade the crystal, so we are limited to the use of solid-gas reactions or to the use of "light", "pressure" or "changes in temperature" to activate the chemical process.

We have chosen to use "light" because the area of photochemistry is already well developed, particularly in solution, but this can readily be extended to the solid-state. The combination of photochemistry and crystallography has led to the development of the science of *photocrystallography*, defined by Philip Coppens, one of the pioneers of the area as "*Photocrystallography is the technique by which the 3-dimensional "metastable" or "excited state" structures of molecules, within a crystalline solid, are determined by crystallographic techniques when the materials are activated by light*" [1].

A factor in the choice of light as the activation source is that there are a wide range of light sources available for the experiments, ranging from inexpensive LEDs to high power pump-probe lasers worth thousands of dollars.

The lifetimes of the species activated by light may range from picoseconds to the metastable (an activated species with an infinite lifetime provided that certain external conditions such as temperature or illumination with light are maintained). The possible outcomes of a photoactivation process are shown in Fig. 1.1. The probing of excited singlet states, which generally have very short lifetimes, perhaps of a few picoseconds remains the domain of time resolved spectroscopy and XAFS, as we shall see in later presentations, while species with triplet lifetimes, of perhaps microseconds, and above, may now be probed crystallographically. In this context, the presence of a heavy metal, with its associated spin-orbit coupling, assists in the relaxation of the selection rules that prohibit intersystem crossing, so long lived triplet states become more accessible. Thus, transition metal complexes have been among the first classes of system to be studied photocrystallographically.

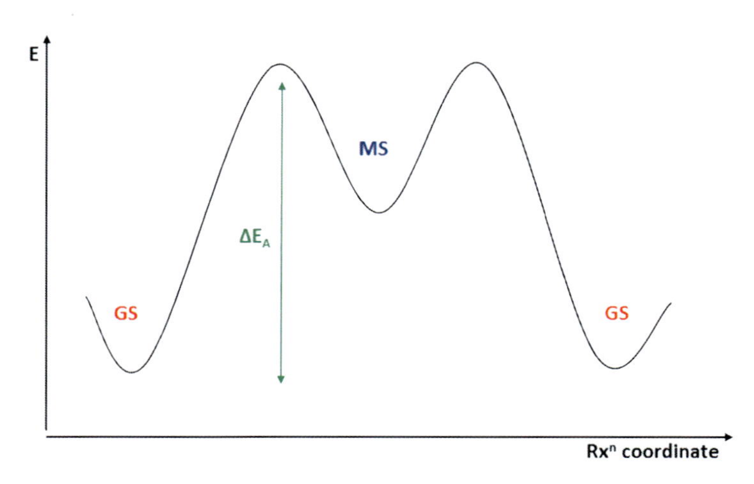

Fig. 1.2 Schematic energy diagram for a reversible metastable system

Experimentally, metastable species are the easiest to study because, upon photoactivation, providing external conditions are maintained, they have infinite lifetimes, and conventional single crystal X-ray diffraction methods can be used for data collection, sometimes called the "*steady state*" methodology. The energy profile illustrated in Fig. 1.2 shows the position of a reversible metastable state where MS represents the local minimum.

1.2.2 Types of Solid-State Processes

There are two classes of solid-state process: (i) those that are reversible, where molecules may be kept in an activated state by continuously pumping energy into the system and, (ii) those that are irreversible, where a process may be started by photoactivation and followed crystallographically until a photostable product is reached. In this discussion we will concentrate on reversible processes leading to the determination of the structure of metastable and "excited state" species but, for completeness we will mention a couple of examples of photocrystallographic studies on irreversible processes.

True homogeneous single-crystal-to-single-crystal reactions, in which there is a 100 % conversion from starting material to product remain rare [2]. It has been suggested that the homogeneous or heterogeneous character of a solid-state photochemical process can be influenced by the method of irradiation, for example in solid-state [2 + 2] cycloaddition reactions [3]. Using wavelengths close to the absorption maxima may favour heterogeneous transformations but using wavelengths that are in the absorption tail can favour homogeneous transformations or, at least, delay crystal degradation [4].

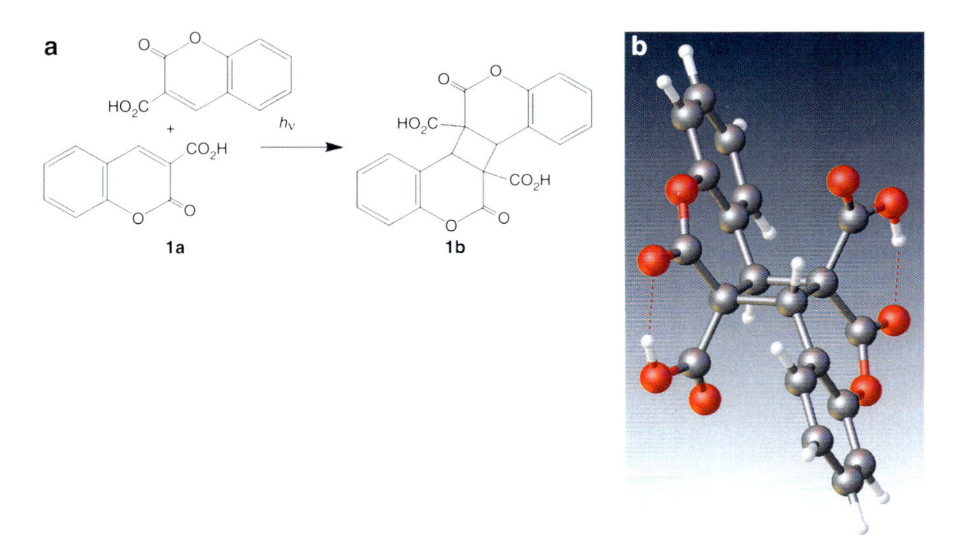

Fig. 1.3 Photochemical [2 + 2] cycloaddition of coumarin-3-carboxylic acid

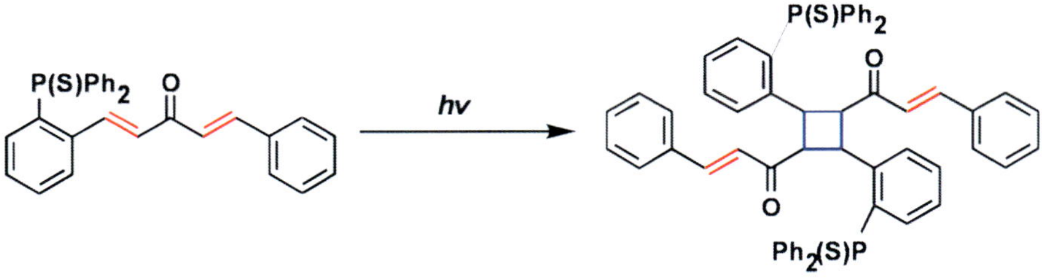

Fig. 1.4 Photocrystallographic [2 + 2] cycloaddition

Photocrystallographic X-ray powder diffraction studies on coumarin-3-carboxylic acid showed that over a 3 h period, when irradiated with a 150 W mercury lamp, there was a regular and significant change to the powder pattern observed. The [2 + 2] cycloaddition was confirmed by a single-crystal experiment (Fig. 1.3), with the transformation occurring without significant loss of crystallinity [5]. More recently, it has been shown that a novel unsymmetrical dibenzylidene acetone (monothiophos-dba), containing two chemically different alkene moieties, undergoes a selective photochemically-induced [2 + 2] cycloaddition solid-state transformation in both the single crystal and the powder, as shown by X-ray diffraction methods (Fig. 1.4) [6]. In the single crystal the rate constant for the cycloaddition reaction was determined $\{k = (2.3 \pm 0.1) \times 10^4\ s^{-1}\}$, with a 69 % conversion attained.

1.2.3 Reversible Metastable Systems

In order for single-crystal photocrystallographic experiments to be successful the structural rearrangement under study must be significant enough that there is a

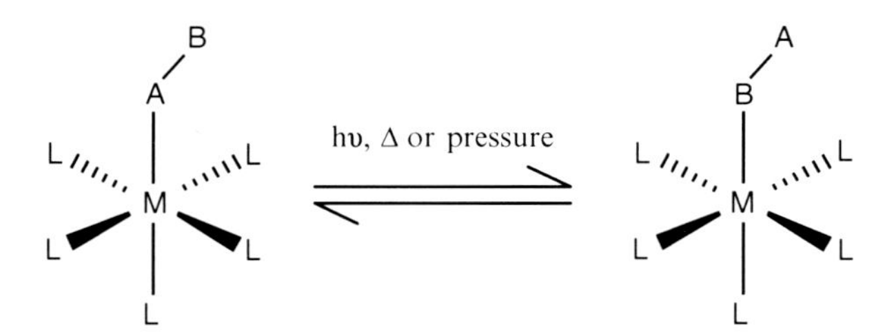

Fig. 1.5 Linkage isomerism: *M* metal, *AB* an arbitrary ligand capable of undergoing (coordinating through A or B linkage isomerism), *L* spectator ligand

noticeable change in the electron density difference map. Materials that undergo solid-state isomerisation provide such changes and are therefore suitable for study by these methods. On the other hand whilst enough movement is required to make results clear, a major structural rearrangement is also not desirable as too much movement would create strain within the crystal and in many cases will lead to destruction the crystal lattice. In light of these considerations the process of linkage isomerisation is ideal for photocrystallographic experiments, as in some cases the change involved is moderate enough that the crystallinity of the sample is not compromised. Linkage isomerism requires a small, ambidentate ligand that can be made to change its mode of bonding to a transition metal centre by external stimulation. A general example of the process is shown in Fig. 1.5.

There are a wide range of di-, tri- and tetra-atomic ligands that can undergo isomerisation, including NO, NO_2, N_2, SO_2, and DMSO, and a large volume of work has been dedicated to the topic. For thermal and photoexcited isomerisations many processes are reversible and can be controlled at low temperature, leading to metastable linkage isomers that are ideal for study using "*steady-state*" methods [1, 7].

Photocrystallographic studies on metastable linkage isomers were pioneered by Coppens who used lasers to photoactivate the crystals [8, 9, 10]. Since then other methods of illumination have been used.

1.2.4 The Experimental Method

For steady-state photocrystallographic experiments a number of key steps are required.

- From the UV absorption spectrum determines the appropriate wavelength for irradiation (generally not the absorption maximum to avoid crystal degradation).
- Calculate the optical penetration depth from the absorption coefficient of the crystal, and pick a crystal of optimum size – taking into account the intensity of the X-ray source as well as the light source.
- On the diffractometer cool the crystal to an appropriate temperature (30–150 K).
- Carry out a standard, accurate low temperature data collection (the ground state structure) which can be used to benchmark the photocrystallographic studies.

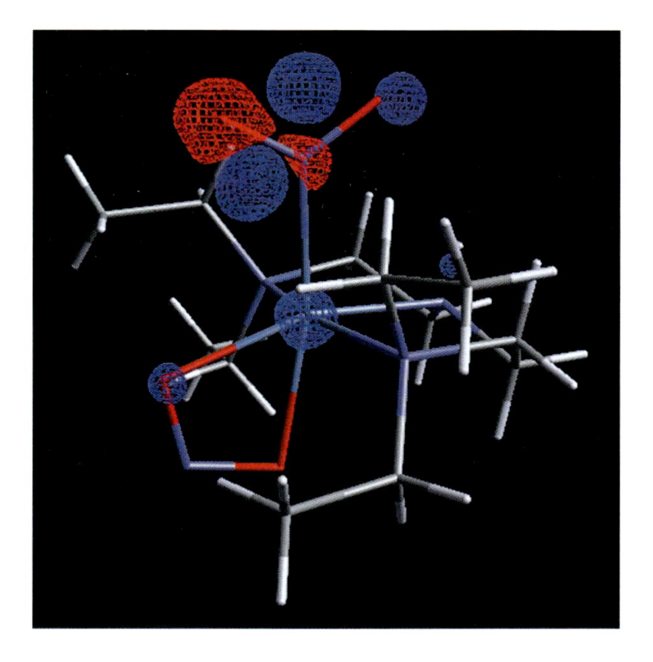

Fig. 1.6 An electron density map showing the regions of increased and reduced electron density for the conversion of a nitro group into a nitrito group upon photoactivation in a single –crystal

- Once the ground state structure has been obtained, irradiate the crystal until an appropriate percentage of molecules have been excited up into the *metastable* state.
- Switch off illumination but maintain the low temperature.
- Collect a full set of accurate X-ray data using the same experimental procedures as for the ground state (combined ground and *metastable* stable structures).
- Use combined data and refine using ground state coordinates – if additional features are observed these normally corresponding to the *metastable* structure. Depending on the level of excitation, subtract ground state structure, leaving *metastable* structure.
- The temperature at which the experiment is run can then be varied in order to maximise the percentage conversion to the *metastable* state structure.

An example of an electron density profile showing the reduction in electron density for the N-bound NO_2 linkage isomer and the growth of the O-bound ONO linkage isomer, in the complex $Ni(Et_4dien)(NO_2)_2$ [11], is illustrated in Fig. 1.6.

While the initial photocrystallographic studies were carried out using lasers it has since proved possible, in many cases, to activate molecules within single-crystals using a set of LEDs of the appropriate wavelength, and a "ring" holding 6 LEDs that fits onto the nozzle of a crystal cooling apparatus has proved effective and versatile (Fig. 1.7), and eliminated many of the safety issues related to using a high power laser [12]. The most spectacular success using this device has been the generation of the metastable nitrito linkage isomer of $Ni(dppe)(NO_2)Cl$, with a reversible 100 % conversion from the ground state nitro isomer [13].

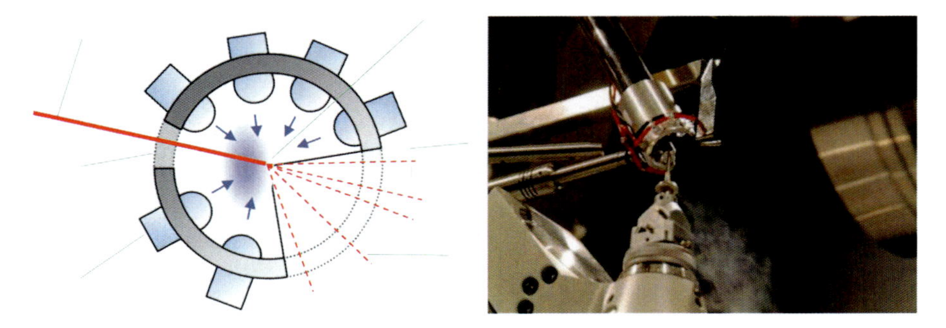

Fig. 1.7 Schematic of the LED ring and it positioned on the diffractometer

1.2.5 Metastable Complexes That Have Been Studied by Photocrystallography

1.2.5.1 Nitrosyl Linkage Isomers

Following spectroscopic work by Hauser et al. [14], neutron diffraction studies by Woike and co-workers determined that two metastable photoactivated existed for sodium nitroprusside: MS1 and MS2 [15]. The exact nature of these photoexcited states was the subject of debate, and when explanations involving electron-transfer models [16] could not explain the surprising stability of MS1 and MS2 at low temperature, the Coppens group proposed an explanation based on Isomerisation of the nitrosyl ligand that better fitted the observations [9].

After irradiation with 488 nm laser light to excite the system to the MS1 state the authors used steady-state crystallographic techniques to show that the ligand reorganises from η^1-NO (nitrosyl) in the ground state to a η^1-ON (isonitrosyl) arrangement (Fig. 1.8). Subsequent irradiation of a crystal with a 1,064 nm laser produced the MS2 state. By interpreting the electron density difference map between the MS2 and ground states, the authors concluded that MS2 is also the result of isomerisation of the nitrosyl ligand, this time to a side-on bound η^2-NO isomer (Fig. 1.8). The MS1 isomerisation was identified by studying the anisotropic temperature parameters for the metastable structure. When the ligand was modelled in the nitrosyl arrangement the thermal parameters strongly differed from those expected for a stable structure, but inversion of the oxygen and nitrogen atoms yielded more physically acceptable parameters indicating that the ligand adopts the isonitrosyl arrangement. Prior to this the η^1-ON arrangement had not been characterised by any crystallographic study, showing the importance of the method.

Coppens has also co-authored other investigations into nitrosyl linkage isomerism, proving that isomerisation to MS1 and MS2 states are not exclusive to the iron complex. Coinciding with the work on sodium nitroprusside, it was reported that the complex $K_2[Ru(NO_2)_4(OH)(NO)]$ is also excited to comparable nitrosyl MS1 and MS2 metastable states, with structural results supporting the proposed η^1-ON isonitrosyl model for MS1 [17]. In addition, in 1997 Coppens again employed steady-state methods to show that the photoinduced metastable state of [Ni(NO)(η^5-Cp*)] involves isomerism to a side-on η^2-NO bound entity similar to the sodium

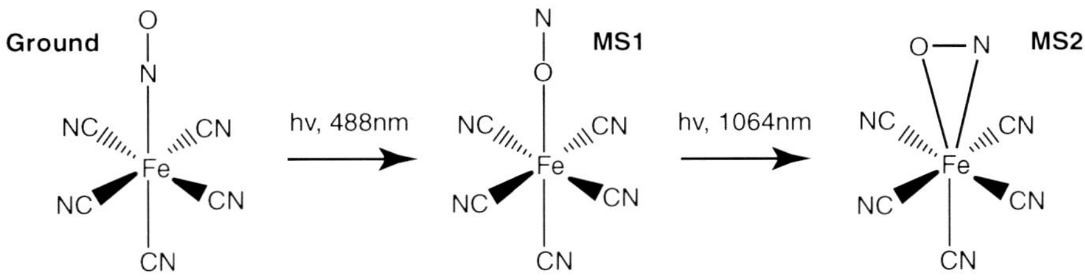

Fig. 1.8 Arrangement of the nitrosyl ligand in the ground and metastable states of sodium nitroprusside

nitroprusside MS2 state [18]. In this work the first η^2-NO entity for a Ni complex was determined by the observation that the Ni-O distance is dramatically reduced in the excited-state species to a value below that expected for a formal bond. Coupled with a Ni-N bond distance still in the expected range, this was strong evidence that the ligand had rearranged to a side-on η^2 bonding arrangement and excluded an explanation that the ligand might take up a more traditional η^1 bent-nitrosyl arrangement.

1.2.5.2 Double Linkage Isomerism (NO and NO_2)

Complexes containing more than one type of ambidentate ligand open up the possibility for double linkage isomers, where both ligand types might convert to a new binding mode on excitation. The first crystallographic study of double linkage Isomerisation was conducted on $[Ru(bpy)_2(NO)(NO_2)](PF_6)$ by the Coppens group [19]. Excitation with light from an Ar^+ laser was found to cause oxygen transfer between ligands, supported by a direct swop of the NO and NO_2 sites that could be seen in the crystallographic data. On excitation at 200 K, structural changes supporting only a nitro-nitrito conversion were observed that suggest a metastable isomer with η^1-O-bound nitrite and N-bound nitrosyl ligands (MS200K). However, on cooling to 90 K changes in the structure correspond to a double isomerisation of nitro-nitrito and nitrosyl-isonitrosyl to generate another metastable state MS90 K, containing the η^1-O-bound nitrite and this time an O-bound nitrosyl. Supported by density function theory (DFT) calculations that confirm these species to be local energy minima on the potential energy surface, the authors proposed a mechanism for oxygen transfer between ligands that relies on these isomerisations.

1.2.5.3 Dinitrogen and Sulphur Dioxide Linkage Isomers

Although considered fairly inert, dinitrogen is observed to bind to metal centres and considering that nitrogen has many important biological and industrial applications an understanding of its binding modes is of major importance. The first evidence for a photoinduced η^2-bound state of nitrogen was obtained by Coppens and co-authors using steady-state techniques [20]. Irradiation of $[Os(NH_3)_5(N_2)]^{2+}$ with laser light produced a metastable state whose electron density difference map showed that electron density reduces along the N-N bond for the ligand in the ground state. At the same time an increase is observed along a perpendicular direction, indicating

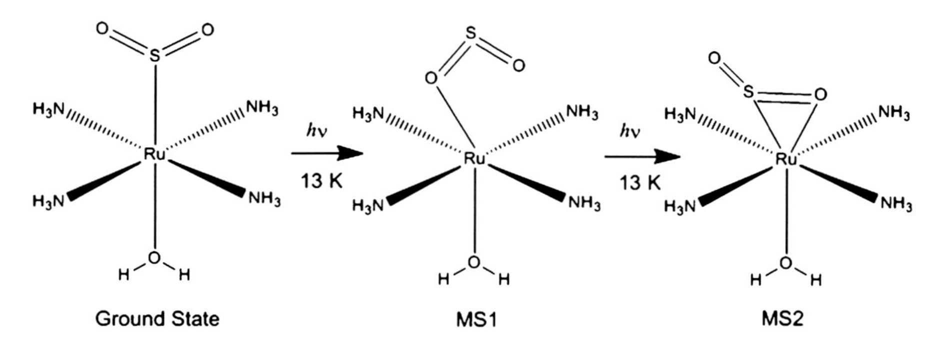

Fig. 1.9 The arrangement of the SO_2 ligand in the ground and metastable states of the $[Ru(NH_3)_4(H_2O)(SO_2)]^{2+}$ cation

rearrangement of the linear η^1-bound ligand to a side-on η^2-arrangement. DFT calculations indicate this arrangement lies at a minimum on the potential energy surface, confirming the likelihood of its existence.

Analogous to the nitrosyl isomerisations, metastable η^1-O-bound (MS1) and side-on η^2-S,O-bound (MS2) arrangements for the sulphur dioxide ligand have also been characterized using steady-state crystallographic methods. Photo-induced conversion of the ground-state η^1-S-bound ligand to the MS2 state was characterized both by early spectroscopic studies [21] and by the Coppens group [22] for a selection of ruthenium sulphur dioxide complexes. Although DFT calculations indicated a second energy minimum supporting the presence of a MS1-type state, evidence of the η^1-O-bound isomer could not be conclusively determined from the data for any of the species. It took a later study by Bowes et al. [23] to conclusively determine the existence of the sulphur dioxide MS1 state for a ruthenium complex.

After irradiation of $[Ru(NH_3)_4(H_2O)(SO_2)]^{2+}$ at 13 K both the MS1 and MS2 states (Fig. 1.9) were identified in the resultant electron density difference map and, with a calculated 36 % excitation level for MS1, the authors could characterise this η^1-O-bound arrangement for the first time in such a complex.

Subsequently Cole et al. have shown that the temperature at which the MS1 state in $[Ru(NH_3)_4(H_2O)(SO_2)]^{2+}$ is present may be dependent on the nature of the counter ion in the crystal structure. At 120 K, for the cation with a camphorsulphonate anion, the MS1 metastable state has been shown to decay to the MS2 state with an estimated half-life of 3.4(8) h and a long-lived population of 2.9(4)% [24]. The same group has also studied the effect of changing the ligand *trans* to the SO_2 group in the complex [25]. Replacement of the *trans* water ligand by pyridine, 3-chloropyridine and 4-chloropyridine leads to the 58(3)% conversion to the MS2 state, for the pyridine case, and the decay process from MS1 to MS2 has been shown to be first order with a non-zero asymptote.

1.2.5.4 Nitro/Nitrito Linkage Isomers

Complexes that undergo photoactivated linkage isomerism between the nitro, η^1-NO_2, and nitrito, η^1-ONO, form in single crystals have been used to investigate the factors that favour the transformation, and both steric and kinetic factors have been taken into account. As a result of studies aimed at improving photoconversion levels the first fully-reversible nitro – nitrito isomerism in the complex [Ni(dppe)(η^1-NO_2)Cl] upon photoactivation with UV LED light at temperatures below 160 K has been reported [13]. Photocrystallographic evidence for the 100 % conversion to the metastable isomer [Ni(dppe)(η^1-ONO)Cl] was supported with data from solid-state Raman spectroscopy experiments. Subsequently, the range of systems that display 100 % reversible nitro – nitrito conversion has been extended to include two analogous complexes involving bulky phosphine auxiliary ligands; [Ni(dppe)(η^1-NO_2)$_2$] and [Ni(dcpe)(η^1-NO_2)$_2$]. Complementing these square-planar phosphino systems, photoisomerisation in a series of octahedral nickel-nitro complexes containing bidentate amine co-ligands has also been investigated [26]. In addition the interesting behaviour of the tridentate amine complex [Ni(Et_4dien)(η^1-NO_2)$_2$] has been investigated. This complex undergoes reversible nitro-nitrito linkage isomerism following exposure to both UV light and heat [11], and similar behaviour has been observed in [Pd(papaverine)(PPh_3)(NO_2)] on exposure to both of these stimuli [27]. These results indicate that the nature of the surrounding crystalline environment has an effect on the photochemical reaction, in-line with previous findings [28].

References

1. Coppens P, Fomitchev DV, Carducci MD, Culp K (1998) Crystallography of molecular excited states. Transition-metal nitrosyl complexes and the study of transient species. J Chem Soc Dalton Trans (6):865–872; Coppens P, Vorontsov II, Graber T, Gembicky M, Kovalevsky AY (2005) The structure of short-lived excited states of molecular complexes by time-resolved X-ray diffraction. Acta Crystallogr Sect A 61:162–172
2. Ohba S, Ito Y (2003) Single-crystal-to-single-crystal photodimerization of 4-chlorocinnamoyl-O, O'-dimethyldopamine. Acta Crystallogr B 59(1):149–155
3. Harada J, Uekusa H, Ohashi Y (1999) X-ray analysis of structural changes in photochromic salicylideneaniline crystals. Solid-state reaction induced by two-photon excitation. J Am Chem Soc 121(24):5809–5810
4. Enkelmann V, Wegner G, Novak K, Wagener KB (1993) Single-crystal-to-single-crystal photodimerization of cinnamic acid. J Am Chem Soc 115(22):10390–10391
5. Mahon MF, Raithby PR, Sparkes HA (2008) Investigation of the factors favouring solid state [2+2] cycloaddition reactions; the [2+2] cycloaddition reaction of coumarin-3-carboxylic acid. Cryst Eng Commun 10(5):573–576
6. Jarvis AG, Sparkes HA, Tallentire SE, Hatcher LE, Warren MR, Raithby PR, Allan DR, Whitwood AC, Cockett MCR, Duckett SB, Clark JL, Fairlamb IJS (2012) Photochemical-mediated solid-state 2+2 -cycloaddition reactions of an unsymmetrical dibenzylidene acetone (monothiophos-dba). Cryst Eng Commun 14(17):5564–5571
7. Cole JM (2011) A new form of analytical chemistry: distinguishing the molecular structure of photo-induced states from ground-states. Analyst 136(3):448–455; Coppens P, Zheng SL, Gembicky M (2008) Static and time-resolved photocrystallographic studies in supramolecular solids. Z Kristall 223(4–5):265–271; Raithby PR (2007) Small-molecule chemical crystallography – from three to four dimensions: a personal perspective. Crystallogr Rev 13:121–142
8. Kalinowski JA, Fournier B, Makal A, Coppens P (2012) The LaueUtil toolkit for Laue photocrystallography. II. Spot finding and integration. J Synchrot Radiat 19:637–646
9. Carducci MD, Pressprich MR, Coppens P (1997) Diffraction studies of photoexcited crystals: metastable nitrosyl-linkage isomers of sodium nitroprusside. J Am Chem Soc 119(11):2669–2678
10. Coppens P, Novozhilova I, Kovalevsky A (2002) Photoinduced linkage isomers of transition-metal nitrosyl compounds and related complexes. Chem Rev 102(4):861–883
11. Hatcher LE, Warren MR, Allan DR, Brayshaw SK, Johnson AL, Fuertes S, Schiffers S, Stevenson AJ, Teat SJ, Woodall CH, Raithby PR (2011) Metastable linkage isomerism in

Ni(Et(4)dien)(NO(2))(2): a combined thermal and photocrystallographic structural investigation of a nitro/nitrito interconversion. Angew Chem Int Ed 50(36):8371–8374
12. Brayshaw SK, Knight JW, Raithby PR, Savarese TL, Schiffers S, Teat SJ, Warren JE, Warren MR (2010) Photocrystallography – design and methodology for the use of a light-emitting diode device. J Appl Crystallogr 43:337–340
13. Warren MR, Brayshaw SK, Johnson AL, Schiffers S, Raithby PR, Easun TL, George MW, Warren JE, Teat SJ (2009) Reversible 100% linkage isomerization in a single-crystal to single-crystal transformation: photocrystallographic identification of the metastable Ni(dppe)(eta(1)-ONO)Cl isomer. Angew Chem Int Ed 48(31):5711–5714
14. Hauser U, Oestreich V, Rohrweck HD (1978) Optical dispersion in transparent molecular systems. 3. New kind of inelastic-scattering of polarized-light by oriented molecules depending upon coherence of radiation-field. Z Phys A-Hadron Nucl 284(1):9–19
15. Zollner H, Krasser W, Woike T, Haussuhl S (1989) The existence of light-induced long-lived metastable states in different Xn Fe(CN)5NO.YH2O crystals, powders and solutions. Chem Phys Lett 161(6):497–501
16. Terrile C, Nascimento OR, Moraes IJ, Castellano EE, Piro OE, Guida JA, Aymonino PJ (1990) On the electronic-structure of metastable nitroprusside ion in Na2 Fe(CN)5NO.2H2O – a comparative single-crystal ESR study. Solid State Commun 73(7):481–486
17. Fomitchev DV, Coppens P (1996) X-ray diffraction analysis of geometry changes upon excitation: the ground-state and metastable-state structures of K_2 $Ru(NO_2)_4(OH)(NO)$. Inorg Chem 35(24):7021–7026
18. Fomitchev DV, Furlani TR, Coppens P (1998) Combined X-ray diffraction and density functional study of [Ni(NO)(eta(5)-Cp*)] in the ground and light-induced metastable states. Inorg Chem 37(7):1519–1526
19. Kovalevsky AY, King G, Bagley KA, Coppens P (2005) Photoinduced oxygen transfer and double-linkage isomerism in a cis-(NO)(NO2) transition-metal complex by photocrystallography. FT-IR spectroscopy and DFT calculations. Chem Eur J 11(24):7254–7264
20. Fomitchev DV, Bagley KA, Coppens P (2000) The first crystallographic evidence for side-on coordination of N_2 to a single metal center in a photoinduced metastable state. J Am Chem Soc 122(3):532–533
21. Johnson DA, Dew VC (1979) Photo-chemical linkage isomerization in coordinated SO_2. Inorg Chem 18(11):3273–3274
22. Kovalevsky AY, Bagley KA, Coppens P (2002) The first photocrystallographic evidence for light-induced metastable linkage isomers of ruthenium sulfur dioxide complexes. J Am Chem Soc 124(31):9241–9248; Kovalevsky AY, Bagley KA, Cole JM, Coppens P (2003) Light-induced metastable linkage isomers of ruthenium sulfur dioxide complexes. Inorg Chem 42(1):140–147
23. Bowes KF, Cole JM, Husheer SLG, Raithby PR, Savarese TL, Sparkes HA, Teat SJ, Warren JE (2006) Photocrystallographic structure determination of a new geometric isomer of [Ru(NH3)(4)(H2O)(eta(1)-OSO)][MeC6H4SO3](2). Chem Commun (23):2448–2450
24. Phillips AE, Cole JM, d'Almeida T, Low KS (2012) Ru–OSO coordination photogenerated at 100 K in tetraammineaqua(sulfur dioxide)ruthenium(II) ($\pm$)-camphorsulfonate. Inorg Chem 51(3):1204–1206
25. Sylvester SO, Cole JM, Waddell PG (2012) Photoconversion bonding mechanism in ruthenium sulfur dioxide linkage photoisomers revealed by in situ diffraction. J Am Chem Soc 134 (29):11860–11863
26. Brayshaw SK, Easun TL, George MW, Griffin AME, Johnson AL, Raithby PR, Savarese TL, Schiffers S, Warren JE, Warren MR, Teat SJ (2012) Photocrystallographic identification of metastable nitrito linkage isomers in a series of nickel(ii) complexes. Dalton Trans 41(1):90–97
27. Bajwa SE, Storr TE, Hatcher LE, Williams TJ, Baumann CG, Whitwood AC, Allan DR, Teat SJ, Raithby PR, Fairlamb IJS (2012) On the appearance of nitrite anion in [PdX(OAc)L2] and [Pd(X)(C^N)L] syntheses (X = OAc or NO2): photocrystallographic identification of metastable Pd([small eta]1-ONO)(C^N)PPh3. Chem Sci 3(5):1656–1661
28. Zheng SL, Gembicky M, Messerschmidt M, Dominiak PM, Coppens P (2006) Effect of the environment on molecular properties: synthesis, structure, and photoluminescence of Cu(I) bis (2,9-dimethyl-1,10-phenanthroline) nanoclusters in eight different supramolecular frameworks. Inorg Chem 45(23):9281–9289

Chapter 2
Raman Crystallography, the Missing Link Between Biochemical Reactions and Crystallography

Paul Carey

The goal of my presentations is to explain the contributions that Raman spectroscopy can make to the understanding of dynamic processes in structural biology. While most of my examples focus on enzymes, the principles we will develop apply to many biomolecular systems. An outline of my talks is:

- Introduction to the Raman effect
- Raman microscopy
- How to obtain selectivity and specificity by using isotopes and Raman difference spectroscopy
- The synergy between Raman and X-Ray crystallography

2.1 Basic Principles

Raman spectroscopy is a light-scattering technique that, for our purposes, relates to the vibrational properties of molecules. Usually physics dictates that inelastic light scattering between a photon and a molecule is a highly improbable event and that Raman scattering is of very low intensity. For example, Raman intensity is orders of magnitude below fluorescence intensity. Fortunately, developments in photonics over the past 20–50 years (starting with lasers) have greatly improved our chances of obtaining detectable Raman spectra. We will begin with the Raman spectrum of water. If we focus a laser beam (continuous radiation in the visible spectrum) into an optical cell containing water and, using a Raman spectrometer, analyze the scattered light at right angles to the focal point, we will see an overwhelming population of elastically scattered photons with exactly the same wavelength

P. Carey (✉)
Department of Biochemistry, School of Medicine,
Case Western Reserve University, Cleveland, OH 44106, USA
e-mail: prc5@case.edu

J.A.K. Howard et al. (eds.), *The Future of Dynamic Structural Science*,
NATO Science for Peace and Security Series A: Chemistry and Biology,
DOI 10.1007/978-94-017-8550-1_2,

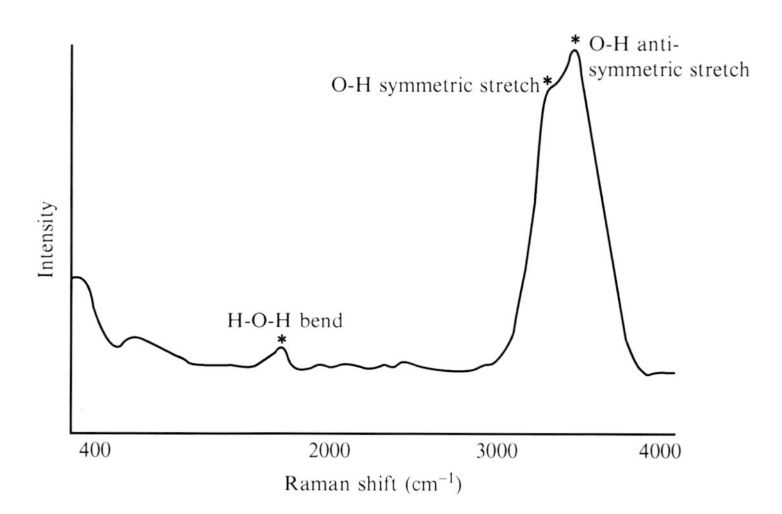

Fig. 2.1 A Raman spectrum of water

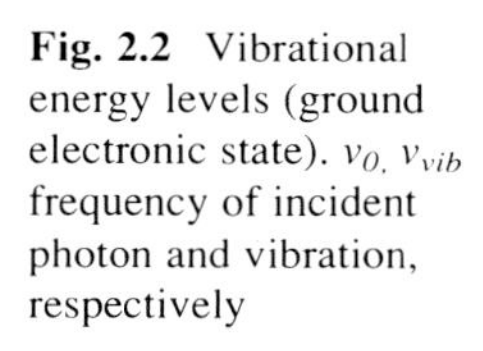
Fig. 2.2 Vibrational energy levels (ground electronic state). ν_0, ν_{vib} frequency of incident photon and vibration, respectively

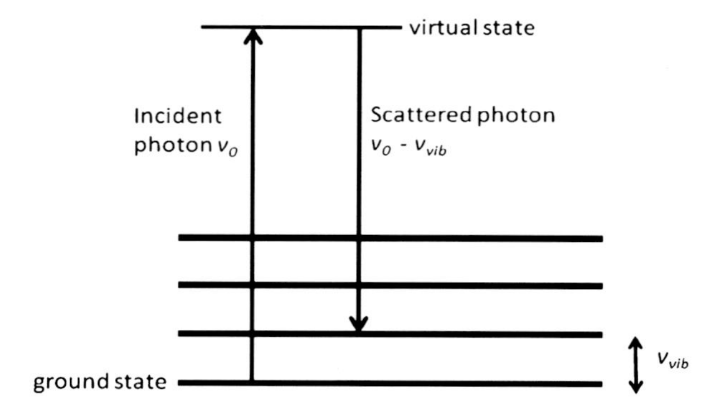

(i.e. energy) as the incident beam. However, a tiny fraction of the photons have exchanged energy with the water molecules, and these give rise to the Raman spectrum. When the intense elastically scattered photons have been removed by optical filters, the Raman spectrometer [1, 2] reveals the Raman spectrum of water, shown in Fig. 2.1.

As part of their quantum chemical make-up, the vibrational energy levels of the water molecules are discrete (i.e. quantized). One vibration involves the stretching of water's O–H bonds and the molecule occupies one of the "rungs" of a vibrational energy level ladder, seen in Fig. 2.2.

Most water molecules sit at the bottom rung – the vibrational ground state. In a Raman scattering event the molecule is excited to its first vibrational excited state and to conserve energy the photon gives up a quantum of energy equal to the vibrational energy spacing. Thus, by measuring the difference in energy between the photon before and after the scattering event, you can measure the vibrational

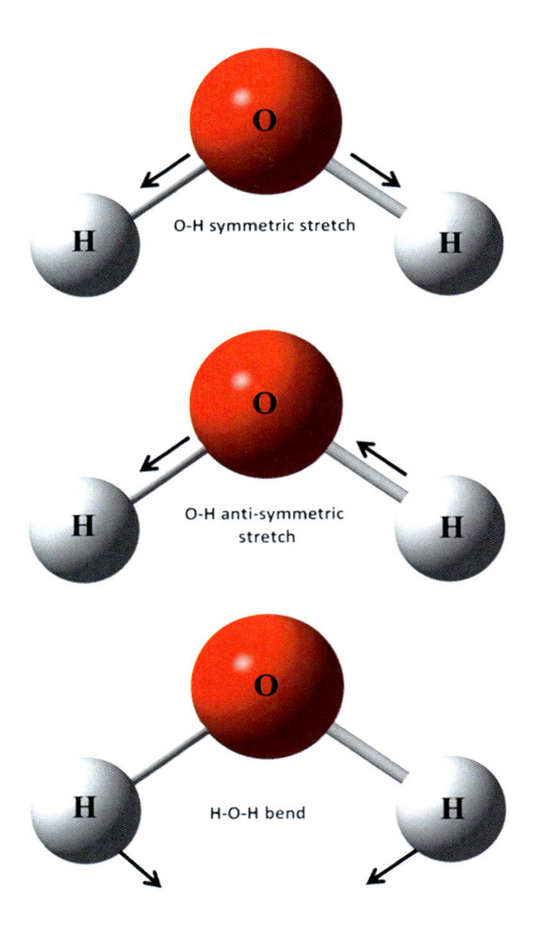

Scheme 2.1 Three modes of water vibrations

energy level spacing of the target molecule. This is the basis of Raman spectroscopy:

Raman measures the energies of molecular vibrations and since these are exquisitely sensitive to molecular identity, conformational state and interactions, the Raman spectrum is a source of chemical information.

A simple rule gives the number of independent vibrations for a given molecule. For a non-linear molecule with N atoms the molecule will have 3N−6 vibrations. Thus, water will have three vibrational modes. These are the symmetric and anti-symmetric stretches and the bending mode shown in Scheme 2.1.

The Raman spectrum of water is shown in Fig. 2.1; as predicted, it has three recognizable "peaks". These are the intense doublet near 3,500 cm^{-1}, that is due to the symmetric and anti-symmetric modes, and a weaker feature near 1,650 cm^{-1}, due to the H-O-H bending motion. The units encountered are usually cm^{-1}, which represents 1/wavelength of the photons. The cm^{-1} unit is directly proportional to the energy spacing of the rungs in the vibrational ladder

Scheme 2.2 Chemical structure of guanosine triphosphate

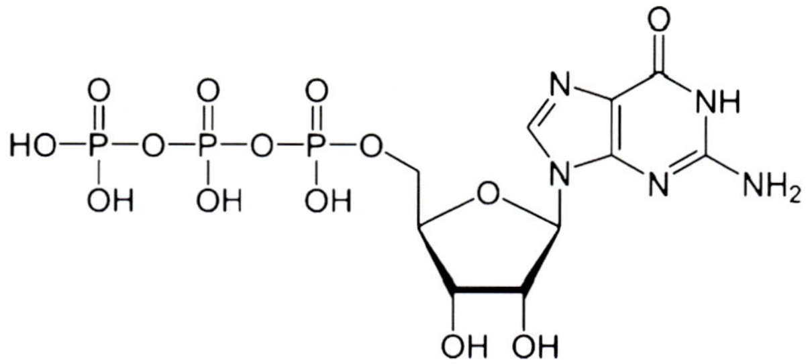

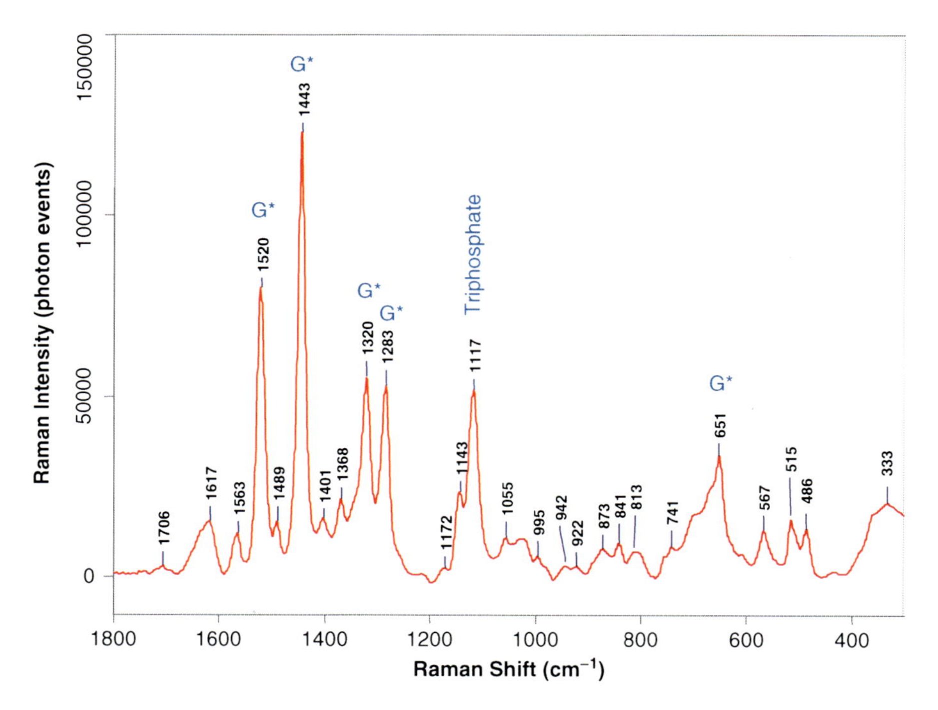

Fig. 2.3 Raman difference spectrum of 50 mM guanosine triphosphate (GTP) in water. G* noted in *blue* represent G ring mode vibrations from ^{13}C and ^{15}N GTP. Triphosphate vibrational mode is also labeled at 1,117 cm^{-1}

The 3N−6 rule means a large biological molecule consisting of thousands of atoms will have three times that number of vibrations. It is challenging to analyze such a vast array of vibrational data! As explained below we use precise Raman difference spectroscopy and the judicious use of isotope labeling to bring the problem to within tractable bounds.

To complete the introduction we will repeat the experiment shown in Fig. 2.1 but now introduce 50 millimolar guanosine triphosphate (GTP) into the optical cell (Scheme 2.2).

The portion of the Raman spectrum between 200 and 1,800 cm^{-1}, the so-called fingerprint region, is shown in Fig. 2.3. You can see many features due to the vibrations of the aqueous GTP. **This illustrates an important rule. The Raman spectrum of water is very weak and does not mask the Raman spectrum of the biological solute**. A big advantage in biochemistry! The Raman spectrum in Fig. 2.3 shows several narrow features due to vibrations of the G ring modes. These are minimally perturbed when the G ring changes its environment and can be used to follow changes in population of the G rings. The triphosphate of GTP has an intense vibration at 1,120 cm^{-1}, and as we will see, this can be used to follow reactions involving the triphosphate moiety. However, before we illustrate GTP reactions, we have to describe the Raman microscope.

2.2 Raman Microscope [3, 4]

A crucial advance in bringing dynamic studies to structural biology was the development of the Raman microscope. This has allowed us to follow enzyme kinetics, with their accompanying conformational changes, in single crystals on the time scale of 10's of seconds. A schematic of a Raman microscope is shown in Fig. 2.4.

Crystals are usually grown in, or transferred to, hanging drops in a 24-well plastic tray of the kind used by X-ray crystallographers. The tray is placed on the microscope stage and the crystal in each drop can be viewed *via* a long focal length objective and a video charge-coupled device (CCD) camera. A laser beam is introduced on the microscope's optical axis using a fiber optic, and the beam focused on the crystal can be viewed on the computer monitor. This provides the degree of experimental control needed since the crystals and focal spot are usually too small (on the 10–100 s of micrometers scale) to be viewed by the naked eye. In the next step the standard illuminating source is blocked and the back-scattered laser light from the focal spot goes back through the microscope, and *via* optical filters and a second fiber optic, is fed into a Raman spectrograph. The Raman

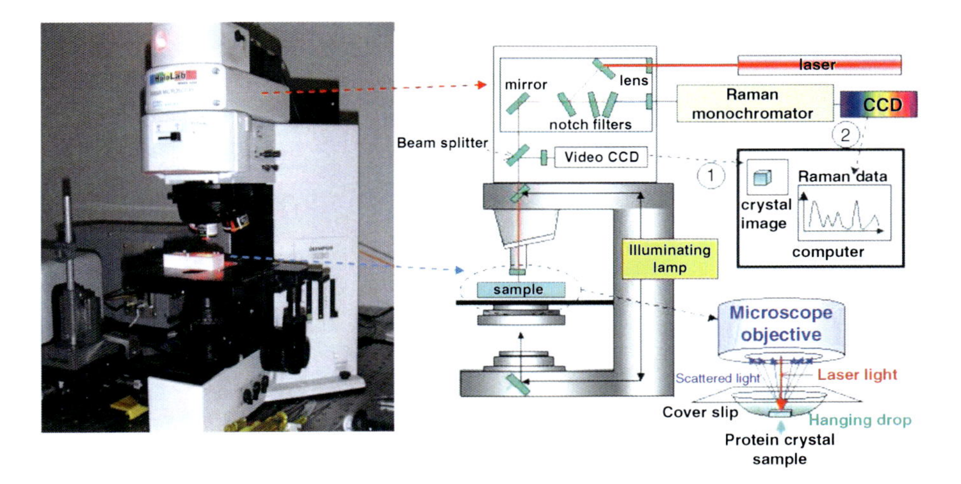

Fig. 2.4 Raman microscope

spectral image at the CCD photon-detector then appears on the computer monitor providing the Raman spectrum associated with the focal spot in the crystal. The focal volume is typically 20 μm in diameter and 30 μm in depth; these conditions determine a minimal crystal dimension of 30 μm for optimal spectral quality. For most systems it takes about 100 s to collect a complete data set, using 100 mW of 647.1 nm Kr^+ laser excitation at the sample. At these power levels the temperature of the crystal increases by about 15 °C when it sits in a 5 μl bath. While undertaking Raman difference spectroscopy, care has to be exercised to perturb the crystal as little as possible when introducing the ligand into the liquid surrounding the crystal. It is important to bear in mind that the intensity of Raman scattering may be dependent on the relative orientations of the crystal axes and the laser beam, a factor that can be used to facilitate the Raman difference method [5], but which can also lead to uncontrolled changes in Raman intensity, if not taken into account [6].

A comparison between the conditions found in solution and in single crystals demonstrates why superior Raman data are obtained from crystals:

- *Concentration*; depending on their molecular weight, proteins are usually in the 2–35 mM concentration range in crystals. In solution it is often difficult to achieve concentrations in the 0.5–1 mM range. Thus, crystals have an immediate concentration advantage up to a factor of 70.
- *Spectral background*; in aqueous solution water contributes a Raman spectrum that essentially forms a continuum in the fingerprint region. This is of weak intensity but becomes an issue at low solute concentrations. In crystals, this background decreases by an order of magnitude relative to the protein spectrum. In addition, aggregated protein in solution leads to strong elastic Rayleigh scattering which "bleeds over" into the Raman regime. Transparent crystals show less Rayleigh scattering. Finally, most proteins have a weak luminescent spectrum that helps obscure the Raman scattering. However, the luminescence is minimal when using red excitation for the Raman spectrum and appears to be partially quenched in crystals compared to solution.
- *Resolution*; in crystals, protein (and ligand) vibrational bandshapes narrow and a better baseline resolution is achieved compared to solution. This we ascribe to the damping of large scale protein vibrational motions in the crystal, due to protein-protein contacts.
- *Ligand concentration*; in solution, for Raman difference studies, protein must be greater than or equal to ligand concentration in order to avoid spectral interference from unbound ligand. A crystal may be soaked with a ligand that is 5 mM or more in the surrounding drop, which reaches 30 mM for the bound ligand in the crystal (if this is the concentration of active sites). Unbound ligand concentration in the crystal is typically 2 mM and gives only weak features at most. Thus, interference from unbound ligand is less of a problem in the crystal compared to solution studies.

Compared to solution studies, Raman crystallography has other advantages that are not related to data quality. These include the ability to undertake competitive inhibition experiments in the crystal, and the ability to quantitate the amount of ligand in the crystal and thus the number of ligands present per protein unit.

2.3 Raman Difference Spectroscopy and the Use of Isotope Labeling

Continuing with the theme of GTP we will see how we are able to follow the Raman spectrum of this molecule as it reacts in the active site of a very complex enzyme. The latter copies a DNA sequence to make the corresponding RNA chain and is called a RNA polymerase (RNAP). The enzyme has ~3,400 amino acids and binds the DNA and RNA sequences seen in Scheme 2.3

As can be seen in Scheme 2.3, in the reaction GTP forms a covalent bond to a cytidine at the 3′ end of the growing RNA chain. The complex has a molecular weight of about 400,000 Daltons. Since there are approximately 40,000 atoms we know the molecule must have approximately 3N or 120,000 vibrations. In order to selectively observe the Raman spectrum of the GTP during the reaction we carry out difference spectroscopy using the Raman microscope. First we obtain the Raman spectrum of a single crystal of RNAP, then we inject GTP into the drop surrounding the crystal. The GTP soaks into the crystal and the reaction occurs at the enzyme's active site. Then using software we subtract the two spectra. This process is illustrated in Fig. 2.5.

As can be seen in Fig. 2.5 the two spectra appear virtually identical, but by careful subtraction and intensity amplification we obtain the difference seen in trace C). The GTP is fully labeled with ^{13}C and ^{15}N isotopes and this shifts the vibrations of the G ring vibrations to unique positions, removing possible confusion with the unlabeled Gs in the DNA and RNA strands seen in Scheme 2.3. Assignments of many of the features in the difference spectrum can be made using the Raman spectrum of aqueous labeled GTP and the literature on the Raman spectra of proteins and nucleic acids. Key assignments are:

- Ring vibrations from the labeled GTP depicted as *G in Fig. 2.5
- Stretching vibrations of the DNA and RNA backbones near 800 cm^{-1}, these are sensitive to backbone conformation
- Amide group modes from the protein near 1,655 cm^{-1}, these monitor protein secondary structure
- Triphosphate vibration from the GTP near 1,120 cm^{-1}

All these features can be followed as a function of time to provide unique kinetic data as the reaction occurs in the crystal. A cartoon of the kinetics for the addition of one GTP to the chain is shown in Fig. 2.6.

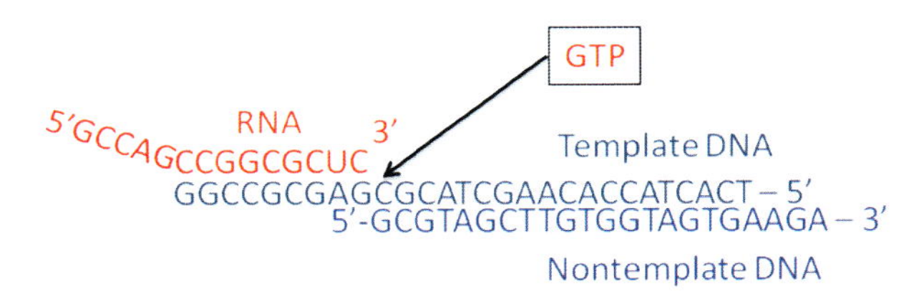

Scheme 2.3 Nucleic acid scaffold composed of a 19 base-pair DNA duplex and 9 base-pair RNA/DNA hybrid.

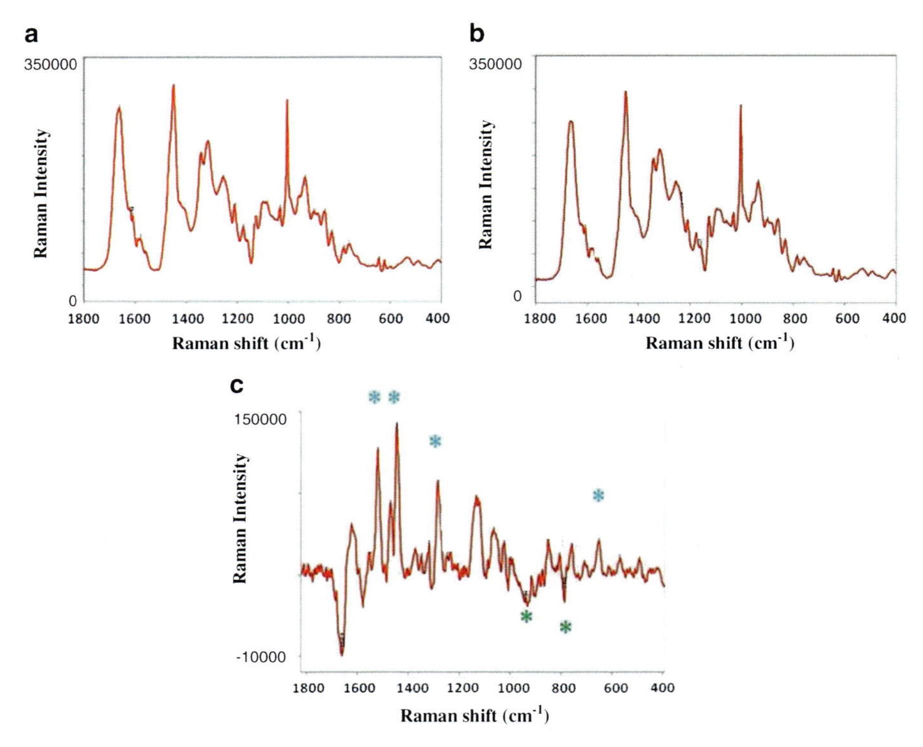

Fig. 2.5 Applying the Raman difference technique. Example: Add $^{13}C/^{15}N$ GTP (GTP*) to a crystal (**a**) and subtract the Raman spectrum of the crystal (without GTP) (**b**); this yields the Raman difference spectrum showing the diffusion of GTP* into the crystal (**c**). Positive Raman peaks are due formation or appearance of species during the *in crystallo* reaction (see G* ring modes labeled in *blue* asterisks). Negative Raman peaks reflect disappearance or perturbation to the system. See *green* asterisks which are mainly due to protein conformational change and nucleic acid backbone vibrations

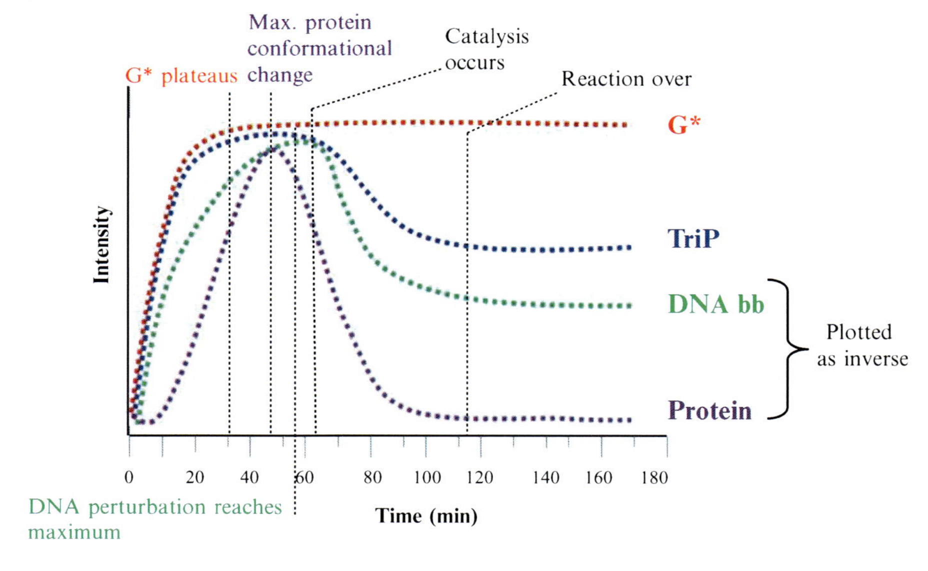

Fig. 2.6 Cartoon of GTP* soak-in

Figure 2.6 shows that GTP soaks into the crystal and the population remains constant after 20 min. The RNA/DNA backbones change over about 1 h and this is accompanied by a large reversible protein conformational change. The population of the triphosphate decreases from 60 to 100 min showing that GTP incorporation is occurring. There is a wealth of mechanistic information that we do not have space to develop here but the salient feature is that it is possible to obtain detailed molecular and kinetic information using the "Raman crystallographic" approach even for a very complex system.

2.4 Pros and Cons of Raman Crystallography

It can be applied to a wide range of protein and nucleic acid crystals. However, in some case coloured co-factors may cause problems if they fluoresce.

Raman is much less operator friendly compared to X-ray or NMR. Considerable patience is needed to acquire data and experience and skill are needed to interpret the data. Interpretation of Raman data can be aided by isotopic substitutions and by quantum mechanical calculations – for which software packages are available.

Raman and X-ray crystallography are a powerful synergistic combination. However, sometimes they diverge!

The intrinsic time scale of Raman scattering is in the femto-second (10^{-15} s) range. Thus, the kind of line broadening effects seen in NMR is usually absent. However, it takes 10s of seconds to acquire a Raman data set from a crystal, limiting the study to slow reactions.

2.4.1 The Synergy Between Raman and X-Ray Crystallography

X-ray structures can be an invaluable aid to interpreting data from Raman crystallographic experiments. In the RNAP experiments outlined above the structural data [7] provide the background for understanding the Raman kinetic data. In turn, the kinetic data shown in the cartoon, Fig. 2.6, tell the X-ray crystallographer the conditions, in particular soak-in times, for trapping the most interesting reaction intermediates. This approach has already borne fruit for an RNAP that catalyzes the initiation step in RNA chain formation [8, 9]. The Raman kinetic data [8] were used to identify the time domain for trapping reaction intermediates as catalysis is occurring [9].

While the importance of this approach cannot be overemphasized, the bad news is that sometimes the X-ray/Raman relationship breaks down. As part of an approach to understand the molecular basis of bacterial resistance to penicillin-like compounds our group has carried out extensive single crystal experiments following the reactions between clinical drugs and bacterial enzymes called β-lactamases. The latter are produced by bacteria to hydrolyze incoming penicillin and render it ineffective in killing the cells.

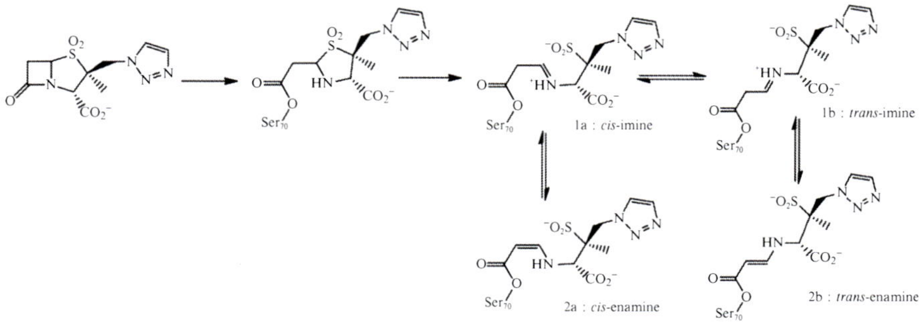

Scheme 2.4 Partial reaction scheme for tazobactam and class A β-lactamase enzymes.

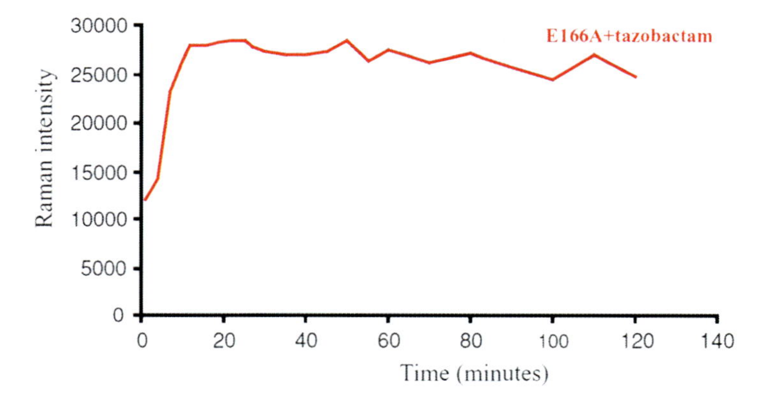

Fig. 2.7 Time dependence of enamine peak height near 1,593 cm^{-1} (normalized to the amide I band) for the E166A crystal with tazobactam soaking in

Tazobactam, a potential β-lactamase inhibitor, is shown in Scheme 2.4, where it undergoes a complex reaction with the enzyme that involves a serine in the active site. Our initial studies [10] used a simple mutant form of the enzyme, E166A, chosen to trap the covalent intermediates shown in Scheme 2.4. The Raman difference spectrum obtained by soaking tazobactam into a single crystal of E166A SHV-1 β-lactamase showed the presence of a *trans*-enamine like species, 2b, in Scheme 2.4 that grew in with time and reached a plateau at about 20 min (Fig. 2.7).

Our crystallographer colleagues reproduced our soak conditions and solved the first "clean" structure of a complex with 100 % occupancy in the active site. Fortunately the Raman interpretation was confirmed and the enzyme complex is indeed a *trans*-enamine [11].

By studying the first round of crystal structures our X-ray colleagues could design a more potent inhibitor with the expectation that could lead to compounds with improved pharmacological properties [12]. However, the news is not always good. Novel compounds have to block the normal form of enzyme (the so-called wild type), lacking the E166A mutation, in bacterial cells. For wild-type enzyme,

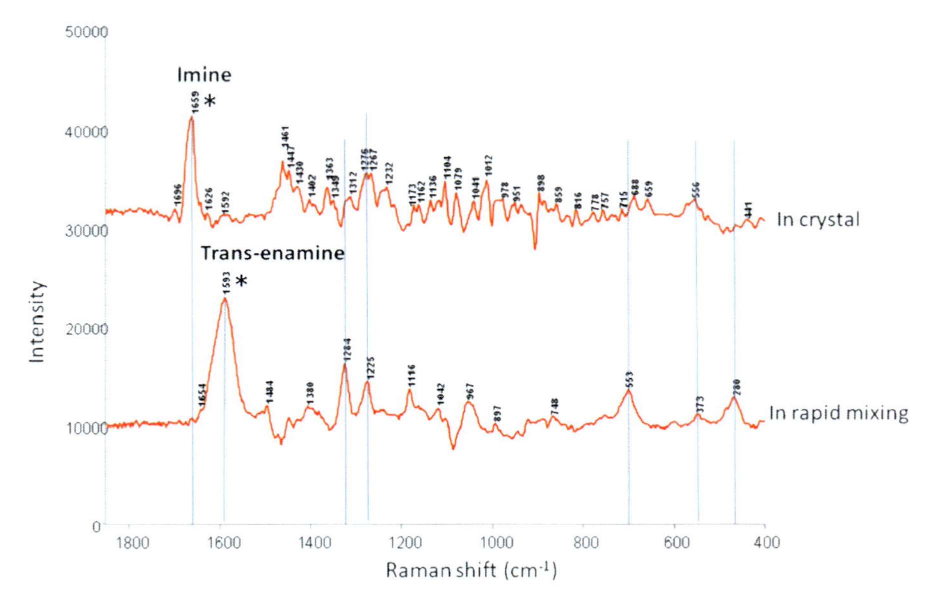

Fig. 2.8 Comparison of tazobactam reacting with CTX-M-9 in single crystal and in solution. Imines and enamines are shown in Scheme 2.4

Raman studies detected some of the enamine and imine species seen in Scheme 2.4 (1a, 1b, 2a and 2b), when tazobactam was soaked into crystals of the WT SHV-1 β-lactamase [13, 14]. However, the crystallographers were not able to trap these intermediates by flash freezing for detailed structural analysis. This is almost certainly because the Raman data are collected at room temperature whereas the for X-ray analysis the crystal is treated with a cryoprotectant and then frozen in liquid nitrogen. We believe that the latter process perturbs the kinetic for acylation and deacylation leading to the rapid hydrolysis of the intermediates.

It is likely that this situation occurs mostly for reaction schemes that are complex involving large conformational changes of the substrate and that contain branch points leading to a perturbation of intermediate populations. However, we have recent data which suggest that an X-ray structure could lead to an erroneous conclusion about which species is inhibiting the enzyme. In this instance the species seen in the crystal is very different to that detected in solution and any modeling to create a better inhibitor based on the crystal structure would not reflect the situation for the reaction in solution.

The example is another β-lactamase reaction involving an enzyme recently discovered in the clinic, namely CTX-M-9. This undergoes the reaction with tazobactam seen in Scheme 2.4.

A recent advance in our lab is that we can rapidly mix enzyme and substrate in aqueous solutions and trap the resulting intermediates by flash freezing. This enables us to obtain the Raman spectra of intermediates at 25 milliseconds and longer after mixing. Figure 2.8 compares the Raman spectra of the reaction in a single crystal of the CTX-M enzyme after reacting with tazobactam, soaked in for 15 min, with the spectrum of a trapped intermediate 25 milliseconds after mixing in solution. The spectra are strikingly disparate. In the crystal the intense band at 1,659 cm^{-1} is due to an imine (e.g., species 1a and 1b in Scheme 2.4) whereas the species seen from the mix in aqueous solution is an enamine (species 2a and 2b in Scheme 2.4), as evidenced by the intense peak at 1,593 cm^{-1}. An X-ray structure of the crystal form would not be a good starting point for drug design since it is likely that *in vivo* CTX-M is blocked by the enamine and not the imine.

Acknowledgements I am grateful to Tao Che, Hossein Torkabadi and Yianna Antonopoulos for assistance in preparing this article. Our research is supported by NIH GM 54072.

References

1. Carey PR (1982) Biochemical applications of Raman and resonance Raman spectroscopies. Academic, New York
2. Schrader B (1995) Infrared and Raman spectroscopy: methods and applications. VCH, Weinheim
3. Carey PR (2006) Raman crystallography and other biochemical applications of Raman microscopy. Annu Rev Phys Chem 57:527–554
4. Gong B, Chen JH, Yajima R, Chen YY, Chase E, Chadalavada DM, Golden BL, Carey PR, Bevilacqua PC (2009) Raman crystallography of RNA. Methods 49:101–111
5. Altose MD, Zheng YG, Dong J, Palfey BA, Carey PR (2001) Comparing protein-ligand interactions in solution and single crystals by Raman spectroscopy. Proc Natl Acad Sci U S A 98:3006–3011
6. Tsuboi M, Thomas GJ (1997) Raman scattering tensors in biological molecules and their assemblies. Appl Spectrosc Rev 32:263–299
7. Vassylyev DG, Vassylyeva MN, Zhang J, Palangat M, Artsimovitch I, Landick R (2007) Structural basis for substrate loading in bacterial RNA polymerase. Nature 448:163–168
8. Chen Y, Basu R, Gleghorn ML, Murakami KS, Carey PR (2011) Time-resolved events on the reaction pathway of transcript initiation by a single-subunit RNA polymerase: Raman crystallographic evidence. J Am Chem Soc 133:12544–12555
9. Basu RS, Murakami KS (2013) Watching the bacteriophage N4 RNA polymerase transcription by time-dependent soak-trigger-freeze X-ray crystallography. J Biol Chem 288(5):3305–3311
10. Helfand MS, Totir MA, Carey MP, Hujer AM, Bonomo RA, Carey PR (2003) Following the reactions of mechanism-based inhibitors with ß-lactamase by Raman crystallography. Biochemistry 42:13386–13392
11. Padayatti PS, Helfand MS, Totir MA, Carey MP, Hujer AM, Carey PR, Bonomo RA, van den Akker F (2004) Tazobactam forms a stoichiometric trans-enamine intermediate in the E166A variant of SHV-1 beta-lactamase: 1.63 A crystal structure. Biochemistry 43:843–848
12. Padayatti PS, Sheri A, Totir MA, Helfand MS, Carey MP, Anderson VE, Carey PR, Bethel CR, Bonomo RA, Buynak JD, van den Akker F (2006) Rational design of a beta-lactamase inhibitor achieved via stabilization of the trans-enamine intermediate: 1.28 A crystal structure of wt SHV-1 complex with a penam sulfone. J Am Chem Soc 128:13235–13242
13. Kalp M, Totir MA, Buynak JD, Carey PR (2009) Different intermediate populations formed by tazobactam, sulbactam, and clavulanate reacting with SHV-1 beta-lactamases: Raman crystallographic evidence. J Am Chem Soc 131:2338–2347
14. Carey PR, Chen Y, Gong B, Kalp M (2011) Kinetic crystallography by Raman microscopy. Biochim Biophys Acta 1814:742–749

Chapter 3
Structure and Dynamics from Time Resolved Absorption and Raman Spectroscopy

Siva Umapathy, Khokan Roy, Surajit Kayal, Nilesh Rai, and Ravi Kumar Venkatraman

3.1 Introduction

Spectroscopy is the study of interaction between electromagnetic radiations with matter. To understand the architecture of atoms and molecules we use many spectroscopic techniques. For example, X-rays are used to determine the crystal structure of molecules, while vibrational spectroscopy is used to study molecular structures. When a photon is incident on a molecule, the electromagnetic field will distort the charge (electron density) cloud around the molecule. In that process it can be absorbed or scattered by the molecule giving rise to different spectroscopic processes. When the frequency of the incident field ν_i, matches with the frequency of the energetics of the molecule ν_m, resonance is established and electromagnetic radiation is absorbed. In UV-Visible spectroscopy the incident radiation promotes the molecule from the ground electronic state to higher electronic states, providing information about the electronic energy levels. The nature of chemical bonding in molecules can be probed by infrared or Raman spectroscopy. The shape of the electronic absorption band is associated with vibrational transitions coupled to the electronic excitation are functions of equilibrium bond length. To understand the vibrational structure in the electronic spectra of molecules we apply the Franck-Condon principle. It states that as the nuclei are so much heavier than electrons, electronic transition takes place much faster than the nuclei can respond [1]. During the electronic excitation the initially stationary nuclei instantly experience a new force field. To adjust within the new force field, the molecule starts to vibrate. The equilibrium separation in the ground state becomes the turning point in the excited electronic state, this is known as the vertical transition. The relative displacement of the electronic potential energy determines the vibrational structure in the electronic spectrum, and the excited state potential is normally more displaced than the ground state potential. The separation

S. Umapathy (✉) • K. Roy • S. Kayal • N. Rai • R.K. Venkatraman
Department of Inorganic and Physical Chemistry, Indian Institute of Science, Bangalore, 560012, India
e-mail: umapathyiisc@gmail.com

J.A.K. Howard et al. (eds.), *The Future of Dynamic Structural Science*, NATO Science for Peace and Security Series A: Chemistry and Biology, DOI 10.1007/978-94-017-8550-1_3,

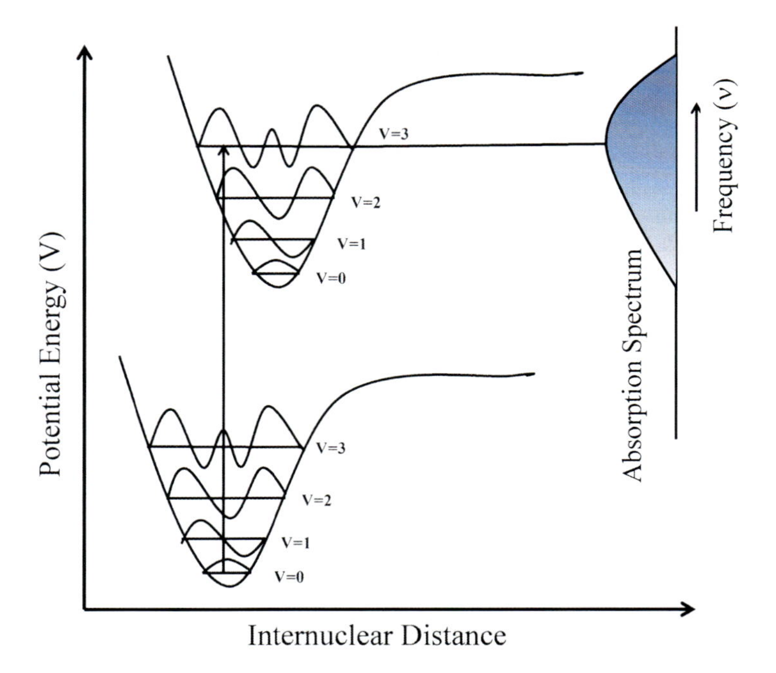

Fig. 3.1 Representation of electronic absorption and Franck-Condon principle

of the vibrational lines depends on the vibrational energies of the excited electronic state. By applying quantum mechanics we can have more insight into the Franck-Condon Principle. In quantum mechanics we say that during electronic transition nuclei retain their initial dynamic state. This dynamic state is represented by wavefunction ψ. So the nuclear wavefunction does not change during electronic transition. At normal temperature most of the molecules are in the lowest vibrational state of the ground state potential energy function. The information about the distribution of molecules can be obtained by Maxwell-Boltzmann distribution formula [2]:

$$\frac{N}{N_o} = e^{-\Delta E/RT}$$

(where N_o, N are the population in the ground state and excited state, ΔE is the energy difference between the two energy states).

The V = 0 vibrational level in the ground electronic state has a maximum in the centre, indicating the region of maximum probability. So after photo-excitation the most probable transition occurs from the V = 0 vibrational level (Fig. 3.1 [3]). The time taken for the electronic transition is of the order of 10^{-15} s, while the time period for a vibration is around 10^{-12} s, which is 1000 times slower than the electronic transition. As a consequence the internuclear distance does not change during the absorption of light, hence the transition process can be represented by the vertical transition which is parallel to the potential energy axis, originating from the lower potential curve to the upper curve. This essentially says that it is difficult to convert rapidly electronic energy into vibrational kinetic energy and the most probable transitions between different electronic and vibrational levels are those for which the momentum and position of the nuclei do not change very much. The quantitative form of the Franck-Condon principle can be obtained by applying quantum mechanics through the transition dipole moment. The transition dipole moment operator is a sum over all nuclei and electrons in the molecule:

$$\vec{\mu} = -e\sum_i r_i + e\sum_I Z_i \, \boldsymbol{R}_I$$

(where the vectors are the distances from the centre of the charge of the molecule).

The intensity of the transition is proportional to the square modulus, $|\boldsymbol{\mu}|^2$, of the transition dipole moment. The overall state of the molecule consists of an electronic part, ψ_e, and a vibrational part, ψ_v. So according to Born-Oppenheimer approximation, the transition dipole moment factorizes as follows [4]:

$$\begin{aligned}\boldsymbol{\mu}_{\mathrm{fi}} &= \int \psi^*_{s_f} \, \psi^*_{v_f} \left\{-e\sum_i r_i + e\sum_I Z_i \boldsymbol{R}_I\right\} \psi_{s_i} \, \psi_{v_i} d\tau \\ &= -e\sum_i \int \psi^*_{s_f} \, r_i \, \psi_{s_i} d\tau_s \int \psi^*_{v_f} \, \psi_{v_i} d\tau_v + e\sum_I Z_i \int \psi^*_{s_f} \, \psi_{s_i} d\tau_s \int \psi^*_{v_f} \, \boldsymbol{R}_I \, \psi_{v_i} d\tau_v\end{aligned}$$

The second term on the right hand side is zero, due to the orthogonality of two different electronic state. So the expression for the transition dipole moment becomes

$$\begin{aligned}\boldsymbol{\mu}_{\mathrm{fi}} &= -e\sum_i \int \psi^*_{s_f} \, r_i \, \psi_{s_i} d\tau_s \int \psi^*_{v_f} \, \psi_{v_i} d\tau_v \\ &= \mu_{S_{f\varepsilon v}} S(v_{\mathrm{f}}, v_{\mathrm{j}})\end{aligned}$$

(where the matrix element $\mu_{S_{f\varepsilon v}}$ is the electronic-transition dipole moment operator. S $(v_{\mathrm{f}}, v_{\mathrm{i}})$ is the overlap integral between the vibrational states ψ_{v_i} in the lower electronic state and the ψ_{v_f} in the final electronic state).

From the above equation we can see that intensity of a transition depends on the overlap integral also known as Franck-Condon Integral.

From the classical electromagnetic theory, the polarization induced in the matter due to the external field is obtained by the vector sum of the individual induced dipoles divided by the volume, this is called the polarization density. The relation between the incident field strength and the polarization density (linear approximation) of the matter is given by the following equation,

$$\overline{P}(v) = \{\varepsilon_r(v) - 1\}\varepsilon_0 \overline{E}(v); \quad \varepsilon_r = \varepsilon_r^{'} + i\varepsilon_r^{''}$$

(where $\overline{P}$ – Polarization density of matter, ε_r – Relative permittivity/Dielectric constant of the matter, ε_0 – Permittivity of the vacuum, $\overline{E}$ – Electric field).

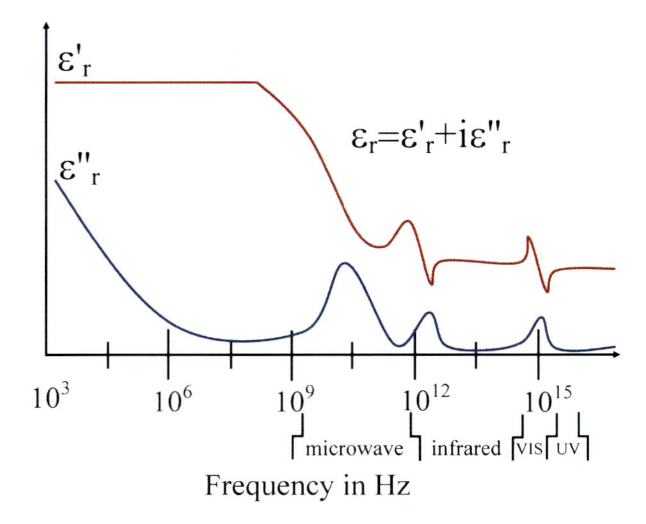

Fig. 3.2 Real and imaginary dielectric constant as a function of frequency

The dielectric constant is a function of frequency of the electromagnetic radiation and it is a complex function. The real function $\varepsilon_r^{'}$ gives dispersion while the imaginary function $\varepsilon_r^{''}$ gives the absorption of the matter as a function of frequency as shown in Fig. 3.2. At different region of electromagnetic spectrum like Microwave, IR and UV-VIS, different molecular motions like orientation, distortion and electronic polarization respectively are in resonance [5]. In general, electronic transitions of most of the organic molecules fall in the UV-VIS region of electromagnetic spectrum. For a transition to be allowed there should be an oscillating dipole moment corresponding to that of the molecular motion in resonance with the frequency of the incident field of the electromagnetic radiation. This is known as transition dipole moment.

3.1.1 Photophysical Processes [6]

During an electronic transition, the molecules nuclei positions are fixed, this is because, the masses of the nuclei are heavier than those of the electrons; this is known as the Franck-Condon principle. As a result of the Franck-Condon principle, the higher vibrational levels of various modes of the singlet electronic excited state are populated during an electronic transition. The higher vibrational level of the singlet electronic excited states then relaxes back to the ground vibrational level of the electronic excited state with the same multiplicity by Internal Conversion (IC) or to a different multiplicity (triplet) by Inter System Crossing (ISC). Generally emission starts from the lowest electronic excited state to the ground state and is called Kasha's rule [7], albeit there are some exceptions [8]. The lowest singlet

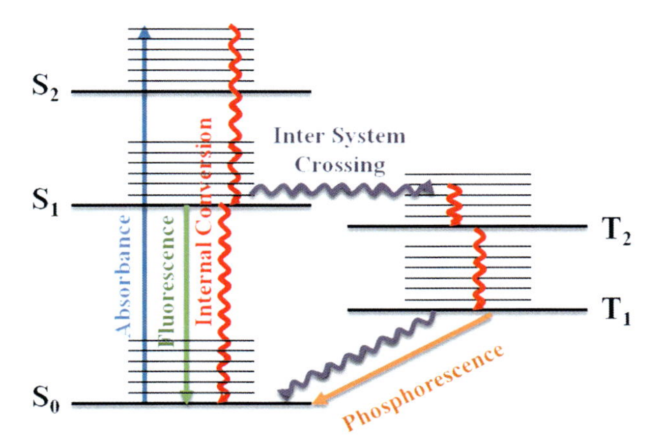

Fig. 3.3 Jablonski diagram

electronic excited state relaxes to ground state, either radiatively by fluorescence or non-radiatively by IC and similarly the triplet electronic excited state relaxes to ground electronic state, either radiatively by phosphorescence or non-radiatively by ISC as shown in the Fig. 3.3.

3.1.2 Photochemical Processes

Despite the photophysical processes, the excited state can take part in chemical reactions, with reactions starting from either singlet or triplet excited state. In general bimolecular photochemical reactions start from triplet states rather than from singlet states, since triplet states are long-lived they undergo more effective collisions with reactants to give products. Photochemical reactions are faster compared to thermal reactions because the activation energy (E_a) needed for the photochemical reactions is much less than that required for thermal reactions and the enthalpy change of the reactions becomes favourable as illustrated in Fig. 3.4, for hydrogen atom abstraction reaction [9]. There are a range of photochemical reactions including, isomerisation along >C=C< bond, Norrish type I and II reactions, etc. [10, 11].

3.1.3 Photoinduced Electron Transfer Reactions (PIET)

Electronic excitation of a molecule increases the Electron Affinity (EA) and decreases the Ionisation Potential (IP) of the molecule as depicted in the Fig. 3.5. Thus molecules in the excited state show enhanced reactivity towards electron transfer reactions (redox reactions) which are known as PIET reactions [12, 13].

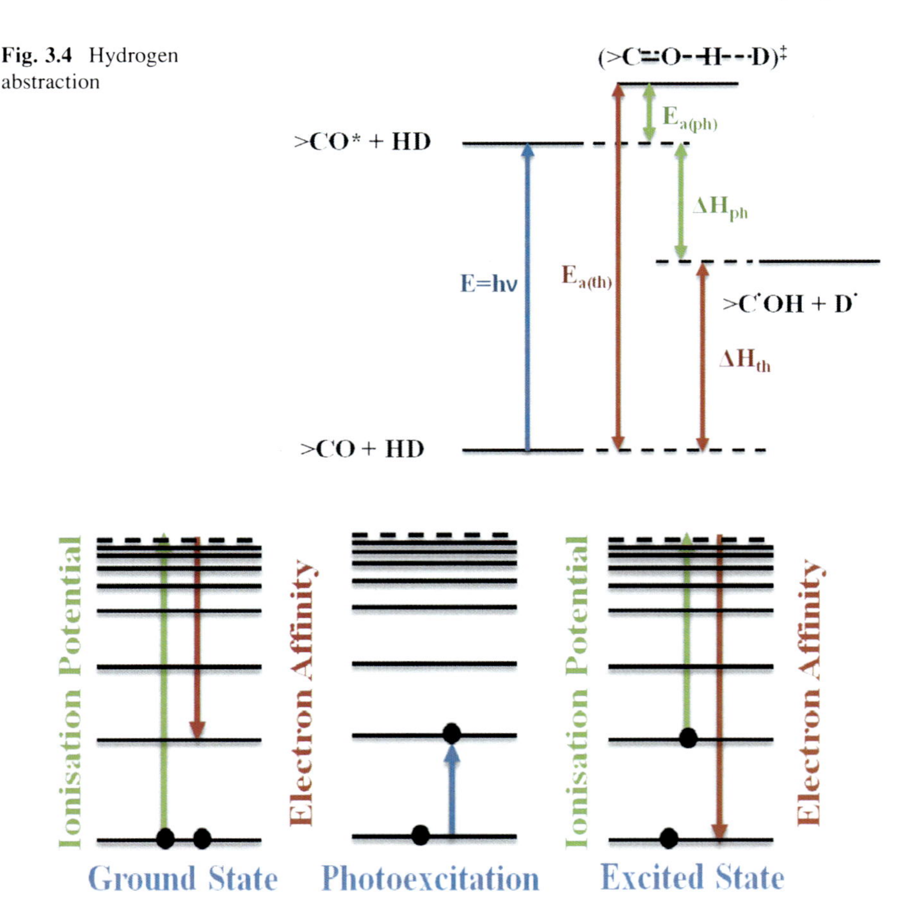

Fig. 3.4 Hydrogen abstraction

Fig. 3.5 IP and EA before and after photoexcitation

3.2 Time Resolved Experiments

In many of the organic chemical reactions, the intermediates and the transient species involved are short-lived and could not be detected by steady state spectroscopy. However time resolved spectroscopy (TRS) can give us insight into the electronic and structural aspects of the intermediates/transient species involved in organic chemical reactions. Time resolved spectroscopy is basically a pump-probe experiment [14], with the pump pulse initiating the reaction while the probe beam, probes the intermediates/transient species formed during the course of the photo-initiated chemical reaction. The pump pulse could initiate chemical reactions by photolysis, radiolysis, temperature-jump, stopped-flow methods, etc. and then the probe comes at various delays from the pump to give the time resolved information of the probed species. In photolysis, both the pump and probe beam are light pulses, typically on the ns-fs time scale. The pulse width of the pump and probe beam are

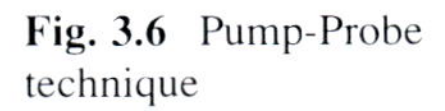

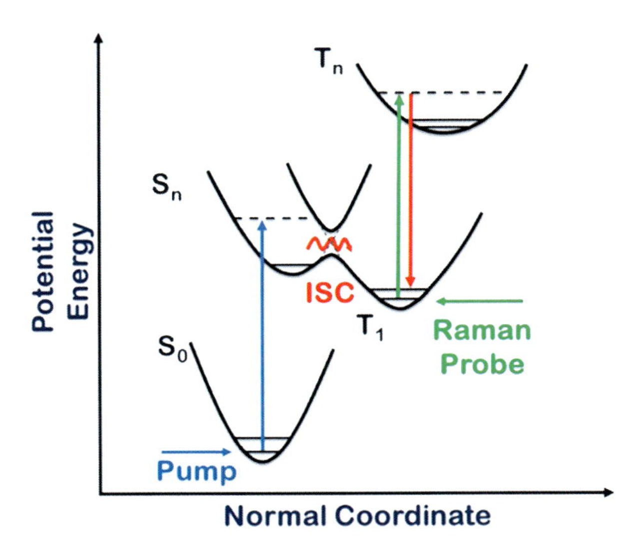

Fig. 3.6 Pump-Probe technique

chosen such that they are narrower than the events that need to be examined. Depending upon the nature of the probe beam and detection method, they can be classified as time resolved UV-VIS and IR absorption, time resolved Raman, time resolved emission spectroscopy, etc.

Time resolved Raman and IR absorption spectroscopy give us insight into the structure of the intermediate/transient species. Time resolved Raman spectroscopy is a scattering technique. Since many of the bimolecular photochemical reactions take place from triplet states, ns-TRS can be used to probe the intermediates/transient species formed during the course of the photochemical processes. ns-LASER flash photolysis (LFP) is an absorption technique used to elucidate the electronic structure of the triplet states or the intermediate species which have a lifetime in the range of few hundreds of ns to μs. ns-time resolved resonance Raman spectroscopy (TR^3s) is also a pump-probe technique, in which the pump beam wavelength must be in resonance with steady-state UV-VIS absorption spectrum to create a singlet electronic excited state population which then undergoes various photophysical processes as shown in Fig. 3.3. In TR^3s the probe beam must be in resonance with a triplet-triplet/intermediate absorption spectrum obtained from LFP. Resonance Raman enhances the Raman activity of certain modes by a factor of up to 10^6 relative to that of the normal Raman. Thus combining LFP with TR^3s can give us both structural and electronic aspects of the triplet state/intermediate species generated by photolysis, a schematic of TR^3s is given in Fig. 3.6.

3.2.1 *Experimental Setup*

The third harmonic (355 nm) from GCR-250, operating at 10 Hz, spectra-physics is used to pump the Optical Parametric Oscillator which can generate wavelengths from 410 to 709 nm with time resolution of a few ns and it is generally used as the Raman probe. Similarly, either second harmonic (266 nm) or third harmonic (355 nm) from INDI LASER, operating at 10 Hz, spectra-physics is generally used as pump. The time delay between pump and probe pulses are controlled by a delay generator from Stanford Research Systems, Inc. The pump and probe are made collinear by using a dichroic mirror (pump wavelength) and a 90° scattering geometry. In an LFP setup, either the third harmonic or second harmonic from a DCR-11 is used as the pump, while a pulsed Xenon arc lamp which is synchronized with a LASER pulse is used as a white light probe. The pump and probe are in perpendicular excitation arrangement and the transmitted white light are dispersed and focused on the exit slit of the Monochromator. The transmitted light intensities are detected by a 5-stage photomultiplier tube (PMT). The signals from the PMT are digitized by an oscilloscope which operates at 500 MSa/s and then stored onto a CPU for further data processing.

3.3 TR^3 Spectroscopic Studies of Transients Involved in the Photochemical Reactions

TR^{3}s is a highly selective and sensitive probe for studying the structure of the intermediates formed during photochemical reactions. Fluoranil (perfluoro-p-benzoquinone) is of interest to photochemists because of its unique structure-reactivity correlation. The photochemistry of fluoranil [15] involves various intermediates, e.g. the excited state complex, ketyl radicals and radical anions, the UV-VIS spectra for each of these are very different. The pump wavelength, 355 nm is in resonance with the ground state absorption spectrum and by choosing the probe wavelength in resonance with the absorption spectrum of the intermediates, TR^{3}s can probe various intermediates formed during the photolysis. The TR^3 spectra of the fluoranil triplet state in CH_3CN and $CHCl_3$ show considerable differences in their spectra which could be due to inversion of the $n\pi^*$ and $\pi\pi^*$ triplet states with solvent polarity [15]. By choosing a probe wavelength at 416 nm, which is in resonance with both ketyl radical and radical anion of fluoranil, it is possible to probe the structure of those intermediates. The TR^3 spectrum with a probe wavelength at 416 nm was compared with the RR and FTIR spectra of the fluoranil radical anion (generated by chemical method) and suggested that the observed TR^{3}s bands are due to ketyl radical.

Another example of the study of reaction intermediates, occurs for the menaquinone (2-methyl-1, 4-napthaquinone) radical anion [16] which is an active intermediate in various biological redox reactions, and understanding the structure of this intermediate will help in elucidating the mechanism of those processes. The TR^{3}s of the napthoquinone and menaquinone triplet state and radical anion intermediates have been carried out, [16] in order to examine the effect of asymmetric substitution of the methyl group on the radical anion structure. The ab initio computational calculation suggests that there is more coupling of the C=C stretching modes with C=O modes. The assignment of the modes of these radical anions helps us to identify the vibrational frequencies observed in Raman studies of photosynthetic systems.

3.3.1 Effects of Solvent on Photoinduced Geminate Ion Recombination Process

The effects of solvent polarity on the fluoranil and durene (tetramethyl-benzene) ion pair generated by photolysis was analysed by TR^3s [17]. A change in the vibrational frequencies with respect to solvent polarity was observed for C=C stretching mode. In polar solvents the most intense band at 1,667 cm^{-1} corresponds to that of the Solvent Separated Ion Pairs (SSIP) and a less intense band is due to the contact ion pair (CIP). In moderate and non- polar solvents (binary solvent mixture), three peaks corresponding to CIP, SSIP and the ketyl radical were observed by curve fitting. As the polarity of the solvent decreases the intensity of the band corresponding to SSIP decreases while that of CIP and ketyl increases. This suggest that in polar solvents, the charges are stabilised by the solvent and form solvated ions while in the non-polar solvents, the ion pairs formed are stabilised through electrostatic attraction with the oppositely charged species.

3.3.2 Solvent Induced Structural Changes in the Excited State

The effects of solvent polarity on the fluoranil triplet excited state were studied by using TR^3s [18], this resulted in Raman spectra which differed in the intensities of certain vibrational frequencies [18]. Based on the experimental data and computational studies, it is concluded that in non-polar solvents the fluoranil triplet excited state has a more distorted structure (chair form) while in polar solvents it achieves a planar structure. The overtone progressions of the ring bending mode, the 418 cm^{-1} peak and alternate intensities in non-polar solvents suggest that there is a double-well potential, i.e. two chair forms of triplet fluoranil which could interconvert via a planar structure.

3.4 Introduction to Stimulated Raman Spectroscopy

The vibrational spectral features of a molecule are closely related to its electronic structure, so we have to know how the vibrational and electronic structures are related. The simplest way to explain the vibrational energy is through a harmonic oscillator where the energy spacing hν, between the energy levels is fixed. But for real molecules the vibrational potential energy is best explained by Morse potential function given by equation

$$V = D_e\left(1 - e^{-\beta q}\right)^2$$

(where D_e is the dissociation energy; β is measure of curvature of the potential).

Both infrared and Raman spectroscopy probe vibrational energy levels in a molecule. IR spectroscopy is an absorption technique while Raman is an inelastic scattering phenomena. If a molecule is irradiated by monochromatic radiation by frequency ω_1, most of the radiation is scattered at frequency ω_1 (Rayleigh scattering), some radiation is scattered inelastically at a frequency $\omega_1 \mp \omega_m$, where ω_m is the vibrational frequency of the molecule. Theoretically the intensity of the Raman band between the energy levels $|i>$ and $|f>$ is given by the following equation

$$I_{fi} = constant.\ I_0(\mathrm{v}_1 \mp \mathrm{v}_{fi})^4 \sum\nolimits_{\rho\sigma} \left[\alpha_{\rho\sigma}\right]_{fi} \left[\alpha_{\rho\sigma}\right]_{fi}^*$$

(where I_0 is the intensity of the incident radiation, $[\alpha_{\rho\sigma}]_{fi}$ is the $\rho\sigma^{th}$ element of the transition polarizability tensor).

3.4.1 *Stimulated Raman Spectroscopy*

As Raman scattering is a weak phenomena, the weak Raman signals are overwhelmed by fluorescence background and the time resolution in normal Raman scattering is of the order of picoseconds. So in order to achieve high time and spectral resolution many non-linear spectroscopic technique has been developed. Stimulated Raman spectroscopy is one of the third order nonlinear spectroscopic techniques that can probe the dynamics of a molecule on an ultrafast time scale. Here three ultrashort pulses interact with the material system to give vibrational information.

3.4.2 *Description of Stimulated Raman Spectroscopy (SRS)*

In spontaneous Raman spectroscopy a monochromatic pump beam ω_P is incident on a sample and Stokes ω_s and anti-Stokes ω_{as} photons are created. Stokes Raman is accompanied by loss in photon energy of the incident pump beam, whereas a gain in the photon energy is observed in anti-Stokes scattering. Stimulated Raman is a third order (four wave mixing) process in which three ultrashort pulses interact with the system giving a fourth signal field, this signal corresponds to the third order

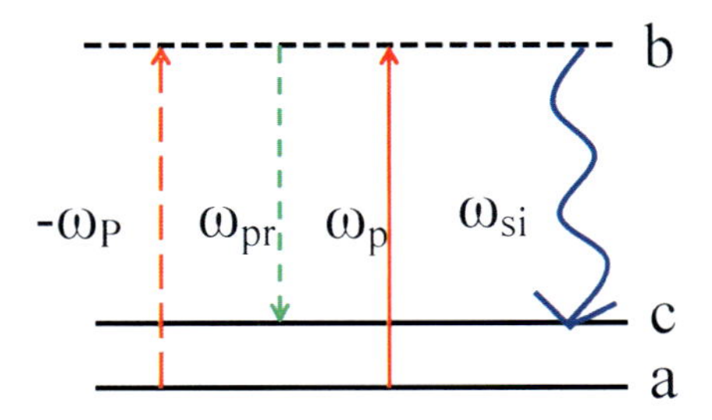

Fig. 3.7 Diagram for stimulated Raman process. *Dotted arrow* represents bra side while *solid line* represents ket side interaction

nonlinear susceptibility (χ^3) of the system. When three different electromagnetic fields interact, there are 108 pairs of possible combinations, the phase matching condition for SRS is given by following equation: [19]

$$\omega_{sig} = -\omega_p + \omega_{pr} + \omega_p$$

Here two electromagnetic fields are coming from the pump pulse and the third field is from probe pulse. To observe any vibrational features the following condition has to be satisfied:

$$\omega_v = \omega_p - \omega_{pr}$$

(where ω_v corresponds to vibrational frequency in the molecular system).

Experimentally SRS is performed with two optical pulses, a Raman pump (ω_P) and a white light continuum which is known as the probe (ω_{pr}). Interaction of the two fields drives the molecular vibrations in the molecule, the third order interaction of the pulses with the molecule can be understood from the energy level diagram (Fig. 3.7). Simultaneous interaction of the pump and probe creates coherence between vibrational levels a band c. In other-words a macroscopic polarization is created which oscillates with frequency ω_v (frequency of vibration in the molecular system).

In femtosecond broadband stimulated Raman spectroscopy [20, 21], the Raman pump field is provided by a narrow bandwidth picosecond pulse and the Raman probe by a femtosecond broadband white light continuum pulse, as broadband white light allows simultaneous detection of a large range of frequencies. SRS is a heterodyned technique which means that signal field is created along the probe direction. Measurement of a probe spectrum with and without the Raman pump followed by calculation of the pump-on:pump off ratio generates a Raman gain spectrum and a Raman loss spectrum depending on the Stokes and anti-Stokes region of the Raman probe with respect to Raman pump (Fig. 3.8). In the case of the loss spectrum the peak intensity is ~1.5 times larger than in the gain spectrum. This is termed as Ultrafast Raman Loss Spectroscopy (URLS) [22, 23], and has been found to have more sensitivity than SRS.

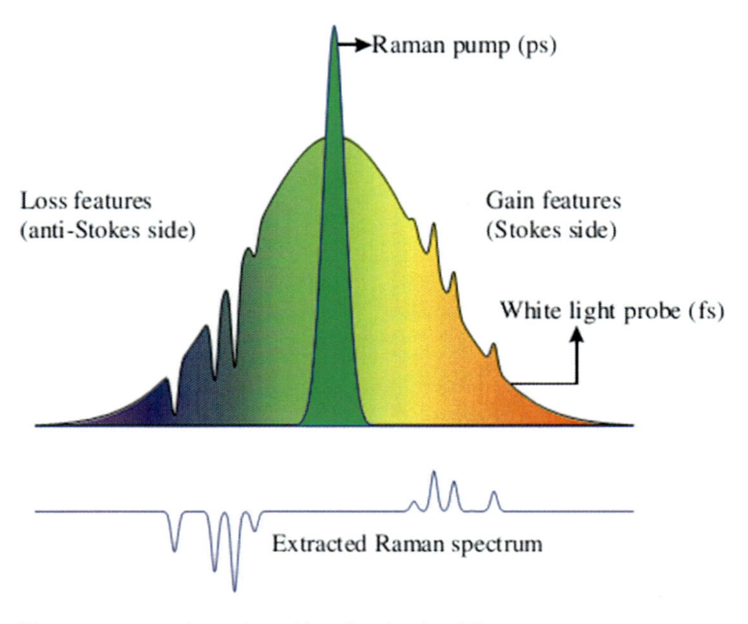

Fig. 3.8 Diagram representing gain and loss in stimulated Raman process

3.4.3 Description of Experimental System

Our laser system consist of a Ti:Sapphire regenerative amplifier (Spitfire, Spectra Physics) seeded by a Mode-Locked Ti:Sapphire Laser (100 fs, 82 MHz, 9 nJ, centered at 790 nm). It is pumped using a Nd:YVO_4 (Millenia) laser. The Ti: Sapphire amplifier generates 100 fs, 2.2 mJ, 1 KHz, centred at 790 nm pulses, 1 mJ is used for pumping an Optical Parametric Amplifier (OPA), the residual is divided into two parts, about 1 % is used for generating the white light continuum probe, while the rest is used to pump TOPAS, which is used as an actinic pulse. The output from the OPA is used to generate Raman Pump, the pico-second Raman Pump Pulse is generated by passing through grating pair in 4f configuration. All three beams are focused onto a 1 mm path length quartz cuvette. The sample was flowing constantly to avoid any degradation of the sample.

3.5 Application: URLS of t-Stilbene in n-BuOH

The ultrafast photophysics of trans-stilbene is well studied, it has an excited state S_1 life time of around 50 ps in n-BuOH, and after photoexcitation in the UV region trans-stilbene isomerizes to the cis form on ultrafast time scale. However there were not many studies on the IVR and vibrational cooling mechanism, by using URLS it is possible to probe the structure of the transient molecule and the dynamics of the individual modes that can give us information about the coupling between the modes and the mechanism of the IVR process.

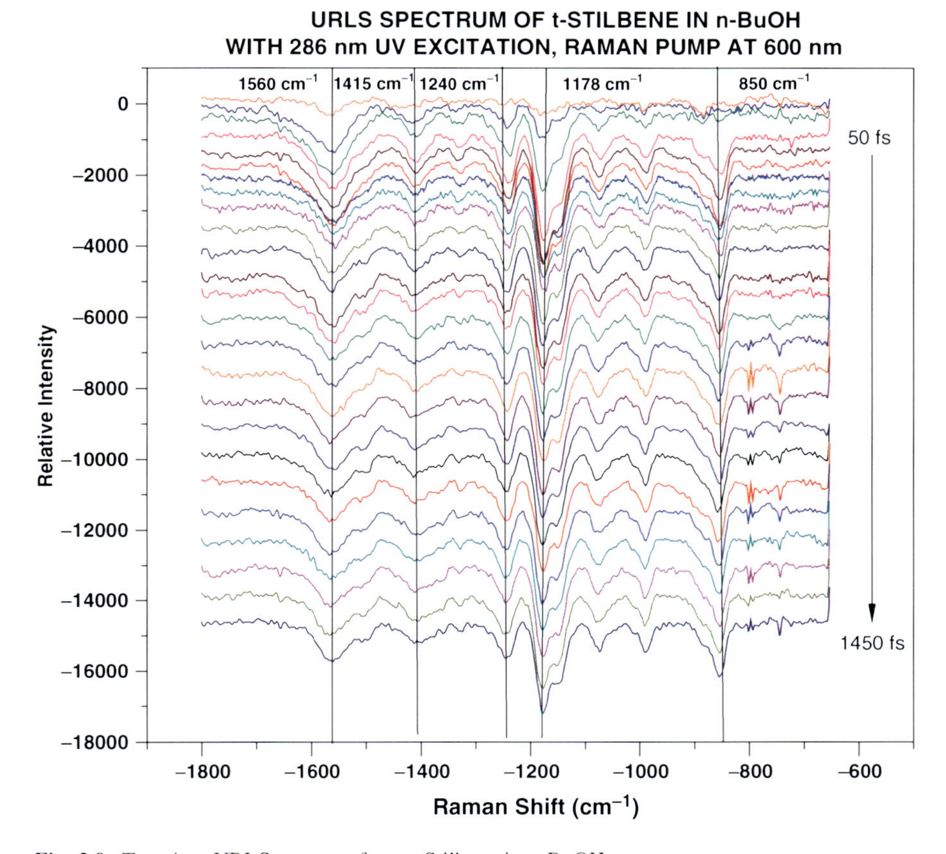

Fig. 3.9 Transient URLS spectra of trans-Stilbene in n-BuOH

The URLS spectrum of trans-stilbene [24, 25] is recorded using the following experimental conditions: a 4 mM solution of trans-Stilbene (Sigma Aldrich) was prepared and the spectra was recorded using a 1 mm cuvette. The centre wavelength of the Raman Pump was 600 nm which is in resonance with the excited state absorption of trans-stilbene, and the energy used for the Raman Pump was 300 nJ with a bandwidth of 0.72 nm. The actinic pump pulse energy was 2 μJ with wavelength centered at 286 nm. The probe beam was dispersed in a spectrometer and detected using a LN_2 cooled CCD detector. All the spectra shown were obtained by subtracting the Raman pump off from Raman pump on condition.

3.5.1 Results

Figure 3.9 shows the transient URLS spectrum of trans-stilbene within 1,500 fs in the 1,800–400 cm^{-1}. Spectra at long delay was also recorded but is not shown here. The peaks >1,300 cm^{-1} corresponds to C=C stretching frequency, the

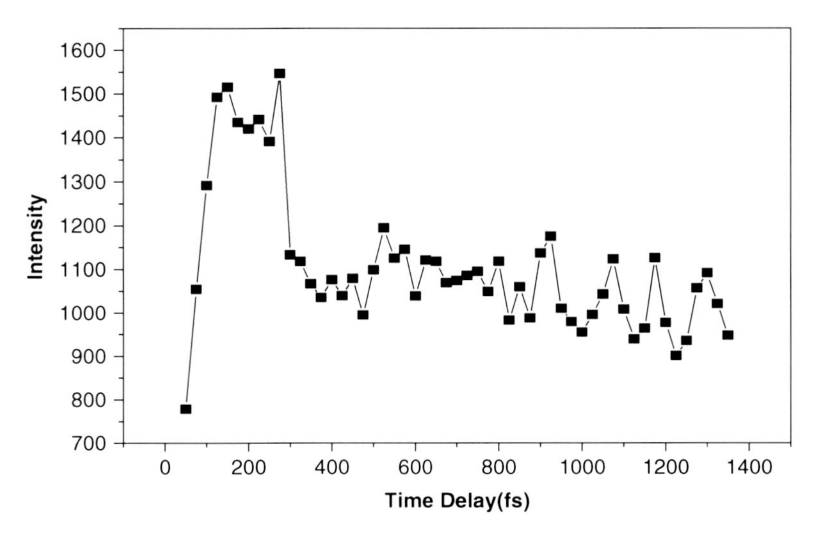

Fig. 3.10 Plot of intensity vs. time for the 1,560 cm^{-1} mode

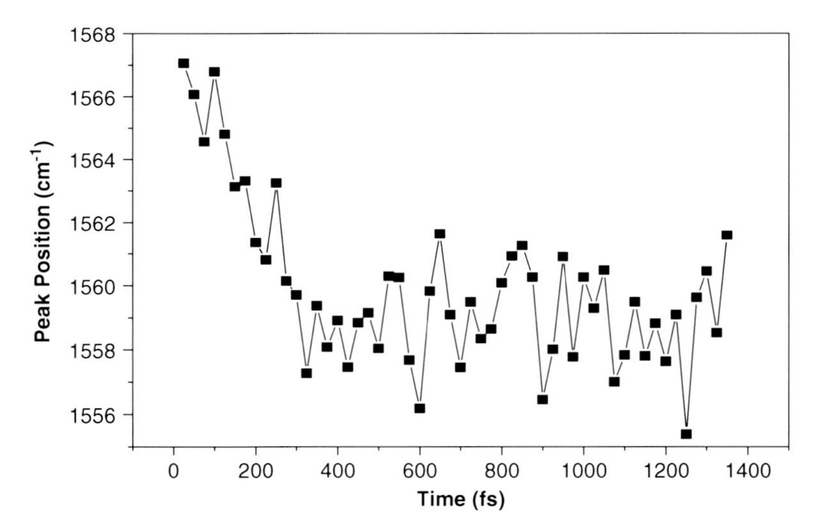

Fig. 3.11 Plot of peak position vs. time for the 1,560 cm^{-1} mode

1,560 cm^{-1} peak corresponds to the ethyl C=C stretching frequency. Figure 3.10 shows the time dependent change in the intensity of the C=C stretching mode, while the change in peak position of this mode with time is plotted in Fig. 3.11.

The variation in the peak intensity of 1,175 cm^{-1} mode, which corresponds to C(ph)-H bending motion, is plotted in Fig. 3.12, while Figs. 3.13 and 3.14 shows the time dependent change in width and position of this mode. These preliminary results suggest that Franck-Condon excited vibrational modes are coupled with

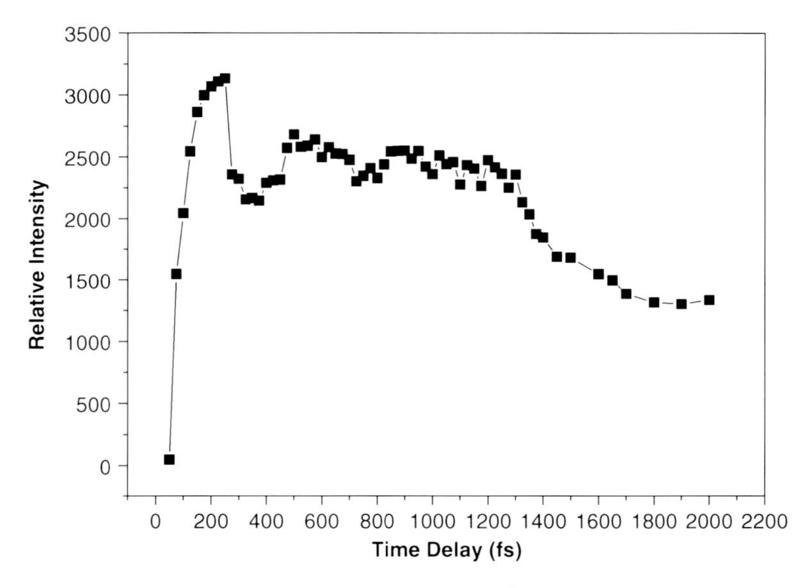

Fig. 3.12 Plot of peak intensity vs. time for the 1,175 cm^{-1} mode

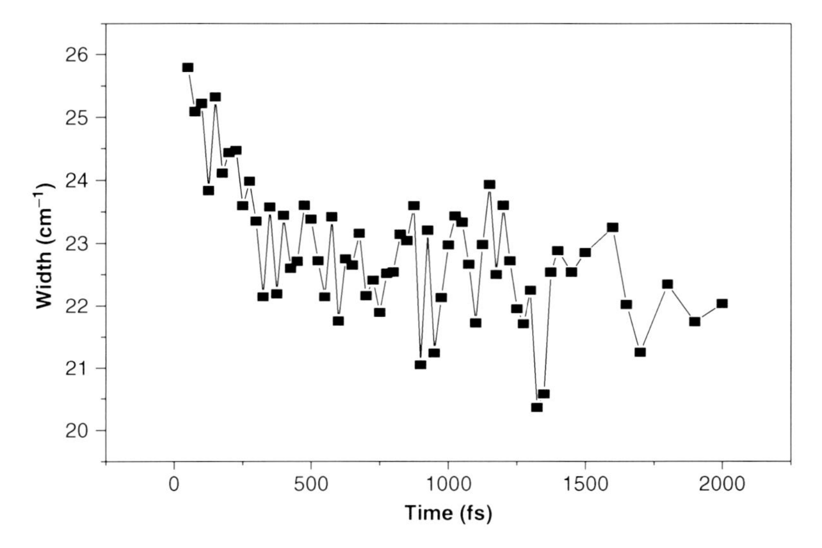

Fig. 3.13 Plot of peak width vs. time for the 1,175 cm^{-1} mode

other low frequency vibrations to induce coherence, which is likely to be solvent dependent. The impacts of these observations are currently being explored. However, these URLS results clearly demonstrate the ability of the technique to study bond specific dynamics on photoexcitation.

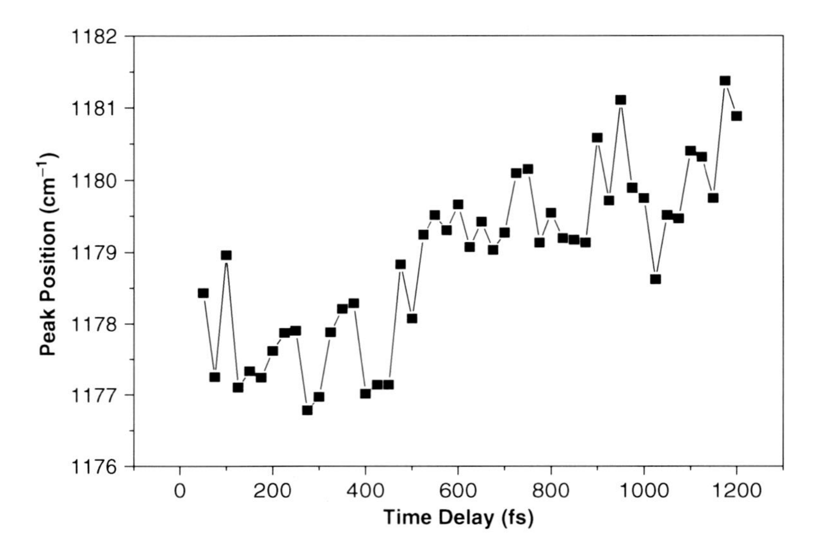

Fig. 3.14 Plot of peak position vs. time for the 1,175 cm^{-1} mode

3.6 Summary

We have demonstrated that time resolved Raman spectroscopy has the potential to explore bond specific changes induced upon photoexcitation of molecular systems. Interestingly, these changes can be correlated to electron density changes across a bond and this could be correlated to the atomic charge density changes observed by crystallographic studies.

Acknowledgement We thank DST, DRDO and the Indian Institute of Science for financial support. SU likes to thank DST for the J C Bose fellowship.

References

1. Turro NJ (1991) Modern molecular photochemistry. University Science Books, Sausalito, California
2. Barrow GM (1988) Introduction to molecular spectroscopy. McGraw-Hill Book Co., London
3. Rohatgi-Mukherjee KK (2006) Fundamentals of photochemistry. New Age International Publishers, Revised 2nd edn, New Delhi
4. Struve WS (1989) Fundamentals of molecular spectroscopy. Wiley, Chichester
5. Atkins P, de Paula J (2007) Atkin's physical chemistry. Oxford University Press, Oxford
6. Turro NJ, Ramamurthy V, Scaiano JC (2010) Modern molecular photochemistry of organic molecules. University Science, Sausalito, California
7. Michael K (1950) Characterisation of electronic transitions in complex molecules. Discuss Faraday Soc 9:14

8. Takao I (2012) Fluorescence and phosphorescence from higher excited states of organic molecule. Chem Rev 112:4541
9. Previtali CM, Scaiano JC (1972) Kinetics of photochemical reactions. Part I. J Chem Soc Perkin 2:1667
10. Coyle JD, Carless HAJ (1972) Photochemistry of carbonyl compounds. Chem Soc Rev 1:465
11. Wagner PJ, Kelso PA, Zepp RG (1972) Type II photoprocesses of phenyl ketones. J Am Chem Soc 94:7480
12. Grabowski ZR, Rotkiewicz K (2003) Structural changes accompanying intramolecular electron transfer: Focus on Twisted Intramolecular Charge-Transfer states and Structures. Chem Rev 103:3899
13. Mattay J (1987) Charge transfer and radical ions in photochemistry. Angew Chem Int Ed Engl 26:825
14. Sahoo SK, Umapathy S, Parker AW (2011) Time resolved resonance Raman spectroscopy: exploring reactive intermediates. Appl Spectrosc 65:1087 and references therein
15. Balakrishnan G, Mohandas P, Umapathy S (2005) Time resolved resonance Raman, Ab Initio Hartree-Fock, and density functional theoretical studies on the transients states of Perfluoro-*p*-Benzoquinone. J Phys Chem A 105:7778 and references therein
16. Balakrishnan G, Mohandas P, Umapathy S (1996) Time-resolved resonance Raman spectroscopic studies on the radical anions of menaquinone and naphthoquinone. J Phys Chem 100:16472 and references therein
17. Mohapatra H, Umapathy S (2009) Influence of solvent on photoinduced electron-transfer reaction: time-resolved resonance Raman study. J Phys Chem A 113:6904 and references therein
18. Balakrishnan G, Sahoo SK, Chowdhury BK, Umapathy S (2010) Understanding solvent effects on structure and reactivity of organic intermediates: a Raman study. Faraday Discuss 145:443
19. Mukamel S (1995) Principles of nonlinear optical spectroscopy. Oxford University Press, New York
20. Kukura P, McCamant DW, Mathies RA (2007) Femtosecond Stimulated Raman Spectroscopy. Annu Rev Phys Chem 58:461
21. Lee SY, Zhang D, McCamant DW, Kukura P, Mathies RA (2004) Theory of femtosecond stimulated Raman spectroscopy. J Chem Phys 121:3632
22. Mallick B, Lakshmanna A, Radhalakshmi V, Umapathy S (2008) Design and development of stimulated Raman spectroscopy apparatus using a femtosecond laser system. Curr Sci 95:1551
23. Umapathy S, Lakshmanna A, Mallick B (2009) Ultrafast Raman Loss Spectroscopy. J Raman Spectrosc 40:235
24. Weigel A, Ernsting NP (2010) Excited Stilbene: Intramolecular vibrational redistribution and solvation studied by femtosecond stimulated Raman spectroscopy. J Phys Chem B 114:7879
25. Hamaguchi H, Kato C, Tasumi M (1983) Observation of the Transient Raman spectra of the S_1 state of trans-stilbene. Chem Phys Lett 100:3

Chapter 4
Time and Space-Resolved Spectroscopy

Kenneth P. Ghiggino

4.1 Introduction

Spectroscopy is the study of the interaction between radiation and matter. While spectroscopy has a long history, the failure of scientists to fully explain the nature of the interactions between radiation and matter played a major role in the development of quantum theory. For example, much of the understanding we have about the electronic structure of molecules and their molecular size and shape, has been learnt from a quantum theory interpretation of experimental spectroscopy.

Molecules have discrete energy states associated with their rotational and vibrational motion and the arrangements of electrons and nuclei. The interaction of the molecule with radiation of the appropriate frequency can result in a change of its energy state with a consequent loss of specific energies from the incident radiation spectrum. This is the basis of *absorption spectroscopy*. Once a molecule has been promoted to an unstable higher energy state it will usually lose this energy in a finite period of time that often leads to emission of a photon and can be studied by *emission spectroscopy*. The frequencies at which molecules absorb and emit radiation (the spectrum) is determined by the molecular energy level spacing and is quite characteristic of the molecular structure and environment of the molecule. Measurement of the lifetimes of the various energy states provides additional information on the dynamics of the processes taking place and the time-scale of the interactions with other species. The simultaneous study of both spectral and temporal information of molecule-radiation interactions is called *time-resolved spectroscopy* and has become an essential tool for many researchers across the biological and physical sciences.

This article will primarily deal with the principles underlying changes in the energy levels associated with the arrangements of electrons in a molecule and

K.P. Ghiggino (✉)
School of Chemistry, University of Melbourne, Melbourne, VIC 3010, Australia
e-mail: ghiggino@unimelb.edu.au

J.A.K. Howard et al. (eds.), *The Future of Dynamic Structural Science*,
NATO Science for Peace and Security Series A: Chemistry and Biology,
DOI 10.1007/978-94-017-8550-1_4,

that involve interactions with radiation in the visible and ultraviolet regions of the electromagnetic spectrum (called electronic spectroscopy). However, it should be kept in mind that changes in other energy states of molecules, such as those associated with rotational and vibrational motion, involve interactions with radiation in other parts of the electromagnetic spectrum and can be studied using similar principles.

4.2 Energy States and Spectroscopy

The energy states and time-scales associated with electronic energy changes in molecules are depicted in the Jablonski diagram shown Fig. 4.1.

The lowest energy configuration of the molecule is usually one where the spins of the electrons are paired (*i.e.* have opposite spin) in a molecular orbital and is referred to as the 'ground' singlet state. Absorption of a photon of sufficient energy will promote an electron to a higher energy molecular orbital called the first 'excited' singlet state. Following photo-excitation the molecules are often formed with excess vibrational energy but in condensed phases this is dissipated rapidly

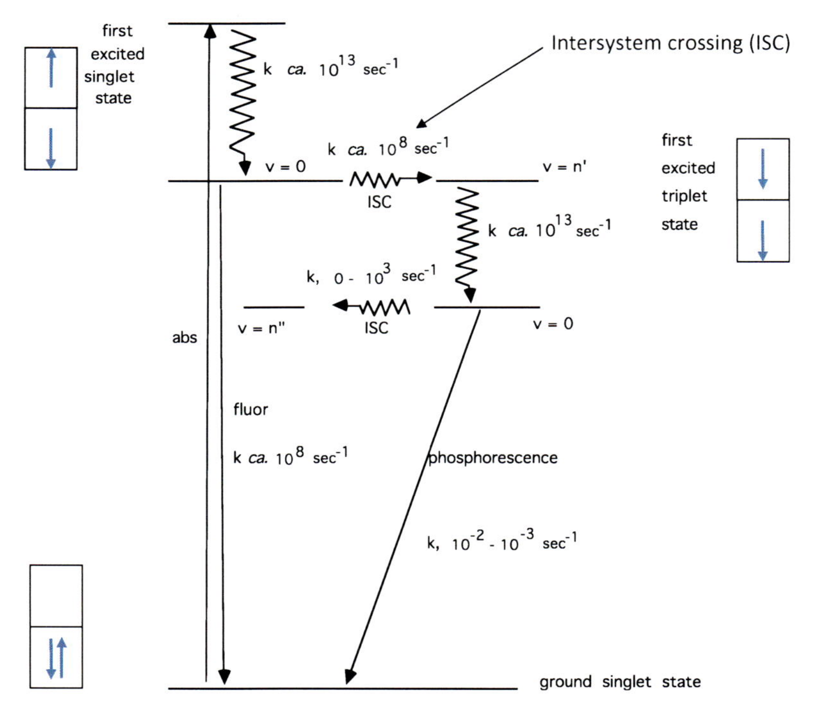

Fig. 4.1 The energy relaxation processes occurring in molecules and approximate time-scales following photo-excitation illustrated using a Jablonski diagram

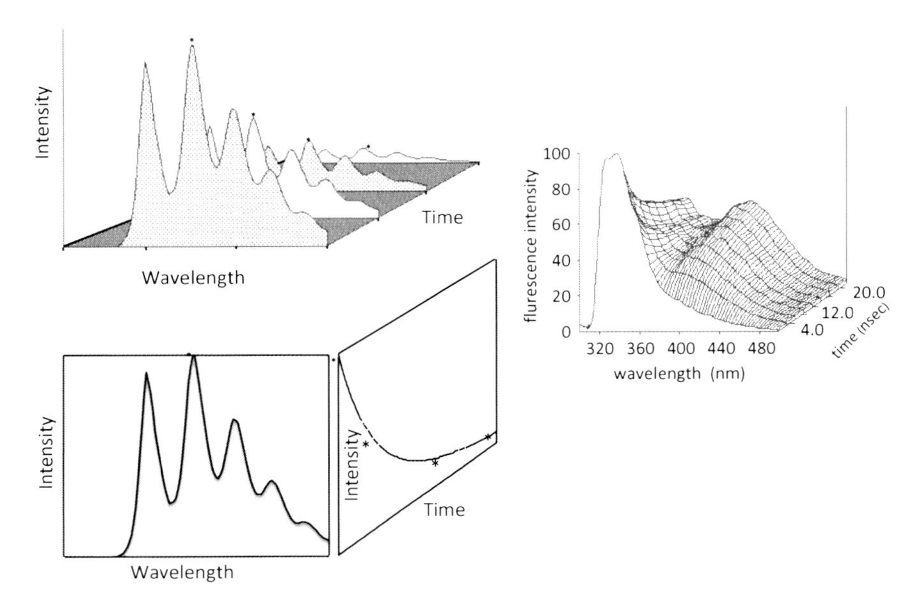

Fig. 4.2 The construction of emission intensity/spectra/time surfaces from fluorescence spectra and fluorescence intensity decay information (*left*), and an experimental time-resolved emission surface for the polymer poly(acenaphthalene) in solution (*right*)

($<10^{-12}$ s) to the surroundings as heat by the process of vibrational relaxation. Once the molecule is in the lowest vibrational energy level of the first singlet excited state, the energy can be lost by the radiative process of fluorescence to return to the ground state or, through an accompanying change in electron spin, undergo the process of intersystem crossing to a triplet excited state where the spins of the two electrons are unpaired (i.e. have the same spin). An alternative process of internal conversion can relax the first excited singlet state to the ground state non-radiatively. The relative magnitude of these processes is determined by the rates for the various steps and these also determine the lifetimes of the electronic energy states involved. Molecules that find themselves in a triplet excited state can return to the ground state by the radiative process of phosphorescence or by a non-radiative intersystem crossing process. Of course competing with these intra-molecular photophysical pathways may be photochemical reactions or bimolecular interactions between the same or different species if the sample concentrations allow interactions during the excited state lifetimes (see [1]).

Each excited electronic state will have a characteristic lifetime (τ) that will be determined by the magnitudes of the various rates for the processes that deplete the state. The time scales for these processes can range from femtoseconds (10^{-15} s) to seconds. Thus, if the initial excitation is by a short light pulse, the intensity of the spectrum will change with time and if multiple states are formed and decay with different rates all this information can be reflected in a multidimensional spectra/ time/intensity surface. The construction of such a surface for emission spectroscopy is shown in Fig. 4.2 together with an experimental time-resolved fluorescence decay surface for the polymer poly(acenaphthalene) in solution. The latter figure shows the temporal decay of fluorescence of the initially excited species with a spectral maximum of approx. 330 nm and the grow-in with time of a new species (an excited state dimer or 'excimer') with a fluorescence spectral maximum at approx. 420 nm. Since each energy state has a characteristic absorption spectrum, a corresponding absorption/wavelength/time surface can also be generated. Such surfaces can provide all the spectral and dynamic information for the species under study.

4.3 Time-Resolved Spectroscopy Techniques

Time-resolved absorption and emission spectroscopy are the two principal techniques that can be applied to obtain information on the spectral and dynamic information of molecules and are described below. However other techniques such as NMR, EPR and radiation scattering methods may also be applied to investigate dynamic behavior depending on the time-scale of the process and the sensitivity required.

The advent of lasers as photo-excitation sources has allowed both high spectral and temporal resolution to be achieved in time-resolved spectroscopy. A limitation on the time-scale of the process that can be studied is the duration of the light pulse that is used to excite the sample. Laser sources are available that can provide light pulses of only a few femtoseconds duration enabling processes that occur on time scales longer than this to be followed. However the time-response of light detectors (such as photodiodes and photomultiplier tubes) and other electronic components can impose limitations so that a variety of clever optical methods have been devised to ensure optimal time-resolution can be achieved.

The principle behind the optical arrangement used in time-resolved 'pump-probe' absorption spectroscopy is shown in Fig. 4.3.

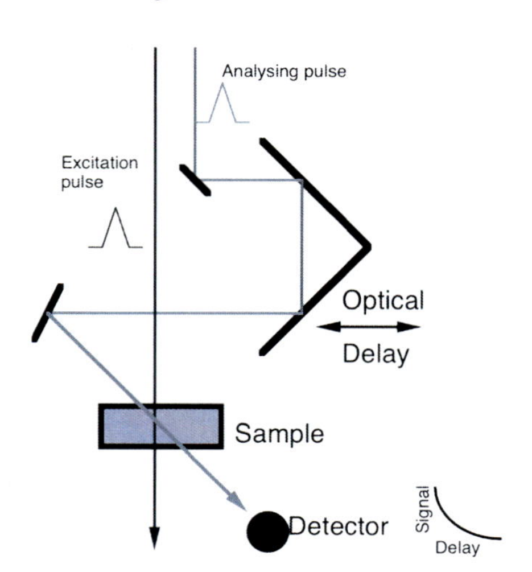

Fig. 4.3 Principle of time-resolved pump-probe absorption spectroscopy using short laser pulse excitation and an optical delay line

A short laser 'pump' pulse photo-excites the sample and the light absorption properties of the excited species formed are investigated by passing through the sample 'probe' light pulses that are delayed in time with respect to the pump pulse. The time delay of the probe pulses is introduced by bouncing the light pulses off a moveable mirror. The accurately known speed of light combined with careful control over the distance the mirror travels using a piezo-driven stage, imposes the time delay at which the absorption spectrum of the transient species is recorded (0.3 mm additional distance travelled by the probe pulse corresponds to a time delay of approximately 1 picosecond (ps)). The intensity of the transmitted light pulses at various optical delays combined with spectral resolution provides the time-resolved absorption spectrum.

For the corresponding time-resolved emission spectrum, again short light pulses are used to photo-excite the sample but in this case the arrival times of photons emitted from the sample with respect to the excitation pulse are measured. A common method employed with high repetition rate lasers is the time-correlated single photon counting technique [2] which uses a 'time-to-amplitude converter' to produce a voltage signal proportional to the time delay of emitted photons for each excitation event recorded. The acquisition of many such events produces a voltage amplitude frequency/time delay histogram that corresponds to the emission decay profile. The time-resolution of this method is restricted to about 20 ps but other techniques such as fluorescence up-conversion and Kerr-gating can achieve resolution approaching the excitation pulse duration.

Once the absorption or emission decay information is collected, it then needs to be analyzed by fitting the data to models based on an appropriate kinetic scheme and the rate information extracted. For the simplest case of a single absorbing or emitting species undergoing intramolecular relaxation to return to its ground state, the decay profile is often best fitted by a first order rate law to give a characteristic excited state lifetime value (τ). More complex kinetics may be followed if the molecule is in multiple environments or undergoes dynamic interactions with itself or other species.

4.4 Applications of Time-Resolved Emission and Absorption Spectroscopy

There are many applications of time-resolved spectroscopy across the biological, chemical and physical sciences that allow the dynamics of light initiated reactions to be followed or the structure and properties of photo-excited molecules and interactions with their environment to be characterized [3–5].

Studies of the photo-induced electron transfer processes that occur in photosynthesis have been of particular interest (*e.g.* [6]). Developing artificial molecular systems that mimic the primary electron transfer steps in natural photosynthesis have also received much attention. An example of the latter studies is depicted in Fig. 4.4.

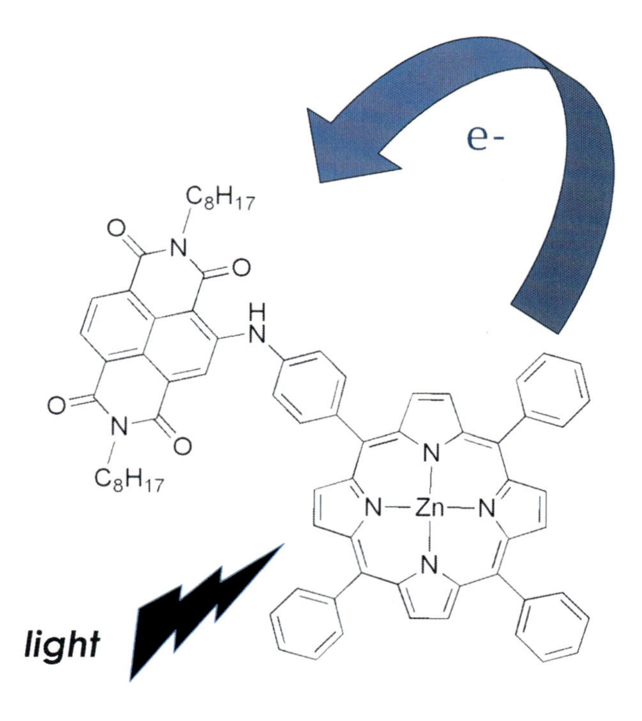

Fig. 4.4 The photo-induced electron transfer process in a linked zinc tetraphenylporphyrin donor and an amino naphthalene diimide electron acceptor

In this molecular system, photo-excitation makes the zinc tetraphenylporphyrin (ZnTPP) a better electron donor and leads to electron transfer to the linked amino naphthalene diimide species (ANDI). The excited states and product species involved in this process can be spectrally identified and the rates for the electron transfer steps determined using time-resolved absorption spectroscopy. In this experiment an ultrafast laser producing light pulses of 50 femtoseconds duration at 415 nm has been used as the excitation source in a 'pump-probe' configuration. The spectral and temporal behaviours of the various species absorbing in solution following photo-excitation are shown in Fig. 4.5.

A detailed analysis of the information in Fig. 4.5 shows that the transient absorption spectra observed arise from a number of species including the first (S_1) and second (S_2) excited singlet states of the initially excited ZnTPP as well as the radical cation ($ZnTPP^+$) and anion ($ANDI^-$) formed following the electron transfer process. Fitting the spectral and temporal data to a suitable kinetic model indicates the electron transfer process from the S_1 state occurs in toluene solvent with a rate of $2 \times 10^{11}\ s^{-1}$ while the electron recombination process has a rate constant of $8 \times 10^9\ s^{-1}$. This information is valuable in designing molecular systems that can optimize the electron separation and recombination steps so that useful electrochemical work can be extracted from the light induced process (see [7]).

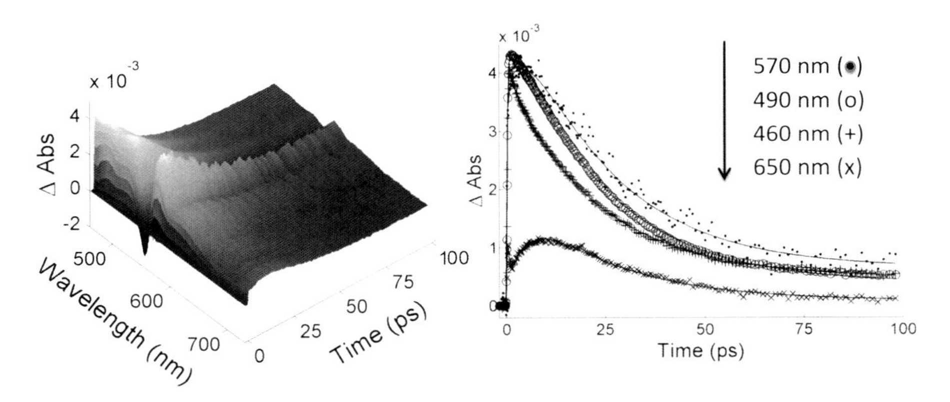

Fig. 4.5 Time–resolved absorption spectral surface (*left*) and transient absorption decay profiles at wavelengths specified (*right*) of a solution of ZnTPP-ANDI in toluene following photo-excitation by a 415 nm laser pulse (after [7])

Some other applications where time-resolved emission and absorption spectroscopy methods can provide valuable insights include:

- Light harvesting processes in polymers and supramolecular assemblies.
- The mechanisms underpinning the operation of both inorganic and organic photovoltaic devices.
- The light induced process involved in vision and photosynthesis.
- The photochemistry of dyes, pigments and biological molecules.
- Ultrafast structural fluctuations of molecules and molecular assemblies.

4.5 Time and Space-Resolved Imaging Spectroscopy

An exciting emerging field in applications of time-resolved spectroscopy is the extension of the technique to small spatial scales. This can involve techniques ranging from time-resolved imaging of samples at diffraction-limited spatial scales (approx. 250 nm) using confocal fluorescence microscopy to super-resolution (sub-diffraction limited) microscopy techniques and time-resolved single molecule methods. Ongoing recent developments allow imaging of dynamic processes at atomic resolution to be possible with pulsed x-ray sources. These techniques and the information that can be obtained from them are discussed separately below:

4.5.1 Time-Resolved Fluorescence in Microscopy

Fluorescence microscopy is a well-established technique that is commonly used to image biological samples and in materials characterization. In normal fluorescence imaging methods (such as those used in conventional and confocal fluorescence

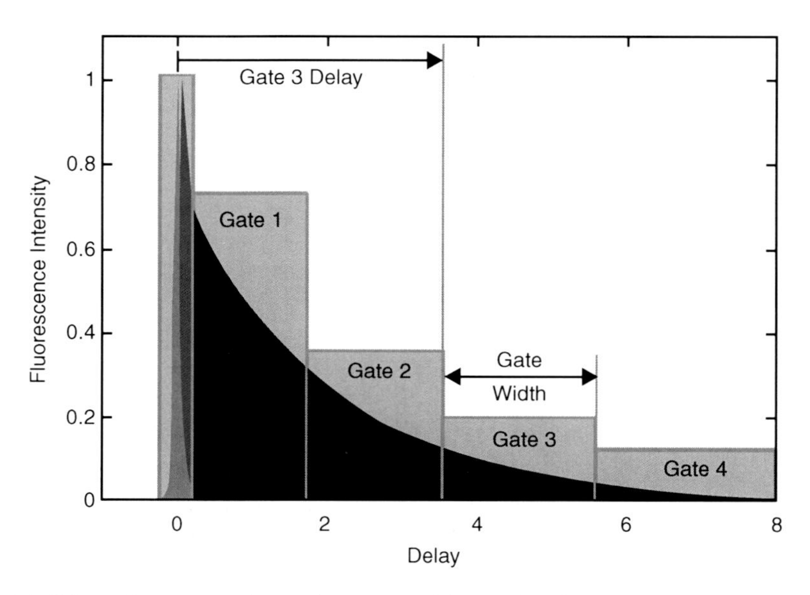

Fig. 4.6 Schematic representation of the method of time-gating the collection of photons during the decay of fluorescence from a sample

microscopes) the images recorded represent the integrated fluorescence intensity over all times from the sample. However, if the temporal decay of fluorescence can be recorded as a function of spatial position, images from different times after photo-excitation can be obtained. This technique of time-resolved fluorescence microscopy has grown rapidly in recent years and can now be considered a mature technology with several commercial instruments available.

A number of methods can be used to undertake time-resolved fluorescence microscopy. These usually involve using a time modulated excitation source or short-pulsed laser excitation and the recording of fluorescence photons emitted at defined times after excitation of the sample placed on the stage under the microscope. Figure 4.6 shows the principle of recording those fluorescence photons detected from the sample only within a certain time 'gate' after the excitation pulse.

Optical box-car mode devices (gated optical intensifiers) capable of being triggered at millions of times/second and with gate widths of <100 ps are commercially available. If such a detection device is integrated with a microscope the outcome is a series of fluorescence images from the sample recorded at different times after the excitation pulse. If there are multiple species with different fluorescence decay times present then the images recorded at short time scales will highlight the shorter-lived fluorescent species while those images recorded at later times will accentuate the longer lifetime components of the image. There are many applications for this technique. For example, the fluorescence lifetimes of certain drugs will depend on the environment where they are located in a biological cell. Fluorescence lifetime imaging of biological cells can reveal the location and nature of the interactions between the drug and the cellular environment.

In materials characterization the morphology and activity of components in a polymer blend can be established. The dynamics of fluorescent components can also be investigated by recording time-resolved fluorescence polarization images that reveal the mobility of the fluorophore on the time-scale of the emission process.

There are various ways to implement time-resolved fluorescence imaging in microscopy apart from the gated optical method discussed above including using time-correlated single photon counting and frequency domain methods (see review by [8]) depending on the time-resolution, cost and quality of the image required. Imaging methods have been extended to allow simultaneous recording of emission/transient absorption and photocurrent images to be undertaken that further expand the information that can be obtained from a given sample.

4.5.2 Super-Resolution Microscopy

The spatial resolution (d) that can be achieved in conventional optical microscopy is determined by the diffraction limit given below:

$$d = \frac{0.61\lambda}{n \sin\theta}$$

where λ is the wavelength of the radiation and $n\sin\theta$ is the numerical aperture of the objective (up to about 1.4). This limits the resolution of optical microscopes to approx. 250 nm. In recent times a number of methods have been developed that are able to circumvent this limit. Near field scanning optical microscopy (NSOM) relies on the coupling of light through a sub-wavelength size aperture (*e.g.* thinly drawn optical fibre) that is brought within a few nanometers of the sample surface. The lateral resolution here is primarily dependent on the size of the aperture and can be of the order of 20 nm. By scanning the aperture tip over the sample, images may be recorded. The technique can be applied to collect spectroscopic information and is particularly popular in the fluorescence mode where it can be combined with time-resolved techniques [9].

Other super-resolution techniques rely on utilising stimulated emission from molecules and two laser pulses (stimulated emission depletion microscopy, STED) or in other cases by using photoswitchable single fluorophore emitters incorporated in the sample where only some are activated and emit at particular times (e.g. stochastic optical reconstruction microscopy, STORM, and photoactivated localization microscopy, PALM). By mapping the central positions of the individual emitters over a period of time a super-resolution image can be reconstructed (see review by [10]). A spatial resolution of 20 nm can be achieved with these techniques. Variants of these techniques and new approaches are under constant development together with the appropriate fluorescent probe molecules to be used [11].

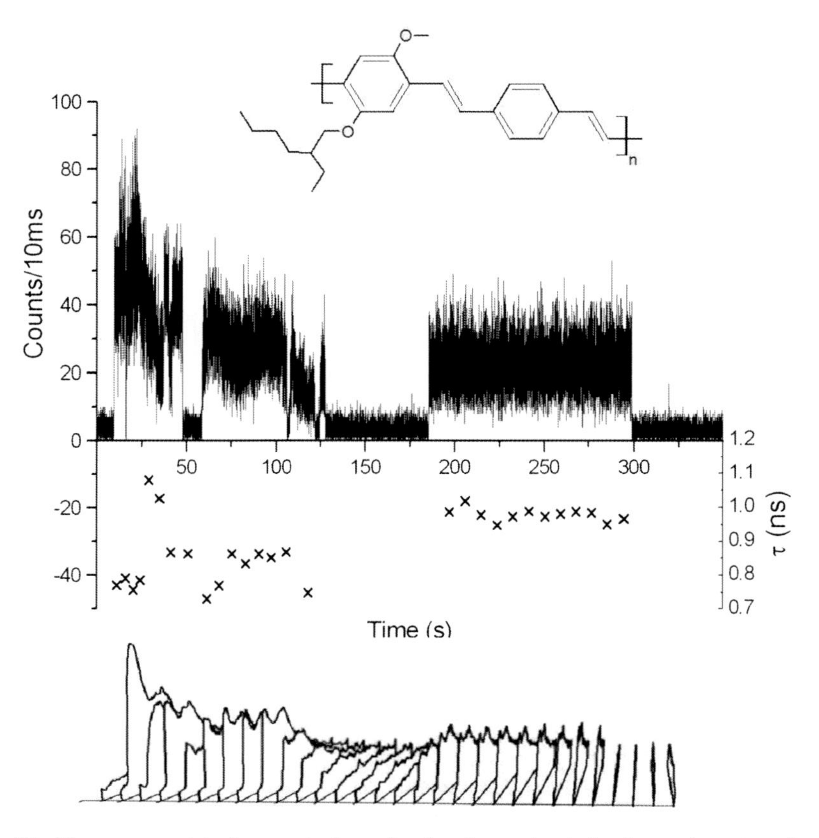

Fig. 4.7 Fluorescence data from a single molecule of a conjugated polymer (structure shown) dispersed in PMMA films. *Upper data set*: fluorescence intensity/time trajectory; *middle data set*: fluorescence lifetimes monitored during trajectory; *lower data set*: fluorescence spectra changes during the trajectory (after [12])

4.5.3 Single Molecule Spectroscopy

If fluorescent molecules are embedded in a solid matrix at a concentration (normally $<10^{-9}$ mol dm^{-3}) such that the molecules are spatially separated by several microns, then fluorescence from the individual molecules can be observed under a fluorescence microscope. Fluorescence spectra, decay times and intensity changes that occur over time for a single emitter can be monitored providing information on the distribution of molecular behaviour that makes up the bulk sample properties. Examples of the application of this method include the study of individual biological molecule behavior in cells and the photophysical behaviour of individual synthetic polymer chains [11]. An example of the information that can be obtained from a single conjugated polymer chain immobilized in a clear plastic (PMMA) matrix is illustrated in Fig. 4.7. By following the fluorescence intensity from the single polymer chain over many seconds, it is apparent that the fluorescence 'blinks' on and off. This can be related to the fluorophore relaxing to a 'dark' state (for example a triplet state of the molecule) or due to quenching by interactions with the substrate. The fluorescence lifetime is also seen to change and there are shifts in the spectra associated with the blinking. Each molecule of the sample may show quite individual spectral and temporal properties depending on its conformation and environment. In these experiments it is important to monitor the behaviour of many individual molecules to characterize the distribution of behaviors. An extension of the technique is single molecule defocused wide-field imaging that can provide information on the nanoscale dynamics of individual molecules [12,13].

4.5.4 X-Ray Based Time-Resolved Structural Studies

X-ray crystallography is a well-established technique to determine crystal structures of both small molecules and biomolecules. However, the availability of pulsed X-ray sources from synchrotrons, free electron lasers and high harmonic generation using high power pulsed lasers has opened up the opportunities for time-resolved X-ray studies. These techniques have been applied to study structural phase transitions, molecular motion and chemical reactions down to femtosecond time-scales [14]. The experimental techniques adopted are a variation of the optical pump-probe experiment. A short laser pulse is used as the pump beam to initiate the process for study and an accurately synchronised in time X-ray probe pulse can probe the structural dynamics that occur (see review by [15]). Both time-resolved X-ray diffraction and absorption (e.g. EXAFS and XANES) can be implemented. An example of the application of this technique is in determining the mechanism of electron transfer in the photochemical transformation of Fe(III) oxalate to Fe(II) oxalate [15]. There is considerable potential for these time-resolved methods with the possibilities to obtain structural information on short-lived excited states and transient species.

Recent work [16] has reviewed progress with the technique of femtosecond nanocrystallography using X-ray lasers for membrane protein structure determination. The use of high power femtosecond pulsed X-rays and a liquid jet of nanocrystals of the protein allows the recording of many single crystal diffraction patterns (*e.g.* Photosystem I) to provide structural resolution at 8.4 Å solely limited by the wavelength of the X-ray beam and the detector. The technique has been extended to undertake time–resolved spectroscopic pump-probe experiments. In this configuration a laser has been used to excite photoactive membrane protein nanocrystals and the X-ray pulses probe the conformational changes that occur upon photo-excitation [16].

There are many exciting possibilities for undertaking a range of time-resolved structural imaging experiments using the innovation of X-ray free electron lasers (XFELs), other excitation sources and appropriate sample handling methods.

References

1. Learmonth RP, Kable SH, Ghiggino KP (2009) Basics of fluorescence. In: Goldys EM (ed) Fluorescence applications in biotechnology and the life sciences. Wiley-Blackwell, Hoboken, pp 1–26
2. O'Connor DV, Phillips D (1984) Time-correlated single photon counting. Academic, London
3. Diels J-C, Rudolph W (2006) Ultrashort laser pulse phenomena: fundamentals, techniques and applications on a femtosecond time-scale, 2nd edn. Academic, San Diego
4. Goldys EM (ed) (2009) Fluorescence applications in biotechnology and the life sciences. Wiley-Blackwell, Hoboken
5. Valeur B, Bereberan-Santos MN (2012) Molecular fluorescence: principles and applications, 2nd edn. Wiley-VCH, Weinheim
6. Berera R, van Grondelle R, Kennis JTM (2009) Ultrafast transient absorption spectroscopy: principles and application to photosynthetic systems. Photosynth Res 101:105–118
7. Robotham B, Lastman KA, Langford SJ, Ghiggino KP (2013) Ultrafast electron transfer in a porphyrin – amino naphthalene diimide dyad. J Photochem Photobiol A Chem 251:167–174
8. Smith TA, Lincoln CN, Bird DK (2009) Time-resolved fluorescence in microscopy. In: Goldys EM (ed) Fluorescence applications in biotechnology and the life sciences. Wiley-Blackwell, Hoboken, pp 195–221
9. Cadby AJ, Dean R, Eliot C, Jones RAL, Fox AM, Lidzey DG (2007) Imaging the fluorescence decay lifetime of a conjugated-polymer blend by using a scanning near-field optical microscope. Adv Mater 19:107–111
10. Hell S (2007) Far field optical nanoscopy. Science 316:1153–1158
11. Sauer M, Hofkens J, Enderlein J (2011) Handbook of fluorescence spectroscopy and imaging – from single molecules to ensembles. Wiley-VCH, Weinheim
12. Ghiggino KP, Bell TDM, Hooley EN (2012) Synthetic polymers for solar harvesting. Faraday Discuss 155:79–88
13. Dedecker P, Muls B, Deres A, Uji-I H, Hotta J, Sliwa M, Soulillion J-P, Mullen K, Enderlein J, Hofkens J (2009) Defocussed wide-field imaging unravels structural and temporal heterogeneity in complex systems. Adv Mater 21:1079–1090
14. Gaffney KJ, Chapman HN (2007) Imaging atomic structure and dynamics with ultrafast x-ray scattering. Science 316:1444–1448
15. Er O, Chen J, Rentzepis P (2012) Ultrafast time-resolved x-ray diffraction, extended x-ray absorption fine structure and x-ray absorption near edge structure. J Appl Phys 112:031101–031116
16. Fromme P, Spence JCH (2011) Femtosecond nanocrystallography using x-ray lasers for membrane structure determination. Curr Opin Struct Biol 21:509–516

Chapter 5
Dynamic X-Ray and Neutron Scattering: From Materials Synthesis and In-Situ Studies to Biology at High Pressure

Paul F. McMillan, Filip Meersman, Fabriza Foglia, Paul Barnes, Simon D.M. Jacques, and Richard Briggs

5.1 Introduction

X-ray and neutron scattering techniques are applied to a very wide range of condensed matter systems and problems ranging from solid state materials to biological organisms in order to study and understand their structures and phase transformations under static and dynamic compression, as well as their synthesis and function under extreme conditions. Here we illustrate the applications of several of these techniques to problems of current scientific and technological interest.

X-ray diffraction provides the standard technique for phase identification and structural studies of crystalline and amorphous materials. It is applied in most

P.F. McMillan (✉) • F. Foglia
Department of Chemistry, University College London, 20 Gordon Street,
London WC1H 0AJ, UK
e-mail: p.f.mcmillan@ucl.ac.uk

F. Meersman
Department of Chemistry, University College London, 20 Gordon Street,
London WC1H 0AJ, UK

Rousselot – Expertise Center, R&D Laboratory, Meulestedekaai 81,9000, Gent, Belgium

P. Barnes
Department of Chemistry, University College London, 20 Gordon Street,
London WC1H 0AJ, UK

Biological Sciences, Birkbeck College, Malet Street, London WC1E7HX, UK

S.D.M. Jacques
The School of Materials, The University of Manchester, Oxford Road, Manchester, M13 9PL, UK

Research Complex at Harwell (RCaH), Rutherford Appleton Laboratory, Harwell Oxford,
Didcot, Oxon OX11 0FA, UK

R. Briggs
Center for Science at Extreme Conditions, School of Physics & Astronomy,
University of Edinburgh, Edinburgh EH9 3JZ, UK

J.A.K. Howard et al. (eds.), *The Future of Dynamic Structural Science*,
NATO Science for Peace and Security Series A: Chemistry and Biology,
DOI 10.1007/978-94-017-8550-1_5,

investigations of materials synthesis and phase transformations in the solid and liquid state. Experiments are enabled both in the laboratory or at high brightness synchrotron X-ray sources with a time resolution that is determined both by detector capabilities and sampling characteristics. Such studies are being used to investigate the structure and function of materials under synthesis or operating conditions, as well as local structures developed in amorphous materials, polymers or biomolecules. The high degree of spatial resolution enabled at synchrotron sources permits experiments under extreme high pressure and high temperature conditions [1, 2], and the emerging availability of free electron laser (FEL) sources combined with streak camera and back-lighting X-ray experiments are leading to new time-resolved studies of materials under dynamic shock and laser ramp conditions [3–7]. Small angle scattering (SAXS) provides information on structure, texture and density fluctuations on the nano- to mesoscale (e.g., 5–50 nm) leading to studies of the organization, formation and function of nanomaterials, including polymers, composites and biologically important systems [8, 9]. X-ray absorption (XAS) spectroscopy combined with analysis of the EXAFS (extended X-ray absorption fine structure) leads to complementary information on local structures, coordination state, ligand binding and oxidation number around the central probe atom.

Recently, a number of these techniques (angle/energy-resolved diffraction, X-ray absorption/fluorescence, total-scattering pair-distribution analysis etc.) is undergoing a renaissance as the technical challenge of conversion to the 2D/3D tomographic data collection mode are being overcome, in order to provide time- and space-resolved maps of the structural organization evolving within functional materials during synthesis or chemical reactions while subject to conditions inside *in situ* environmental cells. This confluence of structural information with time- and spatial-variation is beginning to revolutionise our view of many processes, showing how reactions initiate, spread and differentiate throughout the body of real specimens and objects [10–15].

Neutron scattering studies enabled at reactor or spallation facilities typically have a much lower incident beam flux than that at X-ray sources, so that larger samples and longer collection times are generally involved. However, recent developments in neutron focusing technology combined with new generations of high pressure cells are greatly extending the pressure range that can be studied [16–20]. The special qualities of neutron beams and their interactions with matter make them ideal for penetrating deep inside experimental chambers or reaction environments, as well as for studies of magnetic phenomena, and structures and relaxation dynamics in polymers and biological systems. The use of isotopic contrast techniques leads to unique ways to study these problems. As for X-rays, small angle neutron scattering (SANS) experiments are used to provide information on objects and structures extending from the nano- to the mesoscale [21, 22]. Quasi-elastic neutron scattering (QENS) studies give information on the dynamics of molecular systems on a timescale corresponding to diffusional processes or macromolecular motions [19, 23]. These techniques are being applied to problems ranging from catalysis to the structure and function of polymers and biomolecular systems. Here we describe some of these approaches applied to systems ranging from materials synthesis and *in situ* studies under extreme high-P,T conditions to the structure and function of biological molecules, polymers and whole organisms, using examples drawn from our own research and that of colleagues and collaborators.

5.2 X-Ray and Neutron Scattering

The term X-ray diffraction is typically used to describe "wide angle" (WAXS) scattering studies that probe spatially correlated interatomic separations in crystalline or disordered materials on an approximately 1–20 Å length scale. The methods implemented in the laboratory or at large scale synchrotron or other facilities use the classical methods for data collection and interpretation associated with W.L. and W.H. Bragg, M. von Laue, P. Debye, P. Scherrer, and many others who developed these techniques beginning in the early twentieth century. The experiments give key information on the crystalline or amorphous structures and their phase transformations during synthesis or under extreme pressure, temperature and chemical conditions or gradients, applied either statically or during time-resolved experiments. Much valuable information has been gained on the synthesis and structural properties of solids, liquids, polymers, functional materials and biomolecules in this way.

The WAXS studies are complemented by X-ray absorption spectroscopy (XAS) and EXAFS techniques that give information on local structures, bonding and oxidation states. "Small angle" (SAXS) scattering studies have been developed to investigate structural correlations, density fluctuations and texturing at the nano- to mesoscale, typically up to ~5 nm^{-1}, corresponding to real space separations on the order of 5–50 nm. A main parameter determined in SAXS experiments is the radius of gyration, R_g, that is related to both the size and shape of the macromolecules, nanoparticles or mesoscale fluctuations being studied, although resolved series SAXS peaks are also observed for highly ordered mesoscale arrays of nanoparticles or void assemblies within nanomaterials. Data for less organized materials are typically analyzed using modelling techniques that help us distinguish between different possible degrees of organization. Such studies are used to track the evolution of polymers and biomolecular systems and their structural relaxation during P,T-jump experiments, or the synthesis, processing and function of nanoscale materials and catalytically active surfaces. Analogous information is gained from small angle neutron scattering (SANS) studies. These methodologies are advancing on several frontiers, accompanied by new *in situ* cell designs developed for probing materials under various experimental conditions, X-ray sources and detector technologies permitting 1-, 2- and 3-D scattering and tomographic studies on timescales extending from a few seconds or milliseconds down to femtosecond investigations of structural changes during laser ramp or shock physics experiments. The access to high brilliance synchrotron X-ray sources has also led to the development of new inelastic X-ray scattering (IXS) or X-ray Raman techniques that probe phonon and electronic excitations within materials.

X-ray photons are mainly scattered by electrons so that the contribution to the scattered intensity is determined by the square of the atomic number (Z^2) contributed by each atomic centre. Low-Z elements (H, N or C etc.) thus prove difficult to study without recourse to high intensity synchrotron X-ray sources, and they contribute relatively little to the X-ray diffraction signal, especially when they are combined with heavier elements. By contrast, neutrons are fundamental particles with a rest mass close to that of the proton (1.675×10^{-27} kg). They have no charge and thus can penetrate deeply into materials, including containment apparatus used for high pressure or materials synthesis and *operando* experiments. Also, unlike X-ray photons they possess a property of "spin", that both gives rise to important studies of magnetic properties of materials and also determines the fundamental nature of many neutron scattering experiments. Neutron diffraction experiments lead to wide-angle diffraction and small-angle (SANS) results that are similar to and complementary with WAXS and SAXS studies, although the neutron scattering probes the atomic nuclei rather than the electron density distribution, and is highly sensitive to the isotopic composition. Special characteristics including interactions with the spin state of the sample and the ability to study low-energy excitations make neutron scattering experiments unique.

Incident neutron beams are either produced by nuclear reactors during fission of ^{235}U in enriched uranium fuel rods, or generated by spallation sources during fragmentation of heavy target atoms (e.g. Hg) exposed to high energy proton beams. Neutrons are scattered from atomic nuclei according to their mass and isotopic composition, unlike X-rays that are scattered by electrons surrounding the nuclei. Neutrons have spin 1/2: these interact with different nuclei according to their spin state or nuclear angular momentum. Certain nuclei with equal numbers of protons and neutrons such as $^{12}{}_{6}C$ have zero spin: others such as $^{1}{}_{1}H$ can exist in "spin up" or "spin down" states with respect to a reference axis such as an external magnetic field. That property gives rise to studies of magnetic ordering in materials.

Neutron scattering cross sections (σ) are measured in "barns" (1 barn $= 10^{-28}$ m^2 $= 100$ fm^2. This quantity is defined as the effective area that an element contributes to the scattering (σ_S) or absorption (σ_A) of the incident neutron beam. Both coherent and incoherent scattering processes can occur, with a total scattering cross section $\sigma_S = \sigma_{coh} + \sigma_{inc}$. In coherent scattering the incident neutrons form a wave that maintains its directional and momentum properties throughout the interaction with a diffracting arrangement of nuclei in the sample. During incoherent scattering that occurs by interaction with the nuclear spin states the scattered neutrons lose their temporal and spatial coherency. Both coherent and incoherent scattering cross sections are heavily dependent on the element and isotope studied. For example, incoherent scattering from ^{1}H nuclei is particularly large while that from the $^{2}{}_{1}H$ (i.e., ^{2}D) isotope is much smaller. The overall scattering lengths (*b*: given in 10^{-15} m (i.e., fm) units) can vary widely both in magnitude and sign for different nuclei. For example, for ^{1}H, $b_{coh} = -3.742$ fm and $b_{incoh} = 25.27$ fm, while the scattering lengths for deuterium (^{2}H) are $b_{coh} = 6.671$ fm and $b_{incoh} = 4.04$ fm.

These considerations must be taken into account while designing experiments, and they can be used to develop isotopic contrast studies that provide information on structure and dynamics in materials and biological systems.

In neutron diffraction and SANS experiments no energy exchange is considered. However, neutron beams can be selected or moderated to have energies within the thermal range that can interact with vibrational excitations giving rise to inelastic neutron scattering (INS) techniques. These are applied to study phonon dispersion relations in crystals as well as low frequency excitations such as the "boson" features observed in amorphous solids, liquids and nanomaterials, as well as vibrational dynamics in macromolecules. The energy transfers involved usually range up to a few tens of meV (e.g., 300 cm^{-1}) although higher frequency modes can also be studied in some experiments. The sample volumes for INS experiments are typically larger than those required for diffraction and this leads to limitations in sample environments studied, although results have been reported up to pressures of tens of GPa [20]. Between the elastic and inelastic conditions lies a "quasi-elastic" neutron scattering (QENS) regime occurring at very low energy transfers (i.e., $(h/2\pi)\omega = E_s - E_i < 1$ meV) [23]. This provides new information on the dynamics of macromolecular relaxation and H_2O/D_2O diffusional processes when combined with H/D isotopic contrast experiments. QENS experiments are now being implemented under high pressure conditions up to a few hundred MPa.

5.3 Devices and Detectors Developed for *In Situ* Experiments and Studies

Many variations on diamond anvil cell (DAC), large volume press (LVP) and environmental or reaction cell designs have been developed and applied to carry out *in situ* studies of materials and take full advantage of laboratory and synchrotron-based X-ray facilities and neutron sources for studies under various conditions, including exposure to extreme high-P,T, chemical reactions and structural transformations in polymers and solid state compounds, and structure and function of industrial and biological materials and systems [16, 18, 20, 24–41].

Early X-ray studies recorded the transmission, reflection and diffraction of X-rays by materials *via* blackening of photographic plates or films, mainly by photo-induced formation of Ag nanoparticles from silver salts. These have been largely replaced by image plate technology using photosensistive materials that develop colour centres on exposure to X-rays. The pattern is then read out in scanning mode using a laser to provide a digital image: such detectors are used in synchrotron X-ray diffraction experiments especially at high-P,T beamlines. They do not provide a very high degree of time resolution (on the order of minutes) because of the necessary readout time for each image, and are not typically used for kinetics experiments. CCD cameras with currently up to 4,096 × 4,096 pixels are now used extensively as area detectors for laboratory and synchrotron-based studies. The incident X-rays are absorbed by a thin layer of a phosphor such as Gd_2O_2S:Tb (GOS) or CsI and converted into light signals that are detected by the CCD based on Si photodiode technology. The time resolution is determined by a readout time, that is typically ~1–2 s for commercial systems, although this can now be reduced to <1 s. Both detector types operate between ~1–80 keV making them ideal for synchrotron X-ray scattering experiments carried out with incident beams in this energy range.

Multi-wire detectors with time resolution extending into the millisecond range [42] are also in common use for materials synthesis and studies of functional operation especially at synchrotron sources, as well as P,T-jump studies of polymers and biomaterials [9, 10, 43–46]. Multi-wire and micro-strip gas ionisation detectors contain a grid of fine wires or a patterned metal anode inside a gas-filled chamber that acts as a proportional counter for single photon events. The incident X-rays cause ionisation of the gas particles and electrons are accelerated toward the anode wire by a field of ~100 V/cm and the current is detected by a low noise amplifier. They are typically used as position sensitive detectors for both energy- and angle-dispersive experiments and they can provide submillisecond detection timescales. They have been developed for a wide range of time resolved diffraction, X-ray absorption and tomography studies of materials synthesis and processing. Single wire detectors have been developed for SAXS investigations of polymers and biomolecules following P,T-jump relaxation experiments [8, 47]. For situations where large specimen containers (e.g. industrial processors and image-scanners) and versatile environmental cells (i.e. for imposing specific operating conditions) are involved, tomographic energy-dispersive detector imaging (*TEDDI*) has served the community well [48]. This technique is now undergoing significant and promising enhancements resulting from current advances in detector technology, in particular the orders of magnitude decrease in total image detection-scanning time that can be obtained with the use of fabricated energy-dispersive detector arrays such as the HEXITEC-CdZnTe-array [49, 50].

Laser ramp and shock studies conducted at national laboratories including new free electron laser (FEL) facilities are being combined with time- and spatially-resolved detection techniques to study the dynamic response of materials on a nano- to femtosecond timescale [5–7, 51]. The highest time resolution for X-ray imaging, diffraction, fluorescence and absorption experiments is obtained by X-ray streak cameras that can currently study events down to ~250 fs and below. They provide spatially resolved information on the signal intensity and are used extensively in shock physics experiments [3, 4, 6, 7, 51]. The technology is analogous to that of cathode ray tubes used previously throughout the TV industry. The incident X-rays strike a gold electrode that acts as a photocathode. This is biased at −10 kV with respect to an anode mesh located downstream and launches an ~500 V pulse of electrons towards the 2D detector screen based on photocathode technology (e.g. CsI/or GaAs:P image intensifier) where an optical signal is detected. The electron beam is scanned rapidly by a voltage sweep between deflection plates with a circuit closed by a GaAs photoconductive switch triggered by a Ti:sapphire femtosecond laser operating in the IR range. Such techniques are particularly well adapted to synchrotron X-ray scattering studies of ultra-fast processes where the recording of the events can be linked to the time structure of the electron storage ring, and even faster time-resolved processes are being developed at free electron laser facilities, as well as at large national and international laser facilities where dynamic ramp or shock conditions are combined with backlighting to produce X-ray sources and investigate states of matter under extreme P,T conditions.

In the sections below we illustrate several of these techniques and their application to different types of *in situ* experiments involving phase changes, materials synthesis and functional operation, and also to investigate the dynamical behavior of polymers and biological systems.

5.4 *In Situ* Studies of Sn Melting

Tin lies at a boundary between semiconducting and metallic behaviour between Si, Ge and Pb elements in the periodic table and it has unusual bonding and structural properties. As β-Sn is compressed it transforms into a body-centred structure that exhibits a tetragonal distortion (*bct*). Above P ~40 GPa this is joined by a cubic (*bcc*) polymorph and the two apparently coexist until ~60–70 GPa. We still do not have a fundamental understanding of this unusual behavior.

We also observed unusual behaviour of the melting curve (dT_m/dP) in the 40–70 GPa [52]. Here we took advantage of new techniques developed at the ESRF to combine (a) a well-controlled and mechanically stable environment for laser-heating experiments in the DAC to megabar pressures, with (b) rapid (1–2 s timescale) CCD recording of the X-ray scattering simultaneously with determination of the temperature by optical emission spectrometry, and (c) analysis of the observation of the melting point under high-P,T conditions as the first appearance of a liquid *S(Q)* scattering signature in the X-ray diffraction pattern [53] (Figs. 5.1 and 5.2). The technique was demonstrated for previous melting studies of Pb and Ta [53–55].

Other laser-heated DAC studies have investigated melting phenomena in metals and ceramic compounds using a "laser speckle" technique [56, 57]. Here the diagnostic used to provide evidence of melting is rapid motion observed in the speckle pattern that arises from coherent interference among components of an incident laser beam reflected from the sample surface. Such experiments can provide a rapid and convenient technique for laboratory based melting studies but they can lead to problems of interpretation. The speckle motion could be related to dynamic structural instabilities at the sample surface, or chemical reactions induced by interactions with the pressure transmitting medium. In their synchrotron based XRD experiments for Ta melting at high-P,T in a laser-heated DAC, Dewaele et al. showed the formation of TaC species that could have interfered with determination of the melting relation by the laser speckle method [54]. The X-ray investigations using a first onset of liquid *S(Q)* as a melting criterion resulted in a melting curve in much better agreement with shock physics and theoretical predictions [54, 55, 58].

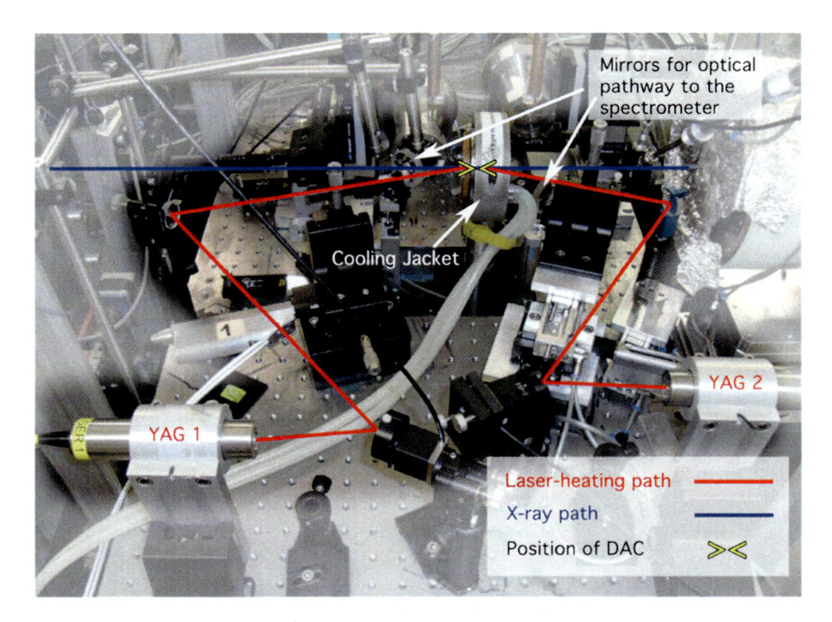

Fig. 5.1 Optics arrangement developed for *in situ* laser heating and melting studies at ID27, European Synchrotron Radiation Facility (Grenoble, France) [53]. The sample is compressed in the DAC inside the cooling jacket at the center of the image and heated using double-sided laser-heating with Nd:YAG lasers (*red* path). The center of the hot spot is coincident with the X-ray beam arriving from the left (*blue* path). X-ray diffraction patterns are collected (CCD detector at right) from the same 2 × 2 μm^2 area as the thermal emission measurements recorded by a spectrometer behind the central area (above the top of this image)

In our work we used the time-resolved XRD technique to determine melting in Sn up to megabar (P > 100 GPa) pressures. Figure 5.2 shows a series of X-ray diffraction patterns collected with increasing temperature, revealing the onset of melting as the first signature of diffuse scattering. Previously the melting curve had been studied up to P ~ 60 GPa using shock techniques combined with MD simulation and modelling (Fig. 5.3). The results indicated a flattening of the melting relation above approximately 40 GPa and it was expected from extrapolation that the maximum T_m value would not exceed ~3,000 K until transformation into the hcp phase was encountered at ~150 GPa. That view appeared to be confirmed by laser-heated DAC experiments conducted up to 68 GPa using the speckle technique to diagnose the onset of melting [57]. However our experimental X-ray results carried out to 105 GPa indicate a quite different picture [52]. Above 70 GPa the melting line rises steeply to attain a value near 5,500 K as the pressure enters the megabar range (Fig. 5.3). These results have implications for the Sn phase diagram as a function of static and dynamic high-P,T conditions.

Mechanical shocks launched into materials by dynamic impact of projectiles in gas gun experiments, explosions or rapid injection of electromagnetic energy provide unique information on the behaviour of matter under extreme compression and high strain rates. Typically mechanical shock experiments follow a trajectory in

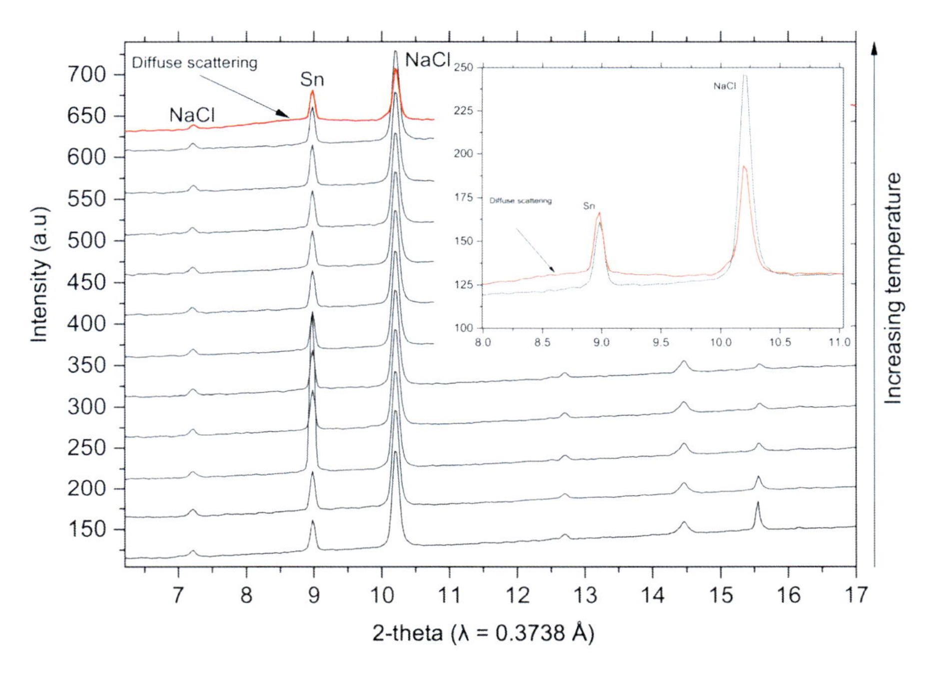

Fig. 5.2 X-ray diffraction patterns of Sn recorded in the DAC up to the melting point at high pressure (red). The data were obtained at ESRF ID27. Data sets are taken ~4 s apart with the temperature increased between each exposure by ramping the laser power steadily as a function of time. The final X-ray pattern at the top reveals the onset of melting by the first observation of diffuse scattering around the most intense Sn reflection near 9° 2θ. The appearance of the diffuse signal from molten Sn is clearly observed (inset in *red*)

P,T space determined by the Hugoniot relation and phase transformations are identified by a marked contrast in material properties. It is possible to investigate other P, T paths by pre-heating or pre-compressing samples to map out important phase relations including the melting curve. Pulsed power techniques using a rapid and massive input of electromagnetic energy into the system can also generate off-Hugoniot states while P,T ramps generated by focusing intense laser beams on a target allow exploration of a wide region in the phase diagram extending into the TeraPascal (TPa) pressure regime with control over the maximum T values obtained. During these experiments it is critically important to examine the structural state of materials subjected to the high dynamic stress and strain rate conditions. X-ray scattering and spectroscopy provide important tools for such studies [5, 7, 51].

5.5 *In Situ* Studies of Semiconductor Clathrate Synthesis

Semiconductor clathrates are tetrahedrally-bonded frameworks based on elements in groups 13–15. The networks define large (4–8 Å) cages that are fully or partly filled by electropositive metal atoms. They have potentially important electronic

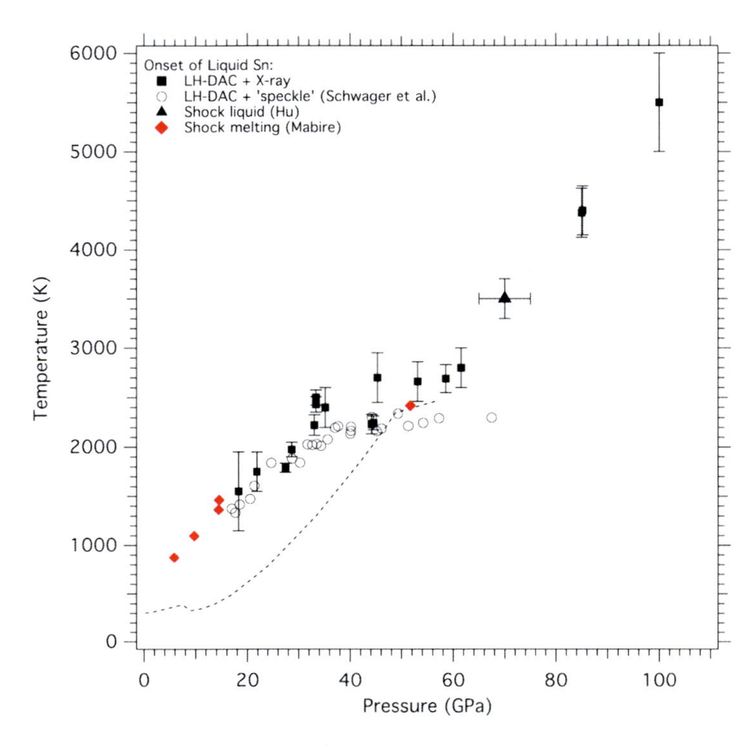

Fig. 5.3 Melting of Sn to P ~1 Mbar. *Black squares* are the present results, *open circles* are LHDAC "speckle" results [57]. Both results agree with previous shock data (*red diamonds*) [59]. The Hugoniot is shown as a *dashed line* [60]

and thermal properties and efforts are under way to understand and control their synthesis and processing. We applied synchrotron-based X-ray diffraction combined with time-resolved multiwire detector technology to study the synthesis mechanism for a major family of semiconductor clathrates [61].

Currently three semiconductor clathrate structure types are known but others are predicted to occur. The type I structure has a primitive cubic unit cell constituted by dodecahedral and tetrakaidecahedral cages occupied by metal guest atoms to give compositions such as Na_8Si_{46}. The type I phase Na_8Si_{46} is typically produced by metastable thermal decomposition under vacuum conditions from the Zintl phase NaSi, by removal of volatile alkali component from the starting solid followed by recrystallisation of the clathrate material. How this process occurs was not known previously but it was generally presumed that because of the large differences in chemical composition and crystal structure of the precursor and product compounds an intermediate amorphous layer should be formed initially at the Zintl phase surface followed by nucleation and growth of the clathrate structure. A second major assumption was that Na atoms or ions remaining within the surface layer after extraction of Na into the vacuum would result in templating formation of the Si clathrate framework structures around the large atomic/ionic species.

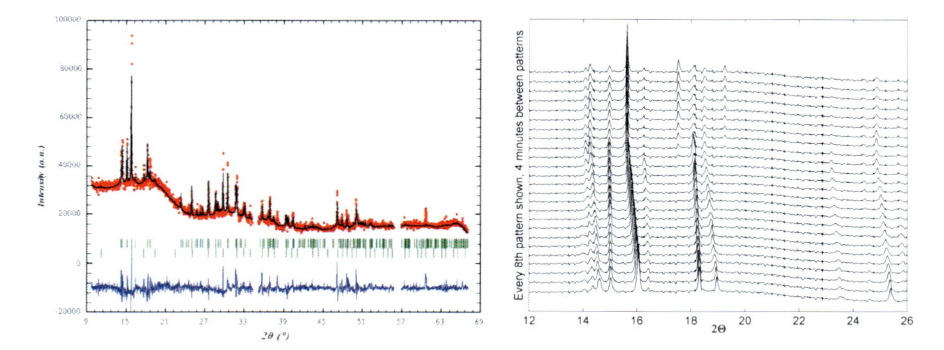

Fig 5.4 In situ XRD experiments during high-T treatment of NaSi Zintl phase and formation of Na8Si46 clathrate. *Left*: A representative X-ray diffraction pattern obtained as part of a time series obtained during heating ramp and synthesis experiments of Na8Si46 clathrate from NaSi Zintl phase. *Right*: a sequence with every 8th 30 s XRD pattern is shown, giving a spacing of 4 min between the displayed patterns over the transformation of NaSi into Na8Si46 clathrate under dynamic vacuum conditions [61, 62]

In order to study the synthesis reaction several experimental requirements had to be met. The starting NaSi phase is air and moisture sensitive and it is derived from a reaction between elemental Na and Si, so the initial samples for the experiment must be prepared and loaded using dry box conditions. Next the reaction takes place under dynamic vacuum (10^{-4} bar) conditions. That required development of new experimental designs for *in situ* studies of the clathrate formation reaction [61, 62]. To initiate the reaction and study the formation process required placing the sample inside a programmable furnace providing temperature ramp rates between 2 and 200 °C/min, up to T ~500–600 °C. These conditions were achieved by designing and constructing a specialised sample environment and furnace arrangement compatible with the synchrotron beamline and X-ray detector. These had been configured during the initial design of the beamline facility to provide maximum flexibility for this type of study. In a typical run the temperature was ramped at 8 °C/min to a target value of 500 °C to observe the initial stages of the reaction, and then held constant. Diffraction data were usually collected every 30 s during the first few hours of an experiment run, then at longer intervals (e.g. 1 min) afterwards. A typical data set along with Rietveld refinements of the NaSi and Na_8Si_{46} phases is illustrated in Fig. 5.4. The broad background signal is due to the glass capillary used to enclose the sample and this was removed before data analysis.

In our work to study this process in detail we took advantage of synchrotron beamline capabilities developed for versatile X-ray scattering combining WAXS and SAXS studies at station 6.2 of the UK Daresbury Synchrotron Radiation Source (SRS), now closed [43, 44, 63]. These capabilities were combined with the parallel development of a new series of multi-wire detectors capable of obtaining high-quality diffraction patterns for a wide range of materials on a 1–10 s timescale, extending down to microseconds in certain cases [43, 44]. To study the Na_8Si_{46} clathrate synthesis process we carried out angle dispersive X-ray diffraction experiments at SRS 6.2 at an incident beam energy 8.856 keV. The diffraction signals were detected using the position sensitive RAPID 2 multi-wire detector providing refineable data over a 60° 2θ range with ~0.06° resolution on a 1–15 s time scale relevant to our synthesis experiments. The detailed mechanisms of clathrate formation were then studied by performing Rietveld and LeBail analyses on each data set.

During the initial temperature ramp we could determine the thermal expansion coefficient of the starting NaSi Zintl phase for the first time. This has a remarkably high value (4.54×10^{-5} °C^{-1}) comparable with ionic conductors such as AgI. Just as the target temperature (500 °C) was attained after 60 min the appearance of characteristic Na_8Si_{46} clathrate peaks could be observed in the pattern and these grew with continued heating. At the point where the clathrate phase emerged the NaSi phase showed un unexpected drop in cell volume associated with loss of Na atoms to the vacuum. The reduction of Na+ to Na species must be accompanied by production of additional Si-Si bonds and this certainly plays a key role in the mechanism of clathrate formation. Interestingly we found no evidence for an intermediate amorphous phase during formation of the clathrate material. Instead we observed an unexpected high degree of structural coherency between the Na_8Si_{46} clathrate and the underlying Zintl substrate, and we could identify particular diffraction peaks that had "joint ownership" between the two structures. The result indicated an unusual, epitaxial-like, clathrate growth process as Na atoms are removed from the NaSi material. The Na atoms at the interface between the $Na_{1-x}Si$ surface and the emerging clathrate act as templates for the growing microporous crystalline solid. Identification of this mechanism suggests new techniques for preparation of semiconductor clathrate materials and interfacing them with semiconductor technology.

5.6 Real Time XRD-CTomography of Catalyst Phase Evolution During Synthesis/Activation

Porous ceramic-supported metals and metal oxides are important heterogeneous catalyst systems for many industrial chemical processes. The nature of the final catalyst dispersion and its spatial distribution within the support material are important factors influencing their ultimate performance. However, such distributions are in general not well understood and, yet further, there is almost no information on how they evolve spatially during the synthesis process.

Dynamic XRD-CT (X-ray Diffraction Computed Tomography) has now been applied [10] to uncover the progression of changing structural distributions of the start-, end- and intermediate-phases formed on a γ-Al_2O_3 support during thermal treatment of the impregnated precursor (Ni-(en)-Cl; en = ethylenediamine) up to activation of the final metal catalyst. XRD-CT requires the use of an intense high energy rapid X-ray detection facility (here, 86.88 keV X-rays on station ID15B of the ESRF synchrotron, Grenoble) collecting a series of transmission "projection" measurements over a schedule of 43 translations (100 μm apart) times 30 rotations (6° apart), corresponding to 1,290 projections for each time-step (~600 s) for impregnated cylindrical supports (diam. = 3 mm by length = 3 mm) during calcination under flowing helium from ambient to 500 °C. The resulting sinograms are

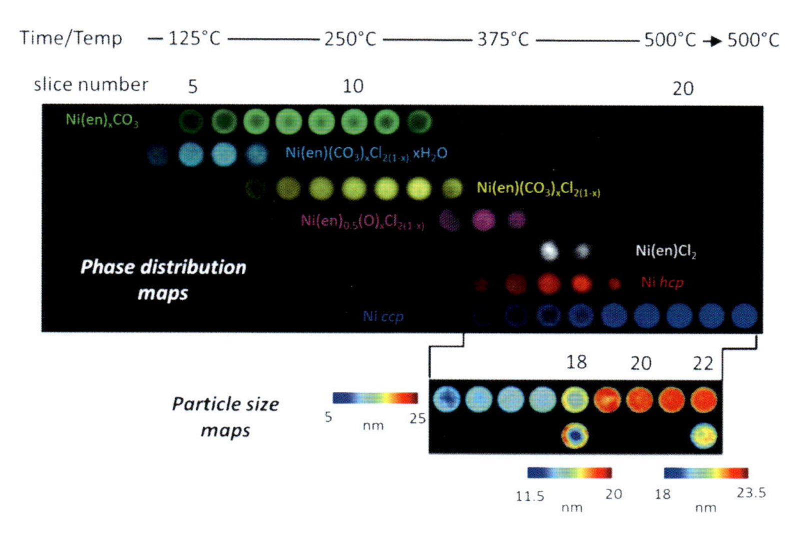

Fig. 5.5 XRD-CT 2D image-maps of features, derived from diffraction patterns obtained during thermal activation of a supported Ni-catalyst, *versus* time and temperature of synthesis. The colour codes for the phase maps are indicated. The particle size maps show the variation in crystallite size (nm) as the growth of the phase proceeds. For time slices 18 and 22, the data have been re-plotted with bespoke colour axes (*right*)

back-projected to form 43 × 43 pixel reconstructed images with contrast based on the selected X-ray diffracted signal selected. By taking representative peaks for each appearing phase, a temperature-series of phase maps can be constructed for one (or more) representative circular slice through the catalyst cylinder.

Figure 5.5 presents a series of such circular slice phase maps for seven phases (out of nine identified) displaying a range of distributions (periphery – 'egg shell'; internal – 'egg yolk') during the reaction, dehydration and decomposition stages leading to the eventual end-product of face-centred-cubic nickel; additional features such as the crystallite size maps, derived from the dependence of diffraction peak width on crystallite size, and the discovery of voids within the alumina support are in effect a welcome bonus. A more detailed analysis of the maps reveals that there are two decomposition pathways to the production of fcc-Ni which result from the two different crystalline phases that form over different locations (1: $Ni(en)_xCO_3$ phase at the periphery as 'egg shell'; 2: $Ni(en)(CO_3)_xCl_{2(1-x)} \cdot xH_2O$ as 'egg-white') and thereafter progress through two different chemical pathways. These observations show that the detailed space/time chemistry involved in real functional catalyst objects is intricate; further they show that a true complete picture of the evolutionary process cannot be gained from conventional single position time-resolved measurements, as is almost-universally practised. Therefore we expect that this technique will have considerable impact on a range of dependent disciplines from functional materials science to chemical engineering and biological materials.

5.7 P,T-Jump SAXS and SANS Studies of Soft Matter and Biomolecules

High pressure studies in biology, soft matter research and biophysics represent a rapidly emerging area with applications ranging from exploring and understanding the deep Earth biosphere and perhaps elsewhere in the universe, including studies of the origins of life [64]. Technological applications extend from food processing and preservation studies to development of bio-derived or -inspired products as well as synthesis and design of nanomaterials [65]. Biophysics studies that combine a wide range of static and dynamic X-ray and neutron scattering techniques can be carried out under high pressure conditions along with a variety of spectroscopy, imaging and thermal analysis measurements of biologically important molecules extending to living systems leading to understanding of their survival and adaptability to extreme conditions. The field is currently undergoing a renaissance with the emergence of new synchrotron and neutron scattering techniques that can be applied to study biological materials both *in situ* and *in vivo* under a wide range of conditions.

Pressure stress affects all levels of cellular physiology including metabolism, membrane physiology, transport, transcription and translation. The biological membrane is one of the most pressure sensitive cellular components. The T- and P-dependent structure and phase behavior of lipid bilayers, differing in chain configuration, headgroup structure and composition is revealed by a combination of thermodynamic and spectroscopic studies along with X-ray and neutron scattering experiments, leading to new understanding of the lateral organization of phase-separated lipid membranes and model raft mixtures, the influence of peptide and protein incorporation on membrane structure and dynamics, and the effects of additives such as ions, cholesterol, and anaesthetics [21, 64, 66].

Synchrotron X-ray scattering as well as SANS experiments allow us to study the time course of lipid phase transitions and to search for possible transient intermediate structures, with a view to unravelling the underlying transition mechanisms. For example, Squires et al. [67] observed a gyroid ($Q^{G}{}_{II}$) to the diamond ($Q^{D}{}_{II}$) bicontinuous cubic mesophase phase transition at 59.5 °C after a P-jump from 60 to 24 MPa. Interestingly, they could also see a transient inverse hexagonal H_{II} phase which is not observed in static experiments (Fig. 5.6). Moreover, the P-jump approach offers distinct advantages over T-jump techniques in that both "up" and "down" experiments can be conducted, the sample pressure reaches equilibrium very quickly, in contrast to T which creates a gradient across the sample [46]. Recently, Brooks and co-workers developed a new SAXS cell that allows both static and dynamic P-experiments in a wide T-range from −20 to +120 °C with P-jumps varying between 0.1 and 500 MPa [47]. The small cell volume (3–11 mm^3) makes adiabatic heating of the sample upon P-variation negligible.

P-jump studies of proteins with SAXS and SANS include the determination of the activation volume of protein folding that provides important information on the folding process and the nature of the transition state [21, 64, 68]. Another important area of research is protein dynamics as this underlies protein activity [69]. The

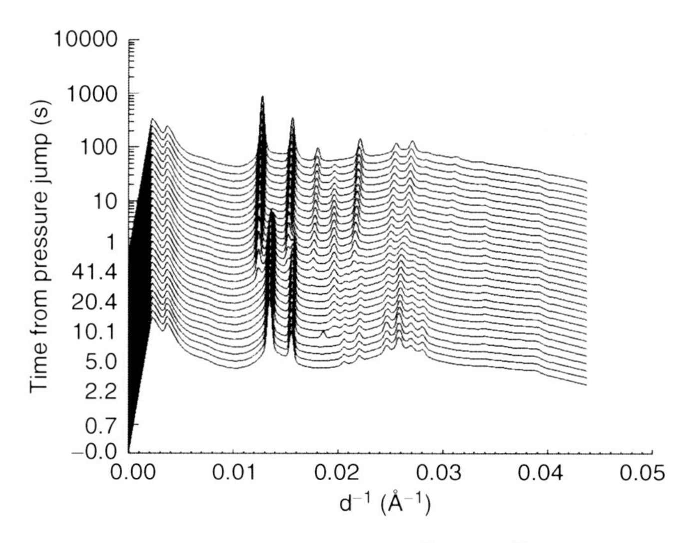

Fig. 5.6 Transformation between cubic mesophases ($Q^{G}II$ and $Q^{D}II$) via an inverse hexagonal intermediate in 2:1 lauric acid: dilauroylphosphatidylcholine at a hydration of 50 % (w/w) (Redrawn from Ref. [67] with data provided by the author). A short-lived species and a longer-lived intermediate peak can be seen at approximately 0.018 and 0.0195 $Å^{-1}$ respectively

dynamics of proteins is often measured by QENS [19, 70, 71]. One major finding so far is a reduction of the internal fluctuations and the diffusion time of the protein with increasing pressure [19]. Similar observations have been made using elastic incoherent neutron scattering that is reported to have some advantages over QENS [72]. Important contributions in methods and cell development have been made by a number of groups [19, 70, 71], although the field is still very much in development.

Kinetic studies are also being applied to whole organisms. In recent work the metabolic activity of *Shewanella oneidensis* MR-1 cells was monitored as a function of time at a given pressure using X-ray near edge structure spectroscopy (XANES) [73, 74]. The results show that the metabolism monitored by Se^{4+} reduction continues at pressures where growth of the microorganisms has already come to a halt. Recent results obtained using quasi-elastic neutron scattering (QENS) to study the transport of water within and through the membrane of cells of this organism are described in the next section.

5.8 Quasi-elastic Neutron Scattering Studies of Organisms at High Pressure

The QENS technique provides a unique method to study the macromolecular dynamics and diffusional processes in polymers and biomolecules extending to whole living organisms contained under high pressure conditions [19, 23, 70, 75–78]. Critical experiments have been developed at neutron scattering facilities using time of flight (TOF) techniques. Pioneering experiments on organisms at high pressure are being carried out at the high resolution TOFTOF instrument at the FRM-II (Forschungs-Neutronenquelle Heinz Maier-Leibnitz) reactor associated with the Technical University in Munich, Germany. An incident neutron beam with wavelength 1.5–16 Å is chopped at speeds between 1,000 and 22,000 rpm and passed through the sample on to a detector bank to achieve a 5 μeV to 5 meV energy resolution and to access the picosecond regime and investigate internal as well as diffusive motions [79]. The TOF flight data are analyzed to separate and study the elastic and quasi-elastic contributions to the scattered spectrum. At this facility experiments are being designed to study the macromolecular dynamics in polymers as well as biologically important molecules by *in situ* QENS under high pressure conditions [19, 70]. We have begun to apply these techniques to investigate the transport of H_2O across cell membranes of organisms that survive and evolve at depths of 10 km or more, providing new insights into the cell dynamics of hyperpiezophile species. To carry out isotope contrast experiments D_2O-exchanged organisms were prepared by feeding them heavy isotopic nutrients at the Deuteration Facility in Grenoble, France (M. Haertlein, T. Forsyth), and transported to FRM-II for QENS experiments [77].

Structure factors $S(2\theta) = \int_{-E_i}^{+\infty} S(2\theta, \omega) d\omega$ are evaluated by integrating the two-dimensional spectra over the energy transfer range (Fig. 5.7). Dynamic studies are enabled by evaluating the incoherent dynamic structure factors after regrouping the profiles in slices at constant $Q(Q = (4\pi/\lambda)\sin\,\theta)$ $(Q = (4\pi/\lambda)\sin\,2\,\theta)$. Each profile at fixed Q is then analyzed by combining Gaussian and Lorentzian functions. The first of these refers to the energy resolution of the instrument and experiment, while the Lorentzian profile takes into account the global diffusion processes. As shown in Fig. 5.8 at higher Q values the curves become broadened and several Lorentzian contributions must be taken into account. For a bacterial cell dispersed in a suspension or growth medium the scattering profile takes into account all possible diffusive as well as macromolecular motions and it is strongly affected by the signal related to the bulk water. The simplest way to evaluate the various components is by using different H/D isotopic compositions both inside the cell (obtained by treating cells in deuterated growth media) and in the surrounding liquid, and then subtracting the contributions in various ways to highlight different contributions to the QENS signal. The technique is based on the large difference in the incoherent scattering length σ_{inc} between H and D (σ_{inc}H = 79.74 barn; σ_{inc}D = 2.01 barn). The procedure can be applied to give information on processes including, e.g.,

- dynamics related to the denser layer near the membrane estimated by the subtraction {[(D-cell in H-buffer) – (D-cell in D-buffer)] – (H-buffer)}
- dynamics across the membrane estimated by {[(H-cell in D-buffer) – (D-buffer)] – [(D-cell in H-buffer) – (H-buffer)]}.

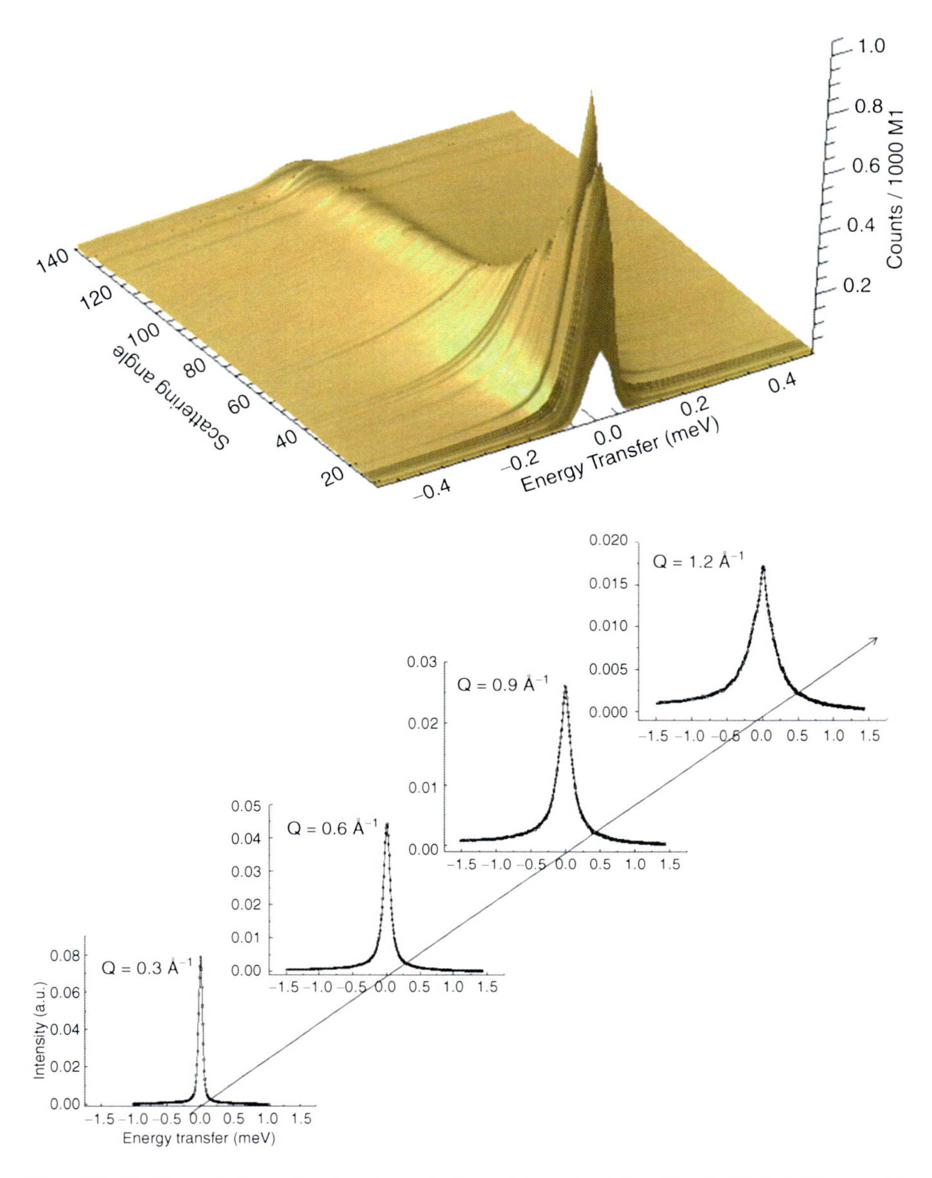

Fig. 5.7 Static and dynamic structure factors determined for D-substituted *Shewanella oneidensis* suspended in H-buffer during QENS experiments at room P. *Left*: $S(Q)$ for all investigated scattering angles; *Right*: Dynamic structure factors recorded at different Q values (0.3, 0.6, 0.9, 1.2 $Å^{-1}$). QENS experiments carried out at TOFTOF (FRM-II)

The diffusion coefficient (D) can be evaluated by considering the half width at half maximum ($HWHM$) of each Lorentzian function. These parameters are related by $HWHM = DQ^2$ assuming a simple jump diffusion model: other models take into account the residence time (τ_o) between jumps [75] (Fig. 5.8). The QENS data

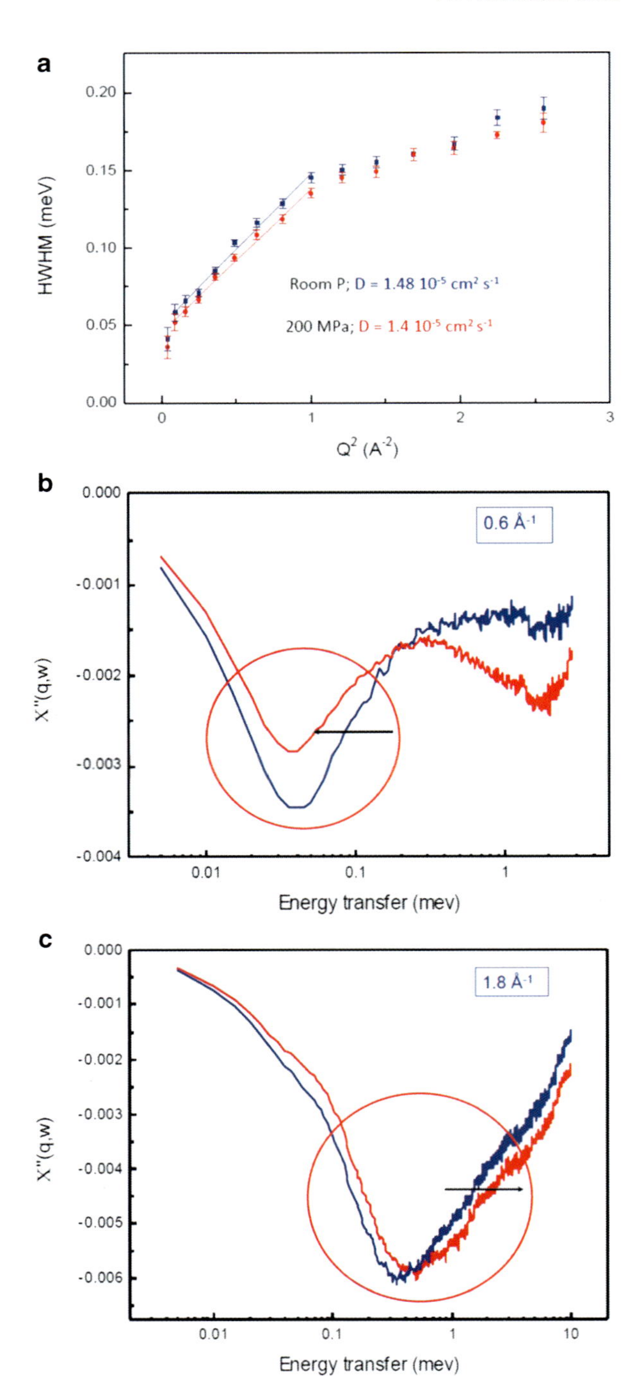

Fig. 5.8 Dynamics of water near the cell membrane of *Shewanella oneidensis* at room P (blue) and 200 **MPa** (*red*) **probed using QENS.** *Left*: Diffusion coefficients calculated from the linear dependency of the HWHM based on low Q values below 1 $Å^{-1}$; *Right*: Dynamic susceptibility data at low and high P recorded at $Q = 0.6\ Å^{-1}$ and $1.8\ Å^{-1}$

contain both fast (confined) and slower (translational) contributions. These can be separated by calculating the dynamic susceptibility ($\chi''(Q, \omega) = S(Q, \omega)/n(Q, \omega)$) to the dynamic structure factor (here n is the Bose factor $n(Q, \omega) \approx k_B T/\hbar\omega$) (Fig. 5.8) [80]. Our new data indicate that the water dynamics slow down at higher pressure, with results that can be compared with literature values for bulk water [78] and for *E. coli* samples at ambient P [77]. Applying the QENS technique at high pressure thus constitutes a new approach to studying the water dynamics associated with living cells under extreme conditions [64].

Acknowledgements Work in PFMs group has been supported by funding from the Wolfson Trust/Royal Society, EPSRC, the Institute of Shock Physics supported by AWE, Leverhulme Trust (UK) and the Deep Life directorate of the Deep Carbon Observatory (Sloan Foundation, USA). PB has been supported by EPSRC and industry. Colleagues and co-workers who contributed to the data shown here include Drs. D. Daisenberger, A. Salamat (high-P,T studies of Sn melting), P. Hutchins (Si clathrate formation), T. Forsyth, M. Haertlein, G. Simeoni, M.-S. Appavou, R. Hazael (high-P biological studies and QENS experiments). We thank Dr. A.M. Squires for providing a high resolution version of his SAXS data.

References

1. Duffy TS (2005) Synchrotron facilities and the study of the Earth's deep interior. Rep Prog Phys 68:1811–1859
2. Liu H, Duffy T, Ehm L, Crichton W, Aoki K (2009) Advances and synergy of high-pressure sciences at synchrotron sources. J Synchrotron Radiat 16:697–698
3. Higginbotham A, Hawreliak JA, Bringa EM, Kimminau G, Park N, Reed E, Remington BA, Wark JS (2012) Molecular dynamics simulations of ramp-compressed copper. Phys Rev B 85:024112
4. Kalantar DH, Belak JF, Collins GW, Colvin JD, Davies HM, Eggert JH, Germann TC, Hawreliak JA, Holian BL, Kadau K, Lomdahl PS, Lorenzana HE, Meyers MA, Rosolankova K, Schneider MS, Sheppard J, Stolken JS, Wark JS (2005) Direct observation of the α-ε transition in shock-compressed iron via nanosecond X-Ray diffraction. Phys Rev Lett 95:075502
5. Rygg JR, Eggert JH, Lazicki AE, Coppari F, Hawreliak JA, Hicks DG, Smith RF, Sorce CM, Uphaus TM, Yaakobi B, Collins GW (2012) Powder diffraction from solids in the terapascal regime. Rev Sci Instrum 83:113904
6. Suggit MJ, Higginbotham A, Hawreliak JA, Mogni G, Kimminau G, Dunne P, Comley AJ, Park N, Remington BA, Wark JS (2012) Nanosecond white-light Laue diffraction measurements of dislocation microstructure in shock-compressed single-crystal copper. Nat Commun 3:1224
7. Wark JS, Whitlock RR, Hauer AA, Swain JE, Salone PJ (1989) Subnanosecond x-ray diffraction from laser-shocked crystals. Phys Rev B 40:5705–5714
8. Conn CE, Ces O, Squires AM, Mulet X, Winter R, Finet SM, Templer RH, Seddon JM (2008) A pressure-jump time-resolved X-ray diffraction study of cubic-cubic transition kinetics in monoolein. Langmuir 24:2331–2340
9. Woenckhaus J, Köhling R, Winter R, Thiyagarajan P, Finet S (2000) High pressure-jump apparatus for kinetic studies of protein folding reactions using the small-angle x-ray scattering technique. Rev Sci Instrum 71:3895–3899
10. Jacques SDM, di Michiel M, Beale AM, Sochi T, O'Brien MG, Espinosa-Alonso L, Weckhuyesen BM, Barnes P (2011) Dynamic X-ray diffraction computed tomography reveals real-time insight into catalyst active phase evolution. Angew Chem Int Ed 50(43):10148–10152

11. Jacques SDM, di Michiel M, Kimber SAJ, Yang X, Cernik RJ, Beale AM, Billinge SJL (2013) Pair distribution function-computed tomography. Nat Commun 4:2536. doi:10.1038/ncomms3536
12. Jacques SDM, Egan CK, Wilson MD, Veale MC, Seller P, Cernik RJ (2013) A laboratory system for element specific hyperspectral X-ray imaging. Analyst 138:755–759
13. Lazzari O, Jacques SDM, Sochi T, Barnes P (2009) Reconstructive colour X-ray imaging. Analyst 134:1802–1807
14. Middelkoop V, Boldrin P, Peel M, Buslaps T, Barnes P, Darr JA, Jacques SDM (2009) Imaging the inside of a continuous nanoceramic synthesizer under supercritical water conditions using high-energy synchrotron X-radiation. Chem Mater 21:2430–2435
15. O'Brien MG, Jacques SDM, di Michiel M, Barnes P, Weckhuyesen BM, Beale AM (2012) Active phase evolution in single Ni/Al2O3 methanation catalyst bodies studied in real time using combined u-XRD-CT and μ-absorption-CT. Chem Sci 3:509–523
16. Bailey IF (2003) A review of sample environments in neutron scattering. Z Kristallogr 218:84–95
17. Goncharenko I, Loubeyre P (2005) Neutron and X-ray diffraction study of the broken symmetry phase transition in solid deuterium. Nature 435:1206–1209
18. Wilding MC, Benmore CJ (2006) Structure of glasses and melts. Rev Mineral Geochem 63:275–312
19. Appavou M-S, Gibrat G, Bellissent-Funel MC (2006) Influence of pressure on structure and dynamics of bovine pancreatic trypsin inhibitor (BPTI): small angle and quasi-elastic neutron scattering studies. Biochim Biophys Acta 1764:414–423
20. Klotz S (2012) Techniques in high pressure neutron scattering. CRC Press/Taylor & Francis, Boca Raton
21. Winter R (2002) Synchrotron X-ray and neutron small-angle scattering of lyotropic lipid mesophases, model biomembranes and proteins in solution at high pressure. Biochim Biophys Acta 1595:160–184
22. Shibayama M, Isono K, Okabe S, Karino T, Nagao M (2004) SANS study on pressure-induced phase separation of poly (N-isoproylacrylamide) aqueous solutions and gels. Macromolecules 37:2909–2918
23. Bée M (2003) Localized and long-range diffusion in condensed matter: state of the art of QENS studies and future prospects. Chem Phys 292:121–141
24. Jayaraman A (1983) Diamond anvil cell and high-pressure physical investigations. Rev Mod Phys 55:65–108
25. Khvostanstev LG, Slesarev VN, Brazhkin VV (2004) Toroid type high-pressure device: history and prospects. High Press Res 24:371–383
26. Liu L-G, Bassett W (1986) Elements, oxides, silicates: high-pressure phases with implications for the Earth's interior. Oxford University Press, New York
27. McMillan PF (2002) New materials from high pressure experiments. Nat Mater 1:19–25
28. McMillan PF (2003) Chemistry of materials under extreme high pressure-high temperature conditions. Chem Commun 9(8):919–923
29. McMillan PF (2005) Pressing on: the legacy of P. W. Bridgman. Nat Mater 4:715–718
30. Wang Y, Rivers M, Sutton S, Nishiyama N, Uchida T, Sanehira T (2009) The large-volume high-pressure facility at GSECARS: a "Swiss-army-knife" approach to synchrotron-based experimental studies. Phys Earth Planet Inter 174:270–281
31. Hemley RJ (2010) Percy W. Bridgman's second century. High Press Res 30:581–619
32. Hemley RJ (1999) Reviews in mineralogy. UltraHigh Press Mineral 37:1
33. Bassett WA (2009) Diamond anvil cell, 50th birthday. High Press Res 29:163–186
34. Besson J-M, Hamel G, Grima T, Nlmes RJ, Loveday JS, Hull S, Hausermann D (1992) A large volume pressure cell for high temperatures. High Press Res 8:625–630
35. Eremets M (1996) High pressure experimental methods. Oxford University Press, Oxford
36. Duffy TS, Ohtani E, Rubie DC (2004) New developments in high-pressure mineral physics and applications to the Earth's interior. Elsevier Science, Amsterdam
37. Hemley RJ, Mao H-k (2002) New windows on Earth and planetary interiors. Miner Mag 66:791–811
38. Irifune T (2002) Application of synchrotron radiation and Kawai-type apparatus to various studies in high-pressure mineral physics. Miner Mag 66:769–790

39. Katrusiak A, McMillan PF (2004) High-pressure crystallography, vol 140, NATO science series. II. Mathematics, physics and chemistry. Kluwer, Dordrecht
40. Wilding MC, Wilson M, McMillan PF (2006) Structural studies and polymorphism in amorphous solids and liquids at high pressure. Chem Soc Rev 35:964–986
41. Holzapfel WD, Isaacs NS (1997) High-pressure techniques in chemistry and physics: a practical approach. Oxford University Press, Oxford
42. Berry A, Helsby WI, Parker BT, Hall CJ, Buksh PA, Hill A, Clague N, Hillon M, Corbett G, Clifford P, Tidbury A, Lewis RA, Cernik RJ, Barnes P, Derbyshire GE (2003) The Rapid2 X-ray detection system. Nucl Instrum Methods Phys Res A 513:260–263
43. Cernik RJ, Barnes P, Bushnell-Wye G, Dent AJ (2004) The new materials processing beamline at the SRS Daresbury, MPW6.2. J Synchrotron Radiat 11:163–170
44. Cernik RJ, Berry A, Helsby WI, Parker BT (2003) The Rapid2 X-ray detection system. Nucl Instrum Methods Phys Res A 513:260–263
45. Cernik RJ, Barclay P, Khor KH, O'Neill W (2006) The manufacture of a very high precision x-ray collimator array for rapid tomographic energy dispersive diffraction imaging (TEDDI). Meas Sci Technol 17:1767–1775
46. Seddon JM, Squires AM, Conn CE, Ces O, Heron AJ, Mulet X, Shearman GC, Templer RH (2006) Pressure-jump X-ray studies of liquid crystal transitions in lipids. Philos Trans R Soc A 364:2635–2655
47. Brooks NJ, Gauthe BLLE, Terrill NJ, Rogers SE, Templer RH, Ces O, Seddon JM (2010) Automated high pressure cell for pressure jump x-ray diffraction. Rev Sci Instrum 81:064103
48. Hall CJ, Barnes P, Cockroft JK, Colston SL, Häusermann D, Jacques SDM, Jupe AC, Kunz M (1998) Synchrotron radiation energy-dispersive diffraction tomography. Nucl Instrum Methods Phys Res B 140:253–257
49. Cernik RJ, Khor KH, Hansson C (2008) X-ray colour imaging. J R Soc Interface 5:477–481
50. Seller P, Bell S, Cernik RJ, Christodoulou C, Egan CK, Gaskin JA, Jacques SDM, Pani S, Ramsey BD, Reid C, Sellin PJ, Scuffham JW, Speller RD, Wilson MD, Veale MC (2011) Pixellated Cd(Zn)Te high-energy X-ray instrument. J Instrum 6, C12009
51. Wark JS, Belak JF, Collins GW, Colvin JD, Davies HM, Duchaineau M, Eggert JH, Germann TC, Hawreliak JA, Higginbotham A, Holian BL, Adau K, Kalantar DH, Lomdahl PS, Lorenzana HE, Meyers MA, Remington BA, Rosolankova K, Rudd RE, Schneider MS, Sheppard J, Stolken JS (2006) Picosecond X-ray diffraction from laser-shocked copper and iron. AIP Conf Proc 845:286–291
52. Briggs R, Daisenberger D, Salamat A, Garbarino G, Mezouar M, Wilson M, McMillan PF (2012) Melting of Sn to 1 Mbar. J Phys Conf Ser 377:012035
53. Dewaele A, Mezouar M, Guignot N, Loubeyre P (2007) Melting of lead under high pressure studied using second-scale time-resolved x-ray diffraction. Phys Rev B 76:144106
54. Dewaele A, Mezouar M, Guignot N, Loubeyre P (2010) High melting points of tantalum in a laser-heated diamond anvil cell. Phys Rev Lett 104:255701
55. Taioli S, Cazorla C, Gillan M, Alfè D (2007) Melting curve of tantalum from first principles. Phys Rev B 75:214103
56. Errandonea D, Schwager B, Ditz R, Gessmann C, Boehler R, Ross M (2001) Systematics of transition-metal melting. Phys Rev B 63:132104
57. Schwager B, Ross M, Japel S, Boehler R (2010) Melting of Sn at high pressure: comparisons with Pb. J Chem Phys 133:084501
58. Dai C, Hu J, Tan H (2009) Hugoniot temperatures and melting of tantalum under shock compression determined by optical pyrometry. J Appl Phys 106:043519
59. Mabire C, Hereil P-L (2000) Shock induced polymorphic transition and melting of tin up to 53 GPa (experimental study and modelling). J Phys IV (France) 10:Pr9-749–Pr9-754
60. Bernard S, Maillet JB (2002) First-principles calculation of the melting curve and Hugoniot of tin. Phys Rev B 66:012103
61. Hutchins PT, Leynaud O, O'Dell LA, Smith ME, Barnes P, McMillan PF (2011) Time-resolved in situ synchrotron X-ray diffraction studies of type I silicon clathrate formation. Chem Mater 23:5160–5167

62. Hutchins PT (2008) In situ synthesis studies of silicon clathrates. PhD thesis, University College, London
63. Greaves GN, Catlow CRA, Derbyshire GE, McMahon MI, Nelmes RJ, Van der Laan G (2008) Two million hours of science. Nat Mater 7:827–830
64. Meersman F, Daniel I, Bartlett D, Winter R, Hazael R, McMillan PF (2013) High-pressure biochemistry and biophysics. Rev Miner Geochem 75:607–648
65. Aertsen A, Meersman F, Hendrickx MEG, Vogel RF, Michiels CW (2009) Biotechnology under high pressure: applications and implications. Trends Biotechnol 27:434–441
66. Wlodarczyk A, McMillan PF, Greenfield SA (2006) High pressure effects on anaesthesia and narcosis. Chem Soc Rev 35:890–898
67. Squires AM, Templer RH, Seddon JM, Woenckhaus J, Winter R, Narayanan T, Finet S (2005) Kinetics and mechanism of the interconversion of inverse bicontinuous cubic mesophases. Phys Rev E 72:011502
68. Woenckhaus J, Köhling R, Thiyagarajan P, Littrell KC, Seifert S, Royer CA, Winter R (2001) Pressure-jump small-angle X-ray scattering detected kinetics of staphylococcal nuclease folding. Biophys J 80:1518–1523
69. Daniel RM, Dunn RV, Finney JL, Smith JC (2003) The role of dynamics in enzyme activity. Annu Rev Biophys Biomol Struct 32:69–92
70. Appavou M-S, Busch S, Doster W, Gaspar AM, Unruh T (2011) The influence of 2 kbar pressure on the global and internal dynamics of human hemoglobin observed by quasielastic neutron scattering. Eur J Biophys 40:705–714
71. Ortore MG, Spinozzi F, Mariani P, Pacarioaroni A, Barbosa LRS, Amenitsch H, Steinhart M, Ollovier J, Russo D (2009) Combining structure and dynamics: non-denaturing high-pressure effect on lysozyme in solution. J R Soc Interface 6:S619–S634
72. Filabozzi A, Deriu A, Di Bari MT, Russo D, Croci S, Di Venere A (2010) Elastic incoherent neutron scattering as a probe of high pressure induced changes in protein flexibility. Biochim Biophys Acta 1804:63–67
73. Picard A, Daniel I, Testemale D, Kieffer I, Bleuet P, Cardon H, Oger PM (2011) Monitoring microbial redox transformations of metal and metalloid elements under high pressure using *in situ* X-ray absorption spectroscopy. Geobiology 9:196–204
74. Picard A, Testemale D, Hazemann JL, Daniel I (2012) The influence of high hydrostatic pressure on bacterial dissimilatory iron reduction. Geochim Cosmochim Acta 88:120–129
75. Bée M (1988) Quasi-elastic neutron scattering, principles and applications in solid state chemistry, biology and materials science. Adam Hilger, Bristol
76. Osaka N, Shibayama M, Kikuchi T, Yamamuro O (2009) Quasi-elastic neutron scattering study on water and polymer dynamics in thermo/pressure sensitive polymer solutions. J Phys Chem B 113:12870–12876
77. Jasnin M, Moulin M, Haertlein M, Zaccai G, Tehei M (2008) Down to atomic-scale intracellular water dynamics. EMBO Rep 9:543–547
78. Chen S-H, Teixeira J, Nicklow R (1982) Incoherent quasielastic neutron scattering from water in supercooled regime. Phys Rev A 26:3477–3482
79. Gaspar AM (2007) Methods for analytically estimating the resolution and intensity of neutron time-of flight spectrometers. The case of the TOFTOF spectrometer. arXiv:0710.5319v1 (physics.ins-det)
80. Wuttke J, Ohl M, Goldammer M, Roth S, Scheider U, Lunkenheimer P, Kahn R, Rufflé B, Lechner R, Berg MA (2000) Propylene carbonate reexamined: mode-coupling beta scaling without factorization? Phys Rev E 61:2730–2740

Chapter 6
Dynamics of Mechanochemical Processes

Elena V. Boldyreva

6.1 Scope of Mechanochemistry

The term "mechanochemistry" seems to have been first introduced by Ostwald in the *Textbook of General Chemistry* in 1891, where he considered, in particular, various methods for the stimulation of chemical processes. This term refers to chemical reactions involving reagents in any aggregated state (liquids, solids), although most frequently it is used in relation to solid-state processes and reactions either initiated by any type of mechanical treatment, including hydrostatic loading, or involving reagents, which were activated mechanically. The IUPAC defines a mechanochemical reaction as a "chemical reaction, that is induced by mechanical energy" (see *IUPAC Compendium of Chemical Technology*, 2nd ed. (the Gold Book), Compiled by A.A. McNaught & A. Wilkinson, Blackwell Scientific Publication, Oxford 1997, http://goldbook.iupac.org).

Mechanochemistry as an applied science dates back to prehistoric times, although the achievements of those times were not patented or even documented. Grinding, impacting, shaking, rubbing, rolling, and other types of mechanical treatment were widely used to generate fire, to process minerals, tissues, corns, herbs, meat, to produce and process the first inorganic materials, e.g. ceramics for pottery, metals for armory, construction materials, as well as to make the first pigments and drugs. Initiation of explosions by impact, widely used for military purposes, also belongs to the first examples of applied mechanochemistry in the "pre-publication era".

Mechanochemistry today deals with all possible types of compounds and finds various applications ranging from biology and pharmaceuticals to minerals processing, geology, and catalysis (Fig. 6.1).

There are many different types of mechanical treatment. Treatment can be continuous, as is the case for hydrostatic compression. However, much more

E.V. Boldyreva (✉)
Institute of Solid State Chemistry and Mechanochemistry SB RAS,
Novosibirsk State University, Kutateladze 18, Novosibirsk 630128, Russia
e-mail: eboldyreva@yahoo.com

J.A.K. Howard et al. (eds.), *The Future of Dynamic Structural Science*,
NATO Science for Peace and Security Series A: Chemistry and Biology,
DOI 10.1007/978-94-017-8550-1_6, © Springer Science+Business Media Dordrecht 2014

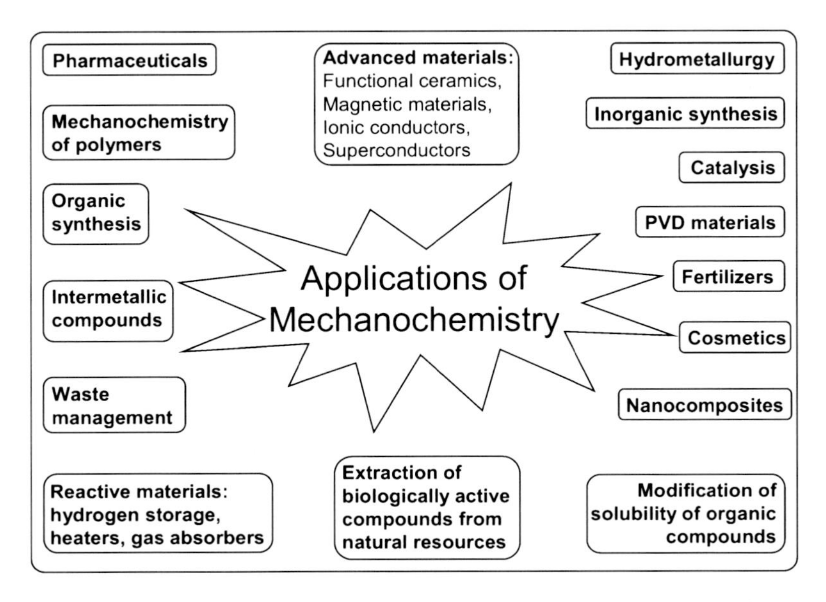

Fig. 6.1 Applications of mechanochemistry

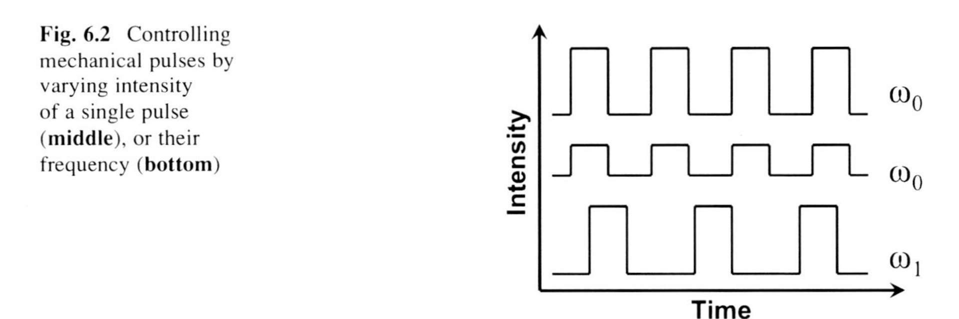

Fig. 6.2 Controlling mechanical pulses by varying intensity of a single pulse (**middle**), or their frequency (**bottom**)

often either mechanical treatment is a single and a very short act (for example, a single impact, indenting, or cutting with a knife), or it consists of pulses, which can be characterized by energy, duration, frequency: depending on the latter, the system has different chances to relax between the pulses (Fig. 6.2).

The outcome of mechanical treatment is, in general, strongly dependent not only on the total mechanical energy input into the sample, but also on the type of treatment. In particular, a continuous hydrostatic compression results in the physical and chemical processes, which are very different from those induced by compression combined with shear. The ratio impact/shear during a treatment can vary in a wide range, depending on the type of a device used (Fig. 6.3), and this also plays a crucial role.

Relaxation of mechanical stress can proceed *via* different pathways (Fig. 6.4), and this is also of primary importance for the resulting chemical changes in the system.

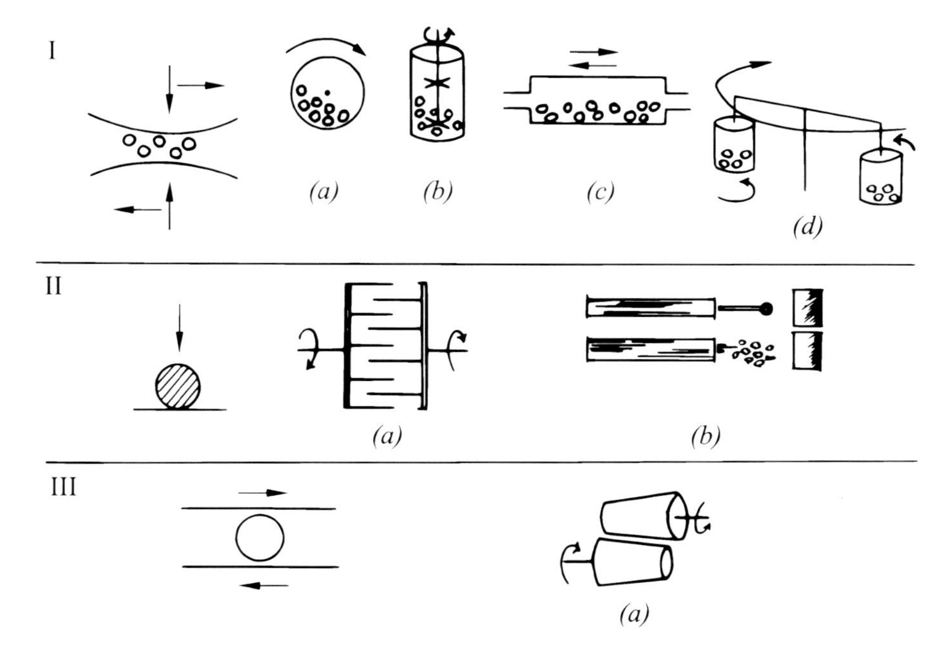

Fig. 6.3 Different types of mechanical treatment: I – shear + impact (*a* – ball mill, *b* – attritor, *c* – vibration mill, *d* – planetary mill); II – impact (*a* – pin mill, *b* – jet-mill); III – shear (*a* – rollers) (Boldyrev [6])

Fig. 6.4 Different routes for stress field relaxation (Boldyrev [6])

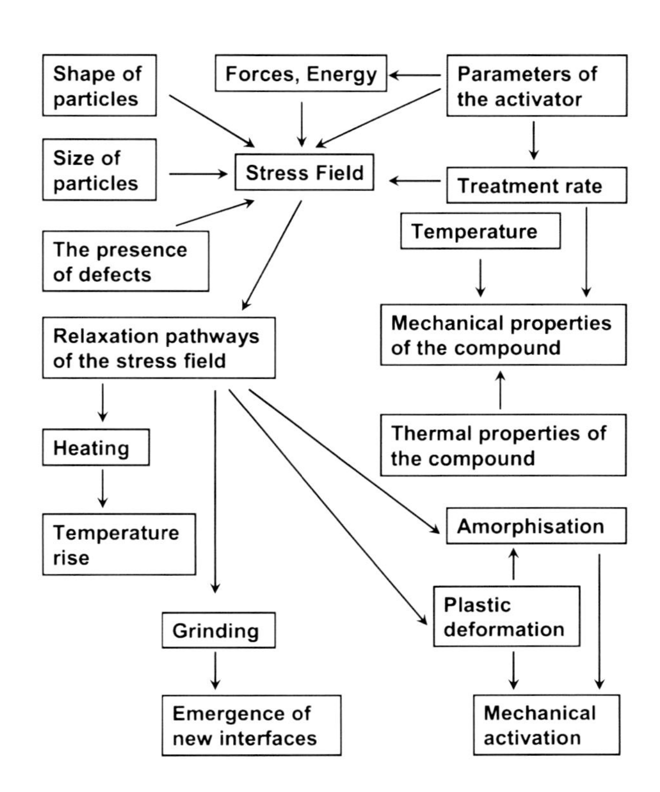

Mechanochemical research includes the following main directions:

1. Experimental and theoretical studies of properties, phase transitions and chemical reactions under well-controlled and well-defined loading conditions (hydrostatic pressure, pressure combined with controlled shear stresses, elastic bending/ uniaxial expansion/uniaxial compression, etc.);
2. Studies of the effects on solid-state properties and reactivity produced by treatment in various mechanical devices (mills, activators, attritors, etc.). The effects of the preliminary treatment on subsequent transformations, and the processes occurring at the moment of mechanical treatment.

One distinguishes between mechanical activation and mechanochemical processes. Mechanical activation means that mechanical treatment (often – preliminary) facilitates a thermal or photochemical reaction. Mechanochemical processes are induced by mechanical action itself (or are supposed to be induced by mechanical action itself).

6.2 Basics of Reactions Involving Solids. The Role of Mechanical Stress in Reactivity of Solids

The main factors which influence reactions involving solid reagent(s) and/or product(s) can be summarized as follows:

1. A solid, by definition, resists any attempts to change its size and shape. Therefore, any phase transition, as well as any chemical reaction, and even merely light absorption, in a solid generate mechanical stress, which can then relax *via* elastic or plastic strain, and, eventually, *via* fragmentation of the solid (see Fig. 6.5 as an example). Even if no external mechanical force is applied to a solid, any transformation in a solid is in some sense mechanochemical. Even if no stress is initially applied to a solid, as soon as a reaction starts, strain and stress are generated, which can then have either a positive, or a negative effect on the further

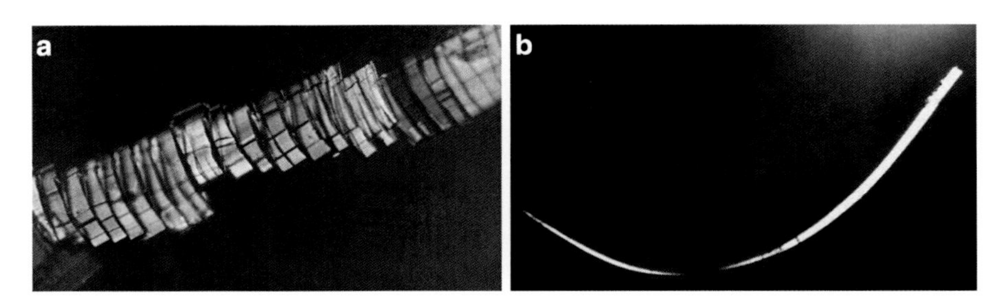

Fig. 6.5 Fragmentation (**a**) and elastic bending (**b**) of single crystals of $[Co(NH_3)_5NO_2]Cl(NO_3)$ on irradiation by visible or UV light

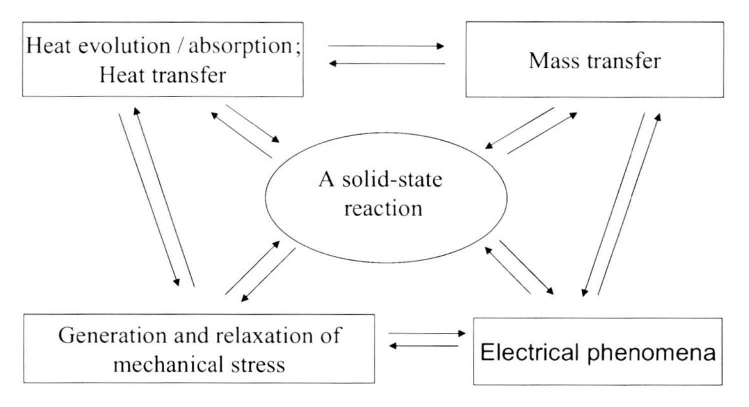

Fig. 6.6 Various phenomena which can account for feed-back in solid-state reactions

reaction course – a positive, or a negative feed-back arise. The problem of feed-back between a chemical reaction and mechanical stress generated by this reaction is one of the central problems of reactivity of solids. The strain of the crystalline environment in response to a chemical transformation is discussed in terms of reaction cavity, or chemical pressure.

2. For all the reactions involving more than one phase (*i.e.* any reactions with the exception of isomerization, dimerization, or polymerization, namely - thermal decomposition, solid + gas, solid + liquid, and solid + solid reactions), bringing reagents into contact and removing the product which could hinder such a contact becomes very important. Reactions occur not in the whole bulk of the sample, but at the interfaces between the phases.
3. As soon as a reaction starts, various phenomena account for changing the conditions for the reaction to continue, so that a positive or a negative feed-back arises. The phenomena can be related not only to generation of mechanical strain and stress, but also to local heating or cooling, to the redistribution of electron and holes at the interface, to the effect of product at the interface on diffusion, to specific interactions (hydrogen bonds, charge-transfer interactions) between components at the interface, to the changes in the concentration of point defects, protons, other chemical species at the interface, to the generation of electrochemical elements, *etc*. (Fig. 6.6). Relative contributions of the different phenomena can be different for different types of solids and for different reactions. For example, for inorganic solids the role of redistribution of electron, holes, point defects at the interfaces can be expected to be larger than for molecular organic crystals. At the same time, hydrogen bonds are important for molecular crystals.

Mechanical action can act on the reaction by improving diffusion, generating strain, structural, electronic and ionic defects, as well as by creating pulses of pressure and temperature. The effect of mechanical treatment on solid-state reactions can be both direct (on the chemical stages and diffusion coefficients, which become different in the mechanical stress field), or indirect (related to the changes in crystal morphology and generation of defects resulting from plastic deformation and fragmentation).

For reactions within a single solid particle, such as isomerization, dimerization, or polymerization, compression, extension, shear of the solid can modify the reaction cavity and the conditions of its relaxation. Mechanical action can break or stretch chemical bonds directly. This can be of importance also for such reactions as decomposition, solid + gas, or solid + liquid. Additional effects can be due to the fact that diffusion in the field of mechanical stress differs significantly from that in a non-stressed solid. For solid + solid reactions, the role of mechanical action is usually in providing better contacts between the reagents because of mixing, decreasing particle size, generating fresh surface for contact, and also by inducing melting and sublimation of the reagent(s).

6.3 Dynamics of Mechanochemical Processes

6.3.1 Probing Individual Bonds: Stretching Bonds in Macromolecules, or Hydrostatic Compression

The "mechanochemistry of single molecules" is a recent direction in the development of fundamental research, dealing with the properties of individual bonds in macromolecules, usually in biopolymers, probed by mechanical action. It finds applications in biochemistry, and also in materials science, when the challenge is to make synthetic polymers with the characteristics approaching or even exceeding those of natural biopolymers. Atomic force microscope (AFM) serves as the main tool both for stretching the bonds and for following the response of the molecules. Dynamics of mechanically-induced molecules within a single large molecule can be followed as a function of the applied force and direction, and this information helps to understand the nature of chemical interactions in the molecule, the factors determining its secondary and tertiary structure, as well as the mechanisms of biochemical reactions.

An alternative technique to stretching molecules using an AFM is squeezing them under hydrostatic conditions. Hydrostatic pressure applied not to single biomolecules, but to crystalline phases can induce anisotropic structural distortion of the same phase, and thus provide valuable information on the properties of bonds and interactions. It can also result in phase transitions or chemical reactions.

A common feature of the two types of very different actions is that they are continuous. Still, kinetic effects can play an important role in structural transformations on hydrostatic compression. These include the duration of keeping a sample under pressure, the details of compression-decompression procedure, the size of particles and the composition of the hydrostatic fluid may have a pronounced effect on phase stability and the products of the transformation (if any).

6.3.2 Processes at the Crack Tip and at a Freshly Cut Surface; Indentation

Another extreme case of mechanical treatment is presented by ultra-fast processes of cracking/cutting/indenting solid samples. To cleave a solid means to break chemical bonds. Chemical processes at the tip of a crack propagating through a crystal were studied as early as the middle of the twentieth century both by model calculations, and experimentally, using single crystals and breaking them under strictly controlled conditions, registering various physical and chemical phenomena *in situ* in real (very short!) time. Several successful attempts were made to measure the temperature directly at the tip of a crack, moving through different materials as a function of crack velocity and deformed zone thickness. Crack propagation can be accompanied by emission of light, X-rays and electrons. One can hear bonds being broken on mechanical destruction (mechanoacoustic effects). One can also analyse the products emitted into the gas phase on cleaving a crystal, and this gives an insight into the elementary processes in the crystal. Structural transformations in the crystalline region through which a crack propagated can be studied *in situ* by X-ray diffraction at a synchrotron radiation source. The first experiments were carried out in Novosibirsk as early as 1980. A crystal of tin was cut by a ruby cutter. A phase transformation from the ambient-pressure phase, β-tin, to the tetragonal form, which is stable at pressures above 100 kbar, was detected in a local region of 0.01 mm^2 under the cutter. On removal of the cutter, the transition was reversible within a few seconds. Similar experiments were carried out with silver single crystals. No metastable modifications were formed, though strong but rapidly vanishing distortions of the initial cubic phase of silver were observed.

Until now, the processes at the tip of propagating cracks have been studied for metals, salts, or polymers. To the best of my knowledge, there have never been similar studies for organic crystals, although anisotropic organic crystals with different types of intra- and intermolecular bonds would be a very interesting system. One could compare structural distortion in the region of cracks running at different velocities through different faces. Such experiments would be important, not only to understand chemical interactions in organic crystals, but also to get an insight into the processes in this systems on grinding.

In model experiments, aimed at achieving a better understanding of the effect of mechanical action of different types on solids one can also use indentation. It has been used to provoke phase transitions and chemical reactions in a number of inorganic solids, and gives valuable information on the mechanical properties and chemical bonding. The only example of a phase transition in a drug crystal induced by indentation is that of the polymorphic transition III – I in single crystals of sulfathiazole. Interestingly enough, the same phase transition can be induced by the ball-milling of a powder sample, but is never complete, since the mechanical treatment in a ball-mill stimulates]the reverse transformation I – III. Recently, some chemical reactions have been induced in organic solids by nano-indentation.

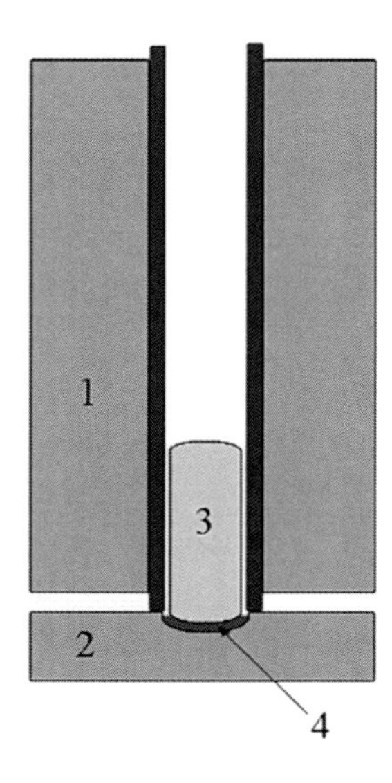

Fig. 6.7 Schematic representation of a model mechanochemical device allowing to control individual mechanical pulses and to follow intermediate products (Tumanov et al. [2])

6.3.3 Mechanical Treatment of Powders in Mills and Other Types of Machines. Mechanical Treatment in Model Devices

Different types of machines produce very different types of mechanical action, with a variable ratio between shear and impact, generating different temperature and pressure pulses. In the middle of the twentieth century there has been much systematic research comparing the efficiency of different types of activators for different types of chemical reactions, phase transitions, amorphization, dispergation. However, these papers seem to be practically unknown to the new generation of mechanochemists which is either not aware of such a problem at all, or starts everything "from scratch" and often "rediscovers the bicycle". At the same time, modern experimental facilities give many more possibilities to study the dynamics of mechanochemical reactions *in situ* using synchrotron radiation, and to follow intermediate reaction products, both in the "real mills" with their complicated mechanical treatment, or in model laboratory devices with well-controlled impact/shear ratio (Fig. 6.7).

6.4 Summary

Using a mill, or grinding a sample manually in a mortar does not automatically mean being a mechanochemist. The processes induced by mechanical treatment are very complex. They can be of thermal or athermal nature, can involve direct bond breaking or be preceded by formation of defects influencing on subsequent thermal reaction. A seemingly solid-state reaction can in fact occur in a fluid phase, because of sublimation, contact melting, or dissolution in traces of liquids present at the surface or in crystal solvents. One of the most important features that makes mechanochemical reactions so special is that the mechanical action is often applied not continuously, so that the system can relax between the mechanical pulses. To study dynamics of mechanochemical processes *in situ*, one needs to achieve a very high time- and space- resolution and combine temperature monitoring with the studies of light, sound, electrons emission, as well as measurements of gasses evolution and X-ray diffraction. Special model devices have an important advantage as compared to standard commercial mechanical apparatus like mills: one can control not only the temperature, the energy, the duration and frequency of mechanical pulses, but also relative contributions of impact, pressure, shear.

Acknowledgements I would like to thank the Course directors for giving me the opportunity to present this lecture and share my knowledge with the audience. I thank A. Polyakova, E. Achkasova, and B. Zakharov for their technical assistance in preparing the manuscript. Research in our group in the field of mechanochemistry was supported through years by Russian Academy of Sciences, Russian Foundation of Fundamental Research and Russian Ministry of Science and Education. The last but not least – I thank Professor Vladimir Boldyrev, the Founding President of the International Mechanochemical Association (affiliated with IUPAC), who has to a large extent shaped what we know as modern mechanochemistry - for his lessons, ideas and very stimulating discussions.

Further Reading

General Monographs and Reviews

1. Baramboim NK (1964) In: Watson WF (ed) Mechanochemistry of polymers. Maclaren Sons, London
2. Thiessen PA, Meyer K, Heinicke G (1967) Grundlagen der Tribochemie. Akademie Verlag, Berlin
3. Fox PG (1975) Mechanically initiated chemical reactions in solids. J Mater Sci 10:340–360
4. Boldyrev VV (1983) Experimental Methods in the Mechanochemistry of Inorganic Solids. In: Herman H (ed) Treatise on materials science and technology, vol 19, Experimental methods, Part B. Academic, New York, pp 186–223
5. Heinicke G (1984) Tribochemistry. Akademie Verlag, Berlin
6. Boldyrev VV (1986) Mechanochemistry of Inorganic Solids. In: Rao CNR (ed) Advances in solid state chemistry. Indian National Science Academy, New Delhi, pp 400–417
7. Tkáčová K (1989) Mechanical activation of minerals. Elsevier, Amsterdam
8. Boldyrev VV (1993) Mechanochemistry and mechanical activation of solids. Solid State Ion 63–65C:537–554
9. Boldyrev VV, Pavlov SV, Poluboyarov VA, Dushkin AV (1995) A comparison of the efficiency of different mechanical activators. Inorg Mater 31(9):1128–1138
10. Shakhtshneider TP, Boldyrev VV (1999) Mechanochemical synthesis and mechanical activation of drugs. In: Boldyreva EV, Boldyrev VV (eds) Reactivity of molecular solids. Wiley, New York, pp 271–311
11. Boldyrev VV, Tkáčová K (2000) Mechanochemistry of solids: past, present, and prospects. J Mater Synth Process 8:121–132

12. Tanaka K, Toda F (2000) Solvent-free organic synthesis. Chem Rev 100:1025–1074
13. Suryanarayana C, Ivanov E, Boldyrev VV (2001) The science and technology of mechanical alloying. Mater Sci Eng A 304–306:151–158
14. Avvakumov E, Senna M, Kosova N (2001) Soft mechanochemical synthesis: a basis for new chemical technologies. Kluwer, Boston
15. Boldyrev VV (2002) Hydrothermal reactions under mechanochemical action. Powder Technol 122:247–254
16. Boldyrev VV (2004) Mechanochemical modification and synthesis of drugs. J Mater Sci 39:5117–5120
17. Beyer MK, Clausen-Schaumann H (2005) Mechanochemistry: the mechanical activation of covalent bonds. Chem Rev 105:2921–2948
18. Trask AV, Jones W (2005) Crystal engineering of organic cocrystals by the solid-state grinding approach. Top Curr Chem 254:41–70
19. Boldyrev VV (2006) Mechanochemistry and mechanical activation of solids. Russ Chem Rev 75(3):177–189
20. Balaž P (2008) Mechanochemistry in nanoscience and minerals engineering. Springer, Berlin/London
21. Boldyrev VV, Avvakumov EG, Boldyreva EV, Buyanov RA et al (2009) Fundamental basics of mechanical activation, mechanosynthesis and mechanochemical technologies. SB RAS Publishing House, Novosibirsk, 343 pp
22. Kaupp G (2009) Mechanochemistry: the varied applications of mechanical bond-breaking. CrystEngCommun 11:388–403
23. Caruso MM, Davis DA, Shen Q, Odom SA, Sottos NR, White SR, Moore JS (2009) Mechanically-induced chemical changes in polymeric materials. Chem Rev 109:5755–5798
24. Boldyreva EV, Boldyrev VV (2010) Mechanochemistry and mechanical activation of solids. Part I. In: Mulas G, Delogu F (eds) Experimental and theoretical studies in modern mechanochemistry. Transworld Research Network, Trivandrum, pp 1–19
25. Boldyrev VV, Boldyreva EV (2010) Mechanochemistry and mechanical activation of solids. Part II. In: Mulas G, Delogu F (eds) Experimental and theoretical studies in modern mechanochemistry. Transworld Research Network, Trivandrum, pp 21–39
26. Sokolov AN, Bučar D-K, Baltrusaitis J, Gu SX, MacGillivray LR (2010) Supramolecular catalysis in the organic solid state through dry grinding. Angew Chem Int Ed 49:4273–4277
27. Friščič T (2010) New opportunities for materials synthesis using mechanochemistry. J Mater Chem 20:7599–7605
28. Balema VP (2011) Mechanical processing – experimental tool or new chemistry? Ceram Trans 224:25–35
29. Stolle A, Szuppa T, Leonhardt SES, Ondruschka B (2011) Ball milling in organic synthesis: solutions and challenges. Chem Soc Rev 40:2317–2329
30. Lomovsky OI, Lomovsky IO (2011) Mechanochemically assisted extraction. In: Lebovka NI, Vorobiev E, Chemat F (eds) Enhancing extraction processes in the food industry. Taylor & Francis, London/New York, pp 361–397
31. Boldyreva EV (2012) Mechanochemistry of organic solids: where are we now? In: Kumar R (ed) Frontiers in mechanochemistry and mechanical alloying. CSIR-National Metallurgical Laboratory, Jamshedpur, pp 17–31
32. Friščić T (2012) Supramolecular concepts and new techniques in mechanochemistry: cocrystals, cages, rotaxanes, open metal-organic frameworks. Chem Soc Rev 41(9):3493–3510
33. Delori A, Frišić T, Jones W (2012) The role of mechanochemistry and supramolecular design in the development of pharmaceutical materials. CrystEngCommun 14(7):2350–2362
34. James SL, Adams CJ, Bolm C, Braga D, Collier P, Friščič T, Grepioni F, Harris KDM, Hyett G, Jones W, Krebs A, Mack J, Maini L, Orpen AG, Parkin IP, Shearouse WC, Steed JW, Waddell DC (2012) Mechanochemistry: opportunities for new and cleaner synthesis. Chem Soc Rev 41(1):413–447
35. Boldyreva EV (2013) Mechanochemistry of inorganic and organic systems: what is similar, what is different? Chem Soc Rev 42:7719–7738

Stretching Bonds in Macromolecules

1. Ribas-Arino J, Marx D (2012) Covalent mechanochemistry. Theoretical concepts and computational tools with applications to molecular nanomechanics. Chem Rev 112 (10):5412–5487
2. Best RB, Paci E, Hummer G, Dudko OK (2008) Pulling direction as a reaction coordinate for the mechanical unfolding of single molecules. J Phys Chem B 112:5968–5976
3. Hyeon C, Thirumalai D (2007) Measuring the energy landscape roughness and the transition state location of biomolecules using single molecule mechanical unfolding experiments. J Phys Condens Matter 19:art. no. 113101
4. Ritort F (2006) Single-molecule experiments in biological physics: methods and applications, J Phys Condens Matter 18:art. no. R01, R531–R583

Hydrostatic Compression

1. Katrusiak A, McMillan P (eds) (2004) High-pressure crystallography. Kluwer, Dordrecht, 567 pp
2. Boldyreva EV, Dera P (eds) (2010) High-pressure crystallography. From novel experimental approaches to applications in cutting-edge technologies. Springer, Dordrecht, 612 pp

Processes at the Crack Tip and at a Freshly Cut Surface

1. Boldyrev VV (1983) Experimental Methods in the Mechanochemistry of Inorganic Solids. In: Herbert H (ed) Treatise on materials science and technology, vol 19, Experimental methods, Part B. Academic, New York, pp 186–223, see Refs to original publications
2. Gilman JJ (1999) Athermal fracture of covalent bonds. Mater Res Soc Symp Proc 539:145–151
3. Gilman JJ (2005) Possible role of dispersion forces in fracture. Philos Mag 85(24):2799–2807

Reactions in Solids and Mechanical Stress

1. Morrison JA, Nakayama K (1963) Reaction of single crystals of potassium bromide with chlorine gas. Trans Faraday Soc 59:2560–2568
2. Boldyrev VV (1973) Topochemistry of thermal decomposition of solids. Russ Chem Rev 7:1161–1183
3. McBride JM (1983) The role of local stress in solid-state radical reactions. Acc Chem Res 16:304–312
4. McBride JM, Segmuller BE, Hollingsworth MD, Mills DE, Weber BA (1986) Mechanical stress and reactivity in organic solids. Science 234(4778):830–835
5. Boldyreva EV et al (1984) Proc Acad Sci USSR 277:893–896
6. Chupakhin AP, Sidel'nikov AA, Boldyrev VV (1987) Control of the reactivity of solids by changing their mechanical properties. React Solids 3:1–19
7. Luty T, Fouret R (1989) On stability of molecular solids "under chemical pressure". J Chem Phys 90:5696–5703

8. Boldyreva EV (1990) Feed-back in solid-state reactions. React Solids 8:269–282
9. Hollingsworth M, McBride JM (1990) In: Volman D, Hammond G, Gollnick K (eds) Advances in photochemistry, vol 15. Wiley-Interscience, New York, p 279
10. Boldyreva EV (1992) The problem of feed-back in solid-state chemistry. J Therm Anal 38:89–97
11. Ohashi Y (ed) (1993) Reactivity of molecular crystals. VCH, Tokyo
12. Luty T, Eckhardt CJ (1995) General theoretical concepts for solid state reactions: quantitative formulation of the reaction cavity, steric compression, and reaction-induced stress using an elastic multipole representation of chemical pressure. J Am Chem Soc 117:2441–2452
13. Boldyreva EV (1997) The concept of the 'reaction cavity': a link between solution and solid-state chemistry. Solid State Ion 101–103:843–849
14. Boldyreva EV, Boldyrev VV (eds) (1999) Reactivity of molecular solids, vol 3, Molecular solid state series. Wiley, Chichester, 328 pp
15. Luty T (2001) Lattice mediation in thermo- and photo-induced reactions; co-operative activation. Mol Cryst Liq Cryst 356:539–548
16. Phillips AE, Cole JM, D'Almeida T, Low KS (2010) Effects of the reaction cavity on metastable optical excitation in ruthenium-sulfur dioxide complexes. Phys Rev B Condens Matter Mater Phys 82:art. no. 155118
17. Bąkowicz J, Turowska-Tyrk I (2012) Photo-induced structural transformations in crystals at high pressure. Part 1. The crystallographic studies of the photochemical reaction at high pressure. J Photochem Photobiol A Chem 232:41–43

Indentation

1. Gilman JJ (1992) Insulator-metal transitions at microindentations. J Mater Res 7(3):535–538
2. Shakhtshneider TP, Boldyrev VV (1993) Phase transformations in sulfathiazole during mechanical treatment. Drug Dev Ind Pharm 19:2055–2067
3. Meier M, John E, Wieckhusen D, Wirth W, Peukert W (2009) Influence of mechanical properties on impact fracture: prediction of the milling behaviour of pharmaceutical powders by nanoindentation. Powder Technol 188(3):301–313
4. Tan JC, Cheetham AK (2011) Mechanical properties of hybrid inorganic-organic framework materials: establishing fundamental structure-property relationships. Chem Soc Rev 40 (2):1059–1080
5. Varughese S, Kiran MSRN, Solanko KA, Bond AD, Ramamurty U, Desiraju GR (2011) Interaction anisotropy and shear instability of aspirin polymorphs established by nanoindentation. Chem Sci 2:2236–2242
6. Chow EHH, Bučar D-K, Jones W (2012) New opportunities in crystal engineering – the role of atomic force microscopy in studies of molecular crystals. Chem Commun 48:9210–9226
7. Varughese S, Kiran MSRN, Ramamurty U, Desiraju GR (2012) Nanoindentation as a probe for mechanically-induced molecular migration in layered organic donor-acceptor complexes. Chem Asian J 7:2118–2125

Temperature Effects

1. Boldyrev VV, Gerasimov KB (1996) On mechanism of new phases formation during mechanical alloying of Ag-Cu, Al-Cu, and Fe-Sn systems. Mater Res Bull 31:1297–1305

2. Urakaev FK, Boldyrev VV (2000) Mechanism and kinetics of mechanochemical processes in comminuting devices. 1. Theory. Powder Technol 107(1–2):93–107
3. Urakaev FK, Boldyrev VV (2000) Mechanism and kinetics of mechanochemical processes in comminuting devices: 2. Applications of the theory. Experiment. Powder Technol 107(3):197–206
4. Zavaliangos A, Galen S, Cunningham J, Winstead D (2008) Temperature evolution during compaction of pharmaceutical powders. J Pharm Sci 97:3291–3304

Monitoring Intermediate Products

1. Boldyrev VV, Pavlov SV, Poluboyarov VA, Dushkin AV (1995) A comparison of the efficiency of different mechanical activators. Inorg Mater 31(9):1128–1138
2. Tumanov IA, Achkasov AF, Boldyreva EV, Boldyrev VV (2011) Following the products of mechanochemical synthesis step after step. CrystEngCommun 13:2213–2216
3. Tumanov IA, Achkasov AF, Boldyreva EV, Boldyrev VV (2012) About the possibilities to detect intermediate stages in mechanochemical synthesis of molecular complexes. Russ J Phys Chem A 86(6):1014–1017
4. Friščić T, Halasz I, Beldon PJ, Belenguer AM, Adams F, Kimber SAJ, Honkimäki V, Dinnebier RE (2013) Real-time and in situ monitoring of mechanochemical milling reactions. Nat Chem 5:66–73

Chapter 7
Measuring and Understanding Ultrafast Phenomena Using X-Rays

Kristoffer Haldrup and Martin Meedom Nielsen

7.1 Introduction

Within the last decade, significant advances in X-ray sources and instrumentation as well as simultaneous developments in analysis methodology has allowed the field of fast- and ultrafast time-resolved X-ray studies of solution-state systems to truly come of age. We here describe some aspects of the physics involved as well as the experimental methodology that have facilitated this development. Building on this foundation, we discuss how the information-poor, but time-resolved (difference) scattering signals can be analyzed in a quantitative model-comparison framework to provide robust information on sub-Ångstrom structural changes taking place on femtosecond to nanosecond time scales. We illustrate this approach by a presentation of recent results from the Centre for Molecular Movies at the Technical University of Denmark.

7.2 X-Ray Scattering and Measuring the Ultrafast

In the first part of the chapter, the main emphasis will be on the interaction of X-rays and matter, notably the interaction between x-rays and electrons, atoms and molecules. Much of the latter is covered in Chapter 1 (pages 1–12) of the classic book, *X-ray diffraction*, by B. E. Warren. The first version was published in 1969, and many updated editions have since been published. Chapter 1 and more of the edition published by Dover in 1990 is freely available on Google Books [1]. Notice that Warren uses cgs units, where the vacuum permittivity of free space is $\epsilon_0 = 1/4\pi$, while we will be using SI units.

K. Haldrup • M.M. Nielsen (✉)
Department of Physics, NEXMAP Section, Centre for Molecular Movies, Technical University of Denmark, Lyngby, Denmark
e-mail: hald@fysik.dtu.dk; mmee@fysik.dtu.dk

J.A.K. Howard et al. (eds.), *The Future of Dynamic Structural Science*, NATO Science for Peace and Security Series A: Chemistry and Biology, DOI 10.1007/978-94-017-8550-1_7,

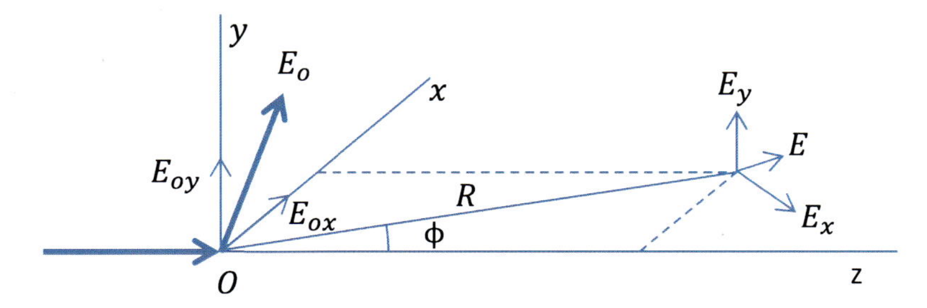

Fig 7.1 An unpolarised beam of X-rays travelling along z is incident upon a free electron at the origin. The electric field vector at O in one particular direction E_0 has components in the x and y directions E_{ox} and E_{oy}. The corresponding electric field vectors of the scattered radiation in the xz plane at a distance R from O an angle ϕ to z are E_x and E_y respectively

7.2.1 Scattering Fundamentals, the Debye Equation

7.2.1.1 Scattering from an Electron

Consider a beam of unpolarised X-rays with electric field vector E and (angular) frequency ω travelling along the z direction which is incident on a free electron positioned at the origin (see Fig. 7.1). The electric field vector in a given direction $\boldsymbol{E}_0$ can be divided into two components E_{0x} and E_{0y}. The (time-dependent, oscillating) component E_{0x} produces a force which acts on the electron and makes it oscillate. In the x direction this force will be eE_{0x} where e is the charge of an electron. If the mass of the electron is m then the equation of motion may be written

$$\ddot{x} = \frac{eE_{0x}}{m}\exp(i\omega t) \tag{7.1}$$

The accelerated (oscillating) electron in turn radiates electromagnetic radiation. At a large distance R from the electron and at an angle ϕ to the z direction the instantaneous values of the electric field vector of the scattered radiation in the xz plane is given by

$$E_x = \frac{-e}{4\pi\epsilon_0 c^2 R}\ddot{x}\left(t - \frac{R}{c}\right)\cos\phi \tag{7.2}$$

where ϵ_0 is the permittivity of free space, c is the speed of light and the term $(t - R/c)$ represents the retarded time.

From Eqs. 7.1 and 7.2 we can write the amplitude of scattered radiation due to the x component as

$$E_x = -\frac{e^2 E_{0x}}{4\pi\epsilon_0 mc^2 R}\cos\phi$$

The equivalent expression for the perpendicular component E_y has no ϕ dependence and we may therefore write the electric field in one direction perpendicular to the direction of motion of the scattered beam as

$$E^2 = \frac{e^4}{(4\pi\epsilon_0 mc^2 R)^2}\left(E_{0y}^2 + E_{0x}^2\cos^2\phi\right) \tag{7.3}$$

The total electric field at the point of observation may be found by averaging over all directions of E_0 in the xy plane to give [Warren, eq. 1.2, p.3]:

$$\langle E^2\rangle = \frac{e^4}{(4\pi\epsilon_0)^2 m^2c^4R^2}\langle E_0^2\rangle\left(\frac{1+\cos^2\phi}{2}\right)$$

The term $\left(\frac{1+\cos^2\phi}{2}\right)$ is known as the polarisation correction for an unpolarised incident X-ray beam. In terms of X-ray beam intensities, which are proportional to the square of the amplitudes of the electric field vector of the radiation, we can therefore write

$$I = I_0\frac{e^4}{(4\pi\epsilon_0)^2 m^2c^4R^2}\left(\frac{1+\cos^2\phi}{2}\right) \tag{7.4}$$

For an atom containing Z free electrons the scattering in the forward direction (scattering angle $= 0$) is Z^2 times the intensity given by Eq. 7.4 since the scattered radiation from each electron in the atom will be in phase at $\phi = 0$. However, as ϕ increases from zero the scattered X-rays from the electrons begin to interfere destructively and the strength of the overall scattering falls off with increasing ϕ. The form of this fall-off is dependent on the charge distribution inside the atom and is represented by an atomic form factor, or scattering factor, f defined as the ratio between the amplitude scattered by an atom and that scattered by a free classical electron when all other conditions remain unchanged.

The proportionality constant between the incident electrical field E_0 and the scattered field emitted by an electron is called the Thomson radius of the free electron:

$$r_0 = \frac{e^2}{4\pi\epsilon_0 mc^2} = 2.8179403267\,(27)\times 10^{-15}\ m \tag{7.5}$$

7.2.2 Atomic and Molecular Form Factors

The scattering vector or the wave vector transfer, see Fig. 7.2, is defined as the difference between the incident wave vector ($\boldsymbol{k}_0$) and the scattered wave vector ($\boldsymbol{k}$):

$$Q = \boldsymbol{k} - \boldsymbol{k}_0 = \frac{4\pi\sin\theta}{\lambda}$$

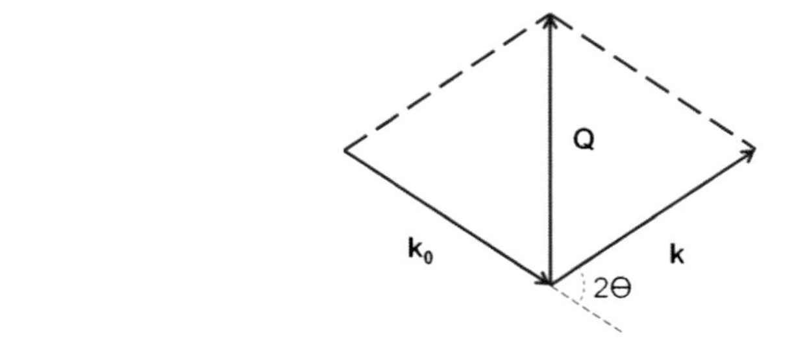

Fig. 7.2 Construction of the scattering vector, Q

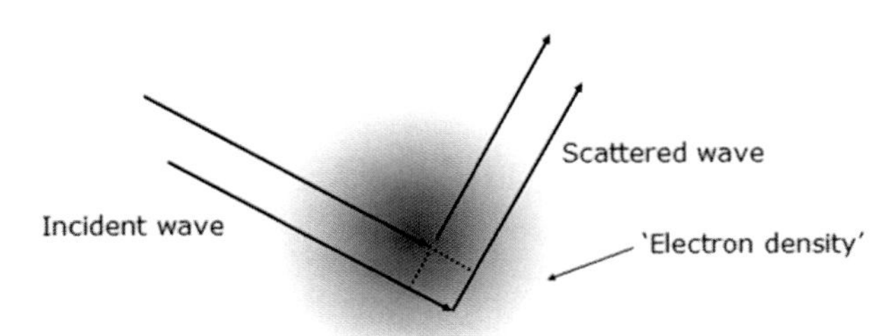

Fig. 7.3 Scattering from the electronic density distribution

7.2.2.1 Scattering from an Atom

The atomic electrons can be viewed as a charge cloud surrounding the nucleus with a density $\rho(\mathbf{r})$. The charge at position **r** is then given as $-e\rho(\boldsymbol{r})d\boldsymbol{r}$, where the integral of $\rho(\boldsymbol{r})$ over all space yields the number of electrons of the atom Z. To evaluate the total scattering amplitude the charge density at $d\boldsymbol{r}$ it must be weighted with the phase factor $e^{i\,\boldsymbol{Q}\cdot\boldsymbol{r}}$ and integrated over $d\boldsymbol{r}$.

The coherent (elastic) scattering amplitude caused by an atom (called the atomic form factor $f^0(\boldsymbol{Q})$ is thus found by integrating the scattering contribution from each infinitesimal volume element of the electron cloud (see Fig. 7.3).

$$f^0(\boldsymbol{Q}) = \int \rho(\boldsymbol{r}) e^{i\boldsymbol{Q}\cdot\boldsymbol{r}} d\boldsymbol{r} \tag{7.6}$$

In the limit that $Q \to 0$, all the different volume elements scatter in phase, so that $f^0(\boldsymbol{Q} = 0) = Z$. As Q increases, the individual volume elements start scattering out of phase and the scattering amplitude decreases as a consequence $f^0(\boldsymbol{Q} \to \infty) = 0$.

The magnitude of the atomic form factor for all common atoms and their respective ions have been determined through measurements and calculations, and are typically approximated through the analytical expression

$$f^0(Q) = \sum_{j=1}^{4} a_j e^{-b_j \left(\frac{Q}{4\pi}\right)^2} + c$$

Where a, b, and c are the so called Cromer-Mann coefficients [2], which are tabulated for a given atom. Examples include:

	C	O	F	S	Cu	Mo
a_1	2.310000	3.048500	3.539200	6.905300	13.33800	3.702500
a_2	1.020000	2.286800	2.641200	5.203400	7.167600	17.23560
a_3	1.588600	1.546300	1.517000	1.437900	5.615800	12.88760
a_4	0.8650000	0.8670000	1.024300	1.586300	1.673500	3.742900
c	0.2156000	0.2508000	0.2776000	0.8669000	1.191000	4.387500
b_1	20.84390	13.27710	10.28250	1.467900	3.582800	0.2772000
b_2	10.20750	5.701100	4.294400	22.21510	0.2470000	1.095800
b_3	0.5687000	0.3239000	0.2615000	0.2536000	11.39660	11.00400
b_4	51.65120	32.90890	26.14760	56.17200	64.81260	61.65840

The intensity of the scattering from an atom is then found as the absolute squared scattering amplitude:

$$I(Q) = r_0^2 \, |f_0(Q)|^2,$$

where r_0 is again the Thomson radius of the free electron.

7.2.2.2 Scattering from a Molecule, the Independent Atom Model

The scattering amplitude from an atom was found by integrating over the phase- and charge-density weighted radiation emitted by each infinitesimal volume element of the electron cloud. The same approach is invoked for the evaluation of the scattering of molecules calculated in the framework known as the Independent Atom Model (IAM). The scattering intensity is found as the scattering from the two form factors, f_1 and f_2, weighted by the phase-difference between photons scattered from each of the electron clouds of the two atoms.

$$A(\boldsymbol{Q}) = f_1 + f_2 e^{i\boldsymbol{Q}\cdot r}$$

Here **r** is the position vector of atom 2, with atom 1 at the origin and where we have suppressed the dependence of f on Q for clarity of presentation. The intensity is given by the absolute square of the amplitude:

$$\begin{aligned} I(\boldsymbol{Q}) &= A(\boldsymbol{Q})A(\boldsymbol{Q})^* = \left(f_1 + f_2 e^{i\boldsymbol{Q}\cdot r}\right)\left(f_1 + f_2 e^{i\boldsymbol{Q}\cdot r}\right)^* \\ &= f_1^2 + f_2^2 + f_1 f_2 e^{i\boldsymbol{Q}\cdot\boldsymbol{r}} + f_1 f_2 e^{-i\boldsymbol{Q}\cdot\boldsymbol{r}} \end{aligned}$$

For spherically symmetric form factors, we have $< e^{i\boldsymbol{Q}\cdot\boldsymbol{r}} >_{orient\ av.} = < e^{-i\boldsymbol{Q}\cdot\boldsymbol{r}} >_{orient\ av.}$ and the expression becomes

$$I(\boldsymbol{Q}) = f_1^2 + f_2^2 + 2 f_1 f_2 < e^{i\boldsymbol{Q}\cdot\boldsymbol{r}} >_{orient\ av.}$$

The orientational average is found by integrating **r** over all possible angles and normalizing:

$$< e^{i\mathbf{Q}\cdot\mathbf{r}} >_{orient\ av.} = \frac{\int e^{iQr\cos\theta} \sin\theta d\theta d\phi}{\int \sin\theta d\theta d\phi} = \frac{2\pi \int_0^{\pi} e^{iQr\cos\theta} \sin\theta d\theta}{4\pi}$$

$$= \frac{\left(-\frac{1}{iQr}\right) \int_{iQr}^{-iQr} e^x dx}{2} = \frac{\sin(Qr)}{Qr}$$

Thus the orientational average of the scattering of two atoms with form factors f_1 and f_2 is dependent on the length of the vector between the two, but not the direction of the vector:

$$I(Q) = f_1^2 + f_2^2 + \frac{2f_1 f_2 \sin(Qr)}{Qr}$$

This expression can be expanded to a set of atoms of any size:

$$< I(Q) >_{orient\ av.} = < \left| \sum_{k=1}^{N} f_k e^{i\mathbf{Q}\cdot\mathbf{r}_k} \right|^2 >_{orient\ av.}$$

$$= |f_1|^2 + |f_2|^2 + 2f_1 f_2 \frac{\sin Qr_{12}}{Qr_{12}} + 2f_1 f_3 \frac{\sin Qr_{13}}{Qr_{13}} + \cdots$$

$$+ 2f_1 f_N \frac{\sin Qr_{1N}}{Qr_{1N}} + 2f_2 f_3 \frac{\sin Qr_{23}}{Qr_{23}} + \cdots + 2f_2 f_N \frac{\sin Qr_{2N}}{Qr_{2N}} \cdots$$

$$+ 2f_{N-1} f_N \frac{\sin Qr_{N-1\ N}}{Qr_{N-1\ N}}$$

Where $\mathbf{r_k}$ is the position vector of each individual atom and r_{kj} is the distance between atoms k and j. This expression yield the scattering intensity produced by the theoretical electron density caused by the constituting atoms if they were independent and arranged in a given (molecular) structure. The expression boils down to:

$$I(Q) = \sum_{i,j,k=1,\,k>j}^{N} |f_i|^2 + f_j f_k \frac{\sin Qr_{jk}}{Qr_{jk}}$$

This is the orientation averaged Debye Function. The intensity is expressed in units of r_0^2. Notice that there are no vector contributions to the Debye function; the input consists only of all distances between the constituent atoms of a molecule, and their respective form factors. From these parameters the function yields the elastic scattering intensity from a randomly oriented molecule as a function of the wave vector transfer.

Fig. 7.4 The CF_4 molecule

7.2.3 *Example CF_4 Molecule*

The molecular structure can be found from the NIST database [3]. The (Cartesian) coordinates of the 4 F atoms and the C atom is (in Ångström) (Fig. 7.4):

```
F  1.3071  0.0000  0.2433
C   1.0932  1.2666  0.6134
F   0.9178  1.3206  1.9375
F   0.0000  1.7310  0.0000
F   2.1471  2.0147  0.2722
```

This gives C-F distances of ~1.34 Å, and F-F distances ~2.18 Å. You can get all the 'exact' bond lengths from the coordinates. Notice that the molecule contains a total of 4 C-F pairs and 6 F-F pairs. Figure 7.5 shows the molecular structure factor for this molecule calculated with the Debye equation.

7.2.4 *X-Ray Scattering Data*

The primary output from the time resolved x-ray scattering (TRXS) experiments conducted, is a 2D scattering pattern, which upon applying the appropriate corrections and performing an azimuthal integration provides 1D scattering curves displaying the X-ray scattering intensity as a function of the scattering angle 2θ. The scattering images are recorded using a position-sensitive, detector. There are a number of different detectors available. Some main groups are listed below, and typical detectors can belong to more than one of the groups above, e.g. the PILATUS is of type (4) and (5), while a MAR CCD 133 is of type (3) and (6).

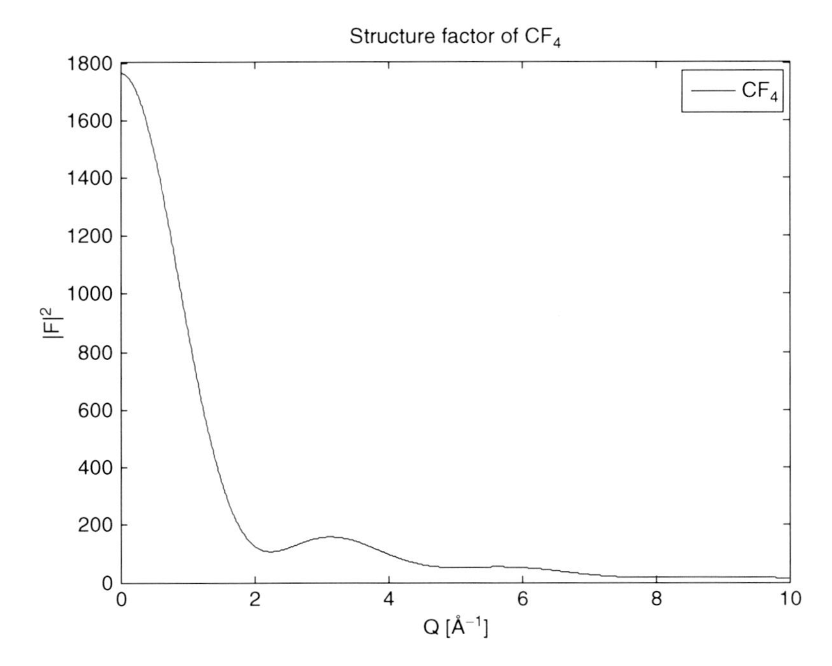

Fig. 7.5 Molecular structure factor of CF_4

(1) Gas detectors, typically with delay line readout
(2) Multi-channel plate detectors
(3) Optically coupled. Often with a fibre optics connecting a (large) fluorescent screen to a (smaller) CCD chip.
(4) Direct detection, where the x-rays are absorbed directly in the semiconductor chip. These devices often include special features such as deep depletion layers to increase efficiency.
(5) Single photon counting area detectors
(6) Integrating detectors

It is possible to device schemes for time resolved scattering that takes advantage of the various detectors' particular strengths. For example an integrating detector can be used to collect many pump-probe events directly on the CCD chip, reducing the demands on data bandwidth, processing and storage, while a single photon counting detector provides very low readout noise [4].

7.2.4.1 Taking Images

The x-ray scattering images are typically recorded in an alternating sequence of laser-on/laser-off images in order to minimize the effects of drift in beamline components and other slowly varying phenomena, such as variations in the

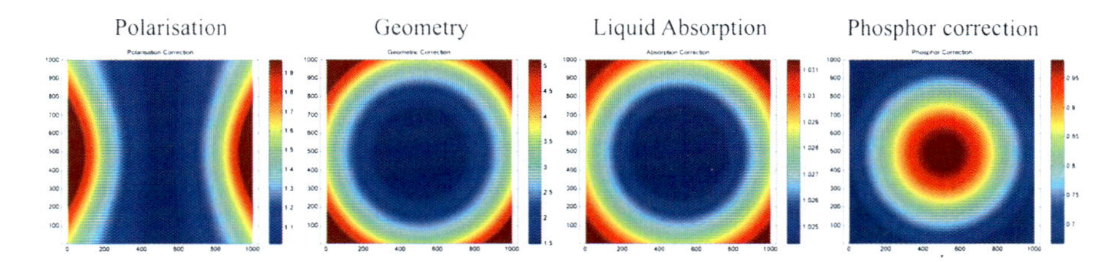

Fig. 7.6 Illustration of the different corrections needed. The liquid absorption is calculated for normal incidence on the liquid sheet

synchrotron X-ray intensity (due to changes in ring current). If possible, the "laser-off" images are often acquired by recording the scattering signal from a "probe-pump" cycle with the laser pump-pulse arriving a long time (e.g. 3 ns) after the X-ray probing event. This helps to minimize long-term side effects arising due to laser exposure.

7.2.4.2 Image Correction and Data Processing

Several corrective steps are applied to the recorded data before any structural analysis can occur. These include notably:

- Flat field corrections (calibrate the response of the pixels to an x-ray photon)
- Dark-current (corrections for the detector response in the absence of any x-rays hitting it).
- Absorption in the phosphor (type 3 detector) or active region of a direct detection chip (type 4). If the x-rays are not completely absorbed in the detector, e.g. when using high energy, the amount absorbed depends on the path length of the x-rays through the detection medium. This will vary as a function of angle of incidence of the x-ray beam on the detector surface.
- Absorption in the sample. This depends on the path length through the sample. E.g. for perpendicular incidence on a liquid sheet, higher angle scattering will have to penetrate further through the sample than low angle scattering. The absorption in the sample can be substantial at lower x-ray energies or in case the x-ray energy is close to an absorption edge.
- Polarization. Due to the horizontal motion of the electrons in undulators, the electromagnetic radiation produced is close to fully horizontally polarized. This necessitates an anisotropic polarization correction of the scattered intensity.
- Space-angle coverage, as the space-angle $d\Omega$ "observed" by each pixel decreases with increasing scattering angle.

Examples of how to calculate corrections factors can be found in the literature [5]. Illustrations of some of the corrections and how they can vary over a 2D detector image are given in Fig. 7.6.

7.2.4.3 Data Reduction and Scaling

The corrected images are subsequently integrated azimuthally and converted to 1D scattering curves, displaying the scattering intensity as a function of the scattering angle 2θ or as function of the scattering vector, q.

A number of effects will cause variations in the scattered intensity from image to image acquired. At synchrotrons, the number of electrons in the storage ring decreases every orbit, and at X-ray Free Electron Lasers, the beam intensity can vary substantially from shot to shot due to the stochastic nature of the emission process. Along with equipment drift and fluctuations in the sample, which is not uncommon for liquid-sheet setups, this implies that scattering curves must be brought to a common, ideally absolute, scale before further processing can take place.

This can be achieved by scaling the scattering curves to the absolute (coherent + incoherent) theoretical scattering intensity from one "liquid state unit cell". This unit cell corresponds to the stoichiometry of the smallest scalable volume of a given solution. The choice of the liquid unit cell as the fundamental unit for scaling, as well as for analysis, enables extraction of relevant analytical parameters, such as molar fractions of excited states, changes in temperature and density of the solution etc. For a 10 mM solution of the PtPOP compound (See below), the unit cell would contain 5,550 water molecules, 2 Pt atoms, 8 P atoms, 20 O atoms, 8 H atoms and 4 K atoms (the counter ions to the PtPOP tetra anion). Scaling is performed over an interval in the high-angle scattering region where the incoherent scattering contribution is dominating, and the scattering signal is relatively insensitive to intensity variations attributed to possible changes in molecular structure.

It is possible to minimize the influence of molecular structural change, by performing the scaling around an isosbestic point – a zero point in the difference scattering signal – identified by performing a fast pre-analysis over a larger interval. For scattering at lower energies, where the maximum recorded scattering vector Q_{Max} is usually smaller, this procedure becomes increasingly important.

7.2.5 *The Difference Scattering Curves*

The difference scattering signal is constructed by subtracting the scattering signal of the non-excited sample solution ($S^{laser\ -\ off}$), from the scattering signal of a laser-excited sample ($S^{laser\ -\ on}$). As previously mentioned, consecutive images are not necessarily recorded with the same intensity and are subject to variations in experimental conditions. These effects can be minimized by collecting laser-on and laser-off images in an alternating fashion, and then use the average of two neighboring laser-off signals for subtraction. In a collection of images labeled according to their recording sequence, the laser-on image n would be bracketed by laser-off images n − 1 and n + 1 respectively, and the difference signal would be found as $\Delta S(2\theta) = S(2\theta)_n^{laser-on} - \frac{S(2\theta)_{n-1}^{laser-off} \mp S(2\theta)_{n\mp 1}^{laser-off}}{2}$,

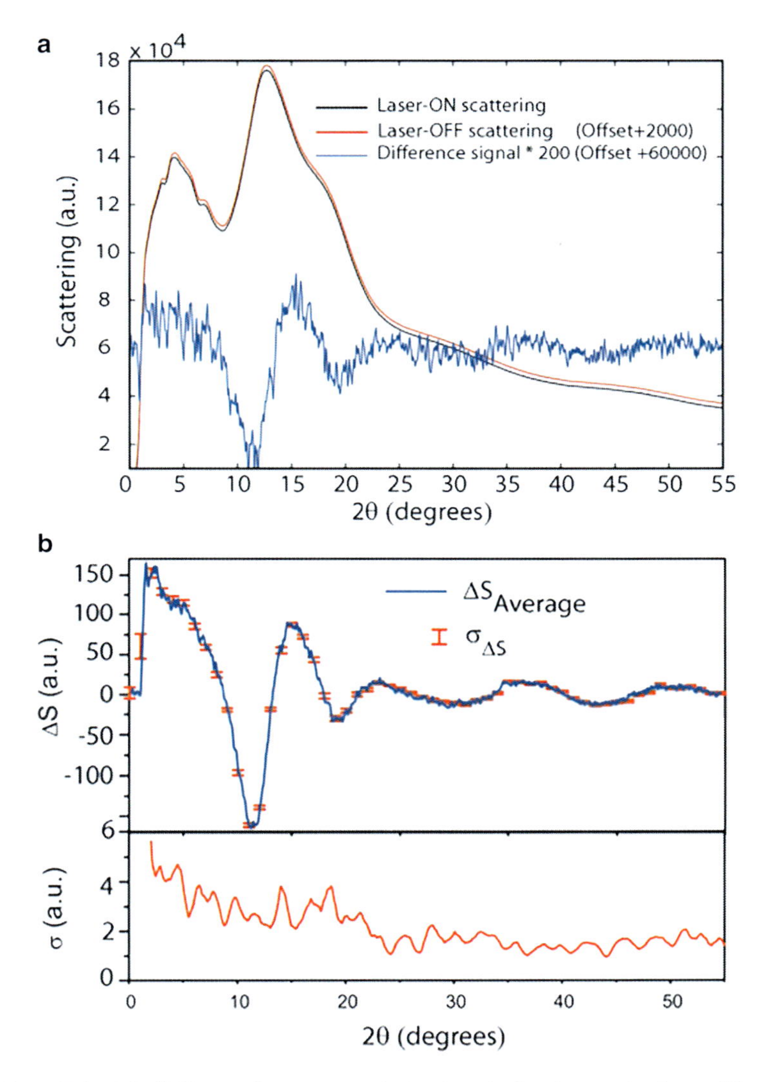

Fig. 7.7 (**a**) Pair of On/Off scattering curves scaled to the calculated scattering from a "liquid unit cell" (see text) and the corresponding difference signal $\Delta S(2\theta)$. The red scattering curve has been offset and the difference curve has been offset and scaled for clarity. (**b**) Average of 45 individual $\Delta S(2\theta)$, obtained in a single experiment, along with the estimated noise level $\sigma(2\theta)$

but more intense beams and faster detectors can allow for different and faster acquisition schemes. Figure 7.7a shows an example of a difference scattering curve based on a single On-Off repetition for a single time delay (100 ps).

In order to reduce noise, it is common to repeat the data acquisition sequence between 10 and 50 times and it has proven crucial for our work to develop a robust method for rejecting outliers in such sets of $\Delta S(2\theta)$ -curves. These outliers often arise due to unstable jet conditions and/or precipitates formed in the investigated solution- state systems and manifest themselves as having e.g. increased or decreased intensity at high- or low scattering angles or through diffraction rings giving rise to sharp peaks in the difference curves. Further effects such as rapid fluctuations in air background scattering or variations in FEL pulses can also contribute to generate outliers. In the course of the data pre-analysis, these are removed based on a point-by-point application of the unbiased Chauvenet criterion [6]. Individual difference curves are excluded from the averaging process if more than 2.5 % of the points on a curve fail this criterion, which has been observed for up to 5–10 % of the curves, depending on the details of the experiment. Figure 7.7b (top) illustrates an average curve based on 45 such repetitions for the same time delay.

The noise level for each point on an averaged $\Delta S(2\theta)$-curve can in principle be estimated from the knowledge of the detector counts on each pixel and applying the laws of error propagation to each subsequent data analysis step. However, this is in practice an unwieldy process, for which reason we use a more robust and straight-forward method, first introduced by Dent et al. in an EXAFS application [7]. It is based on the observation that the information containing part of the signal has low-frequency oscillations on top of which high-frequency noise is superposed. By this method of error-estimation, a low-order polynomial is fitted to the data points in a narrow interval around each data point and the standard deviation estimated from the set of residuals in this interval. It is found, that in the present case an interval length of 20 points and a 2 order polynomial fit are able to accurately capture the highest frequencies contained in $\Delta S(2\theta)$ and the resultant calculated $\sigma(2\theta)$ are shown as vertical bars in Fig. 7.7, bottom and insert. The angle dependence of the noise reflects the complex interplay between scattered intensity, number of pixels and pixel efficiency, the latter related to variations in the beam interaction length with the active material in the detector as a function of angle.

7.2.5.1 Components of the Difference Signal

Having now acquired the scattering signals and performed a series of corrections followed by accurate scaling and calculation of the measured difference signal ΔS_{Meas}, the obvious question is: What gives rise to this signal?

Considering the microscopic processes in play, it is obvious that part of the signal may arise from structural changes in the photo-excited solute

$$\Delta S_{Solute} = S_{On} - S_{Off} = \left[\alpha S_{On} + (1-\alpha) S_{Off}\right] - S_{Off} = \alpha\left(S_{On} - S_{Off}\right) \qquad (7.7)$$

where α is the excitation fraction an S_{Off} and S_{On} is the scattering signal from, respectively, the ground-state structure of the molecule in question and the corresponding excited-state structure.

A second contribution arises from changes in the bulk structure of the solvent. The changes in scattering are due to changes in the three hydrodynamic parameters, density (ρ), temperature (T), and pressure (P). As described in detail in Reference [8] only two of these are needed to accurately describe the related difference signal $\Delta S_{\text{Solvent}}$ and most often one considers the temperature T and density ρ

$$\Delta S_{Solvent} = \Delta T \frac{\partial \Delta S}{\partial T}\bigg|_{\rho} + \Delta \rho \frac{\partial \Delta S}{\partial \rho}\bigg|_{T}.$$

The two components of this signal $\Delta S(\Delta T,Q)$ and $\Delta S(\Delta \rho,Q)$ can be estimated either from Molecular-Dynamics simulations or measured in separate experiments, with the latter showing the best performance [8, 9].

The third contribution to the signal is due to changes in the solvent-solute pair correlation functions. These changes can arise from changes in the number of solvent molecules in the first solvation shell and/or from changes in the coordination of these molecules. While this solvent-shell contribution $\Delta S_{\text{Solute-Solvent}}$ to the difference signal is of very significant scientific interest it is also hard to quantify, and only in recent years have researchers begun to address this contribution directly.

In summary, the acquired difference signal is the sum of three contributions

$$\Delta S_{Measured} = \Delta S_{Solute} + \Delta S_{Solvent} + \Delta S_{Solute-Solvent},$$

with the last contribution most often referred to as the solvent-shell term or the solute-solvent cross term. While important, it is beyond the scope of this text and we will only briefly consider it below.

7.3 Understanding the Data

The difference signal ΔS derived from the raw scattering data as described above is the starting point for the actual analysis. As described in Haldrup et al. [10], this slowly-varying oscillatory curve in Q-space contains all of the experimental information available from the experiment, but this is not sufficient for an *ab-initio* derivation of the structural change due to the laser pump pulse. For simple molecular systems like $C_2H_4I_2$, a Fourier sine transform (thus recovering a real-space representation of (changes in) the electron density) of the difference signal may directly yield clues to the underlying structural changes, although such $\Delta S(R)$ curves must be interpreted with significant care [11]. As an alternative, one may fit a set of simulated difference curves arising from a (potentially very large) number of putative structural changes to the experimental data. From the quality of the fit the (relative) likelihood of any of the proposed structural changes actually explaining the observed data can be evaluated. This model-comparison approach is described in detail in the following sections.

7.3.1 What's in a "Fit"?

They deem a fit acceptable if a graph of the data "looks good". This approach is known as chi-by-eye. Luckily, its practitioners get what they deserve.

Numerical Recipes, W. Press et al. [12]

The conceptual idea of model-fitting is very well described in the "Numerical Recipes" series of reference textbooks [12]. At the end of the day, it comes down to a quantitative way of asking "exactly how well does this particular model explain my data, given that my data are contaminated by this much noise". This question is answered by the χ^2-measure defined as

$$\chi^2 \equiv \sum_{i=1}^{N} \frac{(y_i - y(x_i; p_1 \ldots p_M))^2}{\sigma_i^2} / (N - M)$$

Where the set of y_i's is our measured data at points x_i, $y_i(x_i)$ is the (dependent) data predicted by our model function with parameters $p_1 \ldots p_M$ calculated at the set of points in the (independent) vector x_i, corresponding to e.g. the Q points for which scattering data has been measured. σ_i is the standard deviation calculated at each point by, for instance, the deviation from a locally smooth function [10] or by error propagation. N is the length of x and M is the number of free parameters p_{I-M} in the model. In many cases it is beneficial to transform the $\chi2$-measure to a Likelihood-estimate $L \propto \exp(-\chi^2/2)$ [12] as this may ease the interpretation of the analysis results as discussed below.

7.3.2 Non-structural Fits

The simplest class of fits in time-resolved structural studies arises in cases where one has solid prior knowledge about the possible structural changes occurring in the sample solution. For instance, if both the ground- and excited state structures have been previously determined or if trustworthy calculations of these structures exist, one can simply calculate the expected (solute) difference signal and then assume that the full calculated difference signal ΔS_{Calc} is the linear combination of solute and bulk solvent contributions

$$\Delta S_{Calc} = \alpha \Delta S_{Solute} + \Delta T \left. \frac{\partial \Delta S}{\partial T} \right|_{\rho} + \Delta \rho \left. \frac{\partial \Delta S}{\partial \rho} \right|_{T}$$

In these cases, all the powerful tools of linear algebra can be brought to bear on the problem and finding the best-fit solution that minimizes $|\Delta S_{Meas.} - \Delta S_{Calc}|$ as a function of (α, ΔT, $\Delta \rho$) is often straightforward using tools such as Matlab. This approach was taken in the recent analysis of $Fe(bipy)_3$ data acquired using MHz pump/probe repetition rates [13], and offered clues to how the solvation shell of the molecule changes following photo-excitation, see below.

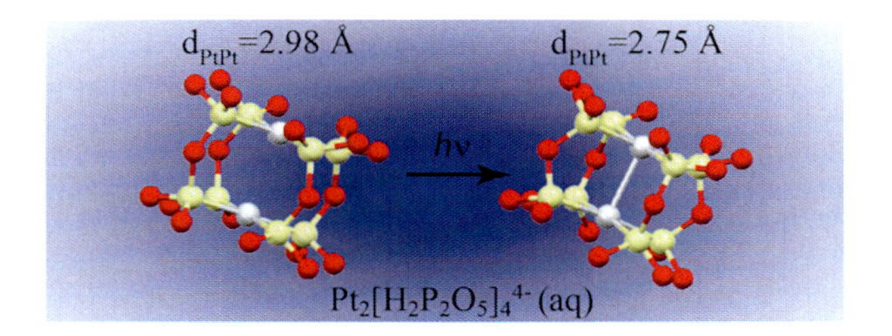

Fig. 7.8 Schematic depiction of the photo-induced bond formation and Pt-Pt contraction in $Pt_2[POP]_4$ in solution

7.3.3 Finding the Excited-State Structure by Fitting Models to the Difference Signal

Whereas non-structural fits as discussed above may at times yield interesting information, one is more often interested in obtaining direct information about the structural changes due to the excitation event. In these cases the structure of the molecule itself becomes a fit parameter. However, as discussed in detail in reference 8 and below, the information-poor nature of the signal in solution-state studies precludes fitting more than a few structural parameters or else run the significant risk of over-fitting and maybe over-interpreting the data.

As a starting point, consider the $Pt_2[POP]_4$ molecule shown in Fig. 7.8. Early optical measurements strongly indicated that the dominant structural change upon photo-excitation is a 0.2–0.3 Å contraction of the Pt atoms along the Pt-Pt axis, but a single article utilizing μs-resolved XAFS indicated that this contraction might be accompanied by a very large (0.5 Å) contraction of the phosphorous end-planes [14]. For our 100 ps time resolved investigations using X-ray scattering [15], we therefore chose to "structurally fit" the Pt-Pt distance (d_{PtPt}) and the P_{plane}/P_{plane} (d_{PpPp}) distance to our experimental data.

On a practical level, this was done by automatically generating a grid of 1,066 molecular structures, systematically varying d_{PtPt} from 2.5 to 3.0 Å and d_{PpPp} from 2.5 to 3.3 Å in steps of 0.02 Å. As a structural starting point we chose a DFT-calculated structure [16] with d_{PtPt} = 2.98 Å based on a quantitative comparison among several DFT- and crystallographically derived structures, benchmarking their expected ground-state scattering signal against a steady-state dataset. As each of the putative excited-state structures will correspond to one particular shape and magnitude of the difference signal, it is possible to quantitatively compare the model results with the experimentally determined difference signal.

$$\chi^2\left(d_{PtPt}, d_{PpPp}, \alpha, \Delta T, \Delta\rho\right) = \sum_Q \frac{\left(\Delta S_{Calc}(Q) - \Delta S_{Meas.}(Q)\right)^2}{\sigma(Q)^2} / (N-5),$$

$$\Delta S_{Calc} = \alpha\Delta S\left(d_{PtPt}, d_{PpPp}, Q\right) + \Delta T \left.\frac{\partial\Delta S(Q)}{\partial T}\right|_\rho + \Delta T \left.\frac{\partial\Delta S(Q)}{\partial\rho}\right|_T$$

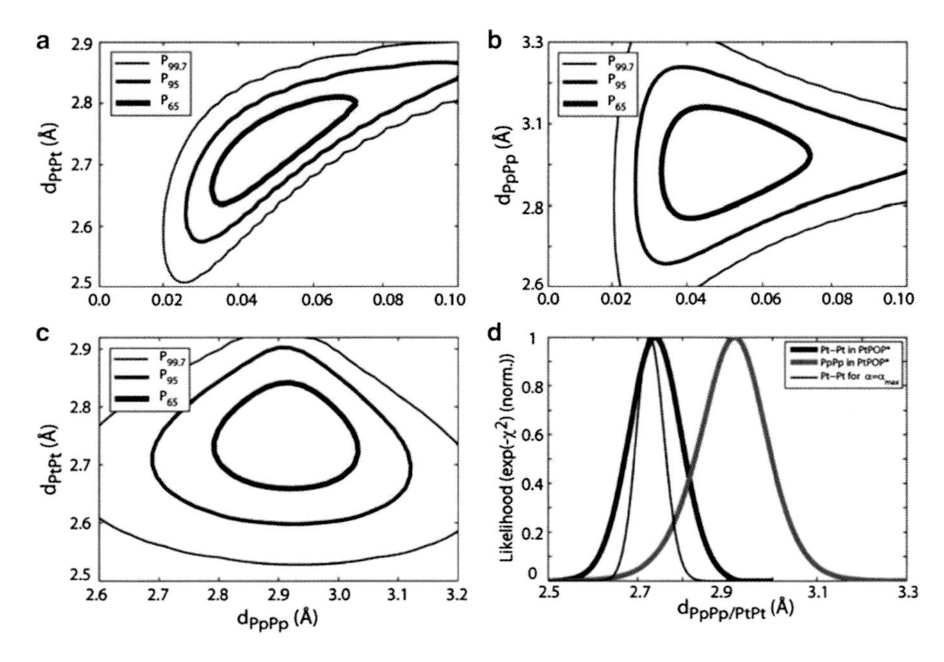

Fig. 7.9 Likelihood distributions from the structural fits of $\Delta S_{PtPOP*}(\Delta t = 100$ ps). The most-likely structural change is observed from Panel D to be a contraction of the Pt atoms of 0.23 Å along the Pt-Pt axis with no contraction of the P end-planes

For simplicity, we now disregard the solvent contributions, although significant information about the energy redistribution in the system may be learnt from, in particular, the temperature increase ΔT. Thus reducing the dimensionality of the problem, the likelihood-estimator becomes a function of three variables, $L(\alpha, d_{PtPt}, d_{PpPp})$.

Figure 7.9 shows the 2D projections of the likelihood-distribution onto the three sets of parameter axis, as well as the 1D projections onto the (combined) d_{PpPp}/d_{PtPt}-axis. From the latter plot it is very evident that the most-likely value for the excited-state Pt-Pt distance is 2.75 Å (0.23 Å contraction) and for the distance between the phosphorous end-planes, the most likely excited-state distance is seen to be 2.92 Å (No significant contraction). The calculated difference signals ΔS_{Calc} for this most-likely structure for the 100 ps and the 1 us time steps are shown super-imposed on the measured difference signals ΔS_{Meas} in Fig. 7.10.

Returning to the 2D contour maps of the likelihood distributions in panels A-C, the contour lines enclose (inner to outer) 65, 95 and 99.7 % of the likelihood distributions, corresponding to 1-,2-, and 3-σ joint confidence intervals of the parameter-pairs in question. For the 1D projections in panel D, the confidence intervals can be similarly constructed to include such particular fractions of the likelihood distribution. This intuitive way of constructing confidence intervals is a pleasing aspect of using the likelihood formalism, and is equivalent to the $\Delta\chi^2$-method where one considers projections of iso-$\Delta\chi^2$ surfaces as discussed in detail in *Numerical Recipes* [12] and from an experimental point of view in Jun et al. [17].

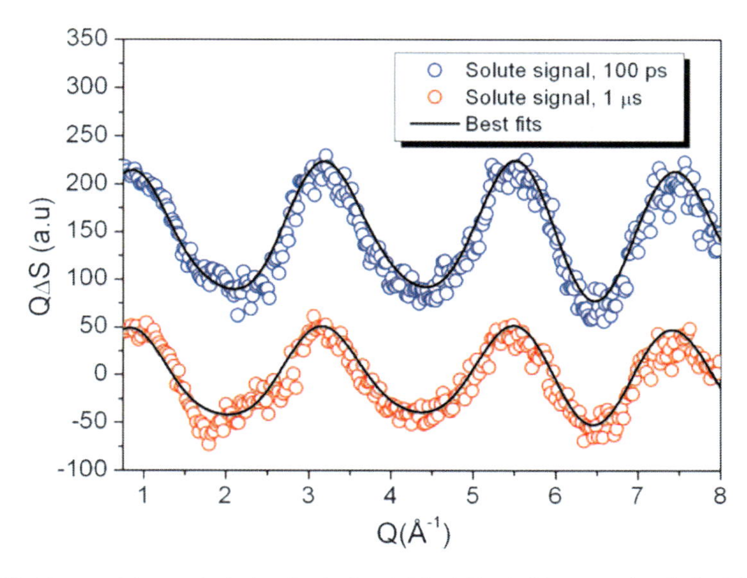

Fig. 7.10 Measured (*open circles*) and calculated (*black lines*) difference signals for PtPOP* at $\Delta t = 100$ ps and 1 μs

At this point it should be mentioned, that simply calculating the expected difference signal for each and every combination of the chosen structural parameters is in some sense a brute-force approach. This is perfectly feasible when the CPU-cost of calculating the difference signal is comparatively low, as it is for medium-sized molecules and using the Debye-formula introduced above. However, for large molecules with many hundreds of atoms or when more sophisticated calculations are needed, the brute-force approach may become prohibitively time consuming. One recently developed method for handling such situations relies on the observation that the scattering signal is a slowly varying function of the structural parameters. Thus, a polynomial interpolation between more sparsely distributed points in the chosen structural space can be utilized to lower computational cost, while still retaining good structural resolution. This has been implemented for XAFS-analysis by G. Smolentsev et al. [18], but the method is readily generalizable to also X-ray Diffuse Scattering.

7.3.4 Fundamental Limitations and Use of Complementary Information

Whereas diffraction patterns with sometimes many thousands of reflections can allow researchers to back-transform the diffraction data and derive the relative position of thousands of atoms in complicated proteins, the information content in diffuse scattering patterns is very much less and does not allow such *ab-initio* structure determination. Still, the shape (and magnitude) of the difference signal contains enough information to allow quantitative model comparisons as discussed above, but the question arises: How many structural parameters can be estimated this way, without over-fitting the data? This is discussed in some detail in [10] and in particular the references therein, and one pertinent result is that an acquired scattering signal with Q-range ΔQ contains a maximum number of "relevant independent points" M given by

$$M = 2\frac{\Delta Q D_{Max}}{\pi}, \Delta Q = Q_{Max} - Q_{Min}$$

Where D_{Max} is the maximum size (in Å) of a particle/molecule considered. The form of this expression makes intuitive sense, as a larger real-space size of the system in question will lead to faster-varying curves in Q-space and it also appears evident that for slowly-varying signals as the ones presented here, one does not obtain more information simply by increasing the number of pixels on the detector or Q-points in the azimuthally integrated signal. The individual ΔS(Q) data points are not truly *independent* measurements as it is, for instance, not equally likely for two neighbouring ΔS(Q) points to have the same or opposite sign. Another way to phrase this is to say that ΔS(Q) is heavily oversampled, as we typically have ~500 ΔS(Q) values, but only 20–30 relevant independent points. From further considerations of these issues [10], it can be demonstrated that at most about ten structural parameters can be estimated from a typical difference signal, and in practice often less. On a hands-on level, this is often made evident when including more (too many) free parameters in the analysis leads to significant increases in the uncertainty estimates for each parameter.

One way to overcome such limitations and to constrain the uncertainty on the investigated structural parameters is to include complementary information in the analysis. Quantitatively, this transforms the projections/sums along p_i in $L(p_i)$-space to transections, as schematically illustrated in Fig. 7.11. Qualitatively this is easily understood, as now not all values of the externally constrained parameter are "equally likely". Instead, all values outside some interval around the measured value of the constrained parameter are set to have zero likelihood.

7.4 Selected Results from Ultrafast Solution-State X-Ray Measurements

7.4.1 Bimetallics

The series of ligand-bridged transition metal compounds commonly denoted M_2[bridge]$_4$, where M = Pt, Ir or Rh has attracted much attention from the photo-chemistry community since the early 1980s [19, 20]. From optical

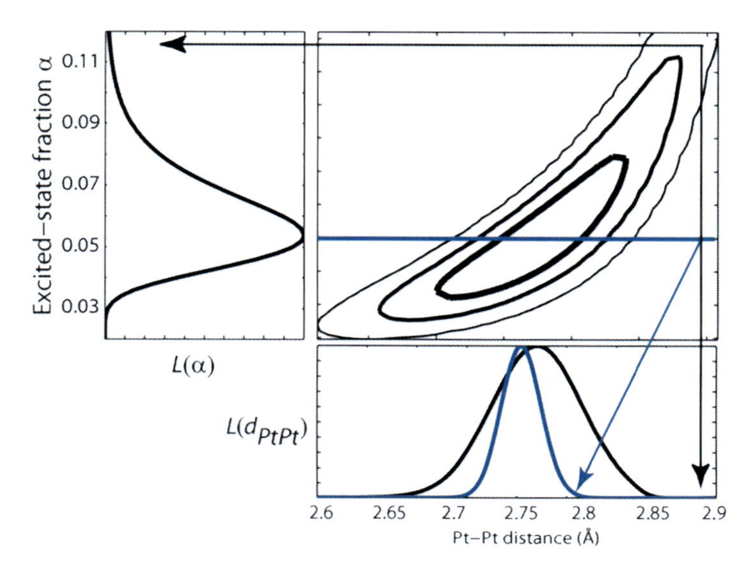

Fig. 7.11 α/d_{PtPt} likelihood distribution. *Black arrows* illustrate sums along the parameter axis (no external knowledge) and *blue line/arrow* represents the situation where e.g. the excitation fraction has been externally determined, leading to a much more narrow likelihood-distribution for the Pt-Pt distance

spectroscopy, it was established that photo-excitation could lead to the promotion of one electron from anti-bonding $d\sigma^*$ orbital to a bonding $p\sigma$ located between the two metal centers. This change in electronic structure was surmised to be accompanied by a significant structural change, a prediction directly confirmed for the same prototypical photoactive compound as discussed above, $Pt_2[POP]_4$, by a time-resolved crystallography study in 2002 [21]. In this pioneering study, low temperature (17 K) was utilized to extend the excited-state lifetime into the regime accessible by the 33 μs X-ray pulses used for those experiments.

In the Centre for Molecular Movies, we have been primarily interested in intramolecular structural dynamics when they take place in solution, the native environment for most chemical reactions and almost all biological ones. In late 2008 we could report the first-ever direct determination of the excited-state structure of PtPOP* in solution [5], using the methods outlined above. In 2009 and 2010 we supplemented these results with structural investigations of the TlPtPOP* [22] and AgPtPOP* [23], short-lived excited-state complexes (exciplexes) formed between the excited state of PtPOP* and metal ions.

One particular complication in the analysis of such time-resolved data sets is which ground-state structure to use in the calculation of the difference-signal Eq. 7.7. Several different structures of PtPOP had been reported in the literature, mainly from single-crystal structure determinations but also from theory. To keep as close a connection to our experiment as possible, we chose one of these published structures based on a direct comparison of calculated scattering with a

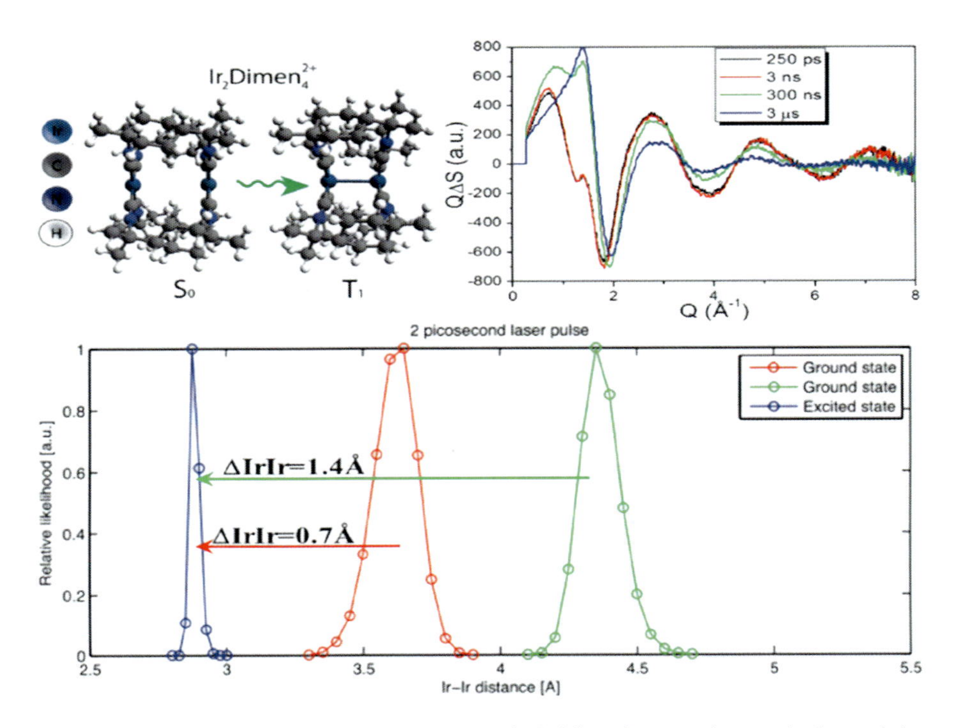

Fig. 7.12 Top panel shows the structural changes in $Ir_2[dimen]_4$ upon photo-escitation and the consequent difference signals. Structural analysis revealed the presence of two ground-state structures, differing significantly in d_{IrIr}. Following photo-excitation to the excited state, both isomers have the same, but much shorter, metal-metal distance

steady-state scattering measurement of a concentrated solution of PtPOP. A recent theoretical study [24] incorporating solvent effects confirmed this to be a good choice of ground state for the PtPOP* and TlPtPOP* investigations, but indicated some discrepancies for AgPtPOP*, possibly due to ground-state association.

Another approach to the challenge of estimating the ground state structure is to include it explicitly in the fitting routine. This was the method of choice in our study [25] of $Ir_2[dimen]_4$ (Fig. 7.12, top) where deformational isomerism leads to the presence of not one but two co-existing ground-state structures in solution, with Ir-Ir distances of 3.6 and 4.3 Å. However, our structural analysis showed that in the excited state both of these structures have the same Ir-Ir distance, 2.90 +/− 0.02 Å as schematically depicted by the likelihood-plots in Fig. 7.12, lower panel.

7.4.2 *$Fe(bipy)_3$ at APS*

In addition to their interesting photochemical properties, the compounds discussed above were also chosen for the very large scattering power of the metal centres Pt (Z = 78) and Ir (Z = 77). In combination with the significant changes in metal ion

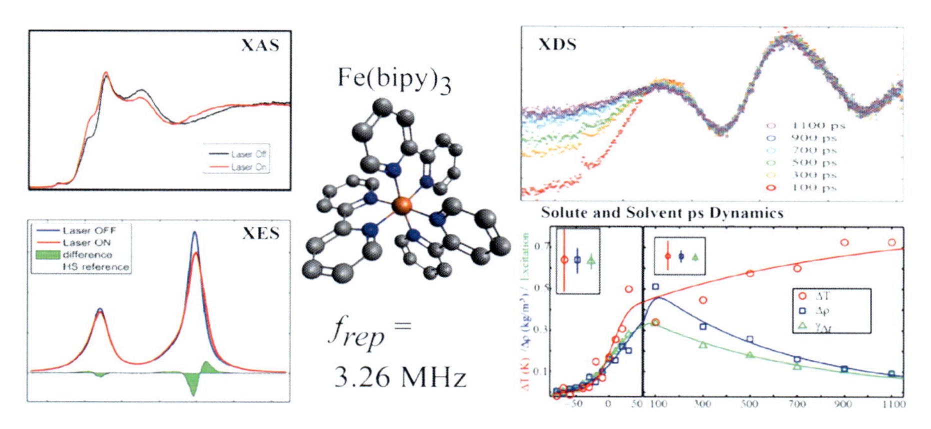

Fig. 7.13 Collage illustrating the three different types of data acquired in one experiment by using MHz pump-probe repetition rates at APS Sector 7. The molecule under study was Fe(bipy)_3 in water, a prototypical spin-crossover compound. By including X-ray Diffuse Scattering we were able to detect an ultrafast change in the solvent density (*lower right, blue* points) upon photo-excitation of the solute. Theory suggests a direct relationship between this macroscopic observation and the microscopic process of structural changes in the solvation shell

positions, this led to very strong difference signals (Fig. 7.12, top), despite the three orders of magnitude decrease in usable X-ray intensity for Time-Resolved studies that is due to the mismatch between the repetition frequency of the synchrotron X-ray pulse train (MHz) and the optical laser pulses (kHz). The recent availability of MHz rate laser systems [26, 27] with sufficient average power in combination with gateable area detectors [28] has allowed us to overcome this limitation for scattering studies and we and our collaborators have recently demonstrated full feasibility of 3.26 MHz repetition rates [13], thus essentially utilizing the full intensity of third generation synchrotron sources for time-resolved scattering studies. Taking advantage of this very large increase in available X-ray flux on the sample, we investigated the prototypical, but only weakly scattering, spin-crossover compound $Fe(bipy)_3$ in aqueous solution at APS Sector 7. Following photo-excitation, this compound exhibits both a large change in electronic structure (from $^1A_1$ to $^5T_2$) as well as significant change in Fe-N bond length. These dynamics have been well characterized using time-resolved X-ray Absorption Spectroscopy as well as conventional optical spectroscopy, but never by using solution-state scattering and only superficially by X-ray Emission Spectroscopy, despite the latter probe's excellent sensitivity to spin states. The relevant interaction cross sections were simply too low for such investigations.

Figure 7.13 shows in schematic form how we combined three different measurement methods (XAS, XES and scattering) in one experimental setup. The S/N obtained for these data sets allowed us to carry out a detailed analysis, and as shown in lower-right plot we were able to determine an unexpected, transient increase in the bulk solvent density. In the analysis of this experiment, we relied on previous, detailed measurements [29] of the structure of HS $Fe(bipy)_3$ and thus we could keep the structure constant, varying only excitation fractions and bulk-solvent parameters (changes in temperature ΔT and density Δρ). Thus rendering the problem as a simple minimum-search $\min\{\chi^2(\alpha, \Delta T, \Delta\rho)\}$ and using the inbuilt minimization methods of Matlab, this led to fast turnover in the analysis process.

References

1. Warren BE (1990) X-ray diffraction. Dover Publications. ISBN: 0-486-66317-5. http://books.google.dk/books?id=wfLBhAbEYAsC&lpg=PP1&dq=warren%20x-ray&hl=da&pg=PP1#v=onepage&q&f=false
2. Cromer DT, Mann JB (1968) X-ray scattering factors computed from numerical Hartree-Fock wave functions. Acta Crystallogr A 24:321–324
3. http://webbook.nist.gov/cgi/cbook.cgi?ID¼C75730
4. Ejdrup T et al (2009) Picosecond time-resolved laser pump/X-ray probe experiments using a gated single-photon-counting area detector. J Synchrotron Radiat 16:387–390
5. Boesecke P, Diat O (1997) Small-angle X-ray scattering at the ESRF high-brilliance beamline. J Appl Crystallogr 30:867–871; Skinner LB et al (2012) Area detector corrections for high quality synchrotron X-ray structure factor measurements. Nucl Instrum Method A 662:61–70; and others
6. Taylor JR (1997) An introduction to error analysis: the study of uncertainties in physical measurements. University Science Books, Sausalito
7. Dent AJ et al (1991) The extraction of signal to noise values in x-ray absorption spectroscopy. Rev Sci Instrum 63:856–858
8. Cammarata M et al (2006) Impulsive solvent heating probed by picosecond x-ray diffraction. J Chem Phys 124:1245041–1245049
9. Kjær K et al (2013) Introducing a standard method for experimental determination of the solvent response in laser pump, X-ray probe time-resolved wide-angle X-ray scattering experiments on systems in solution. Phys Chem Chem Phys 15:15003–15016
10. Haldrup K et al (2010) Analysis of time-resolved X-ray scattering data from solution-state systems. Acta Crystallogr A 66:261–269
11. Ihee H et al (2005) Ultrafast X-ray diffraction of transient molecular structures in solution. Science 309:1223–1227
12. Press W et al (1988) Numerical recipes. Cambridge University Press, Cambridge (also available on the web)
13. Haldrup K et al (2012) Guest – host interactions investigated by time-resolved X-ray spectroscopies and scattering at MHz rates: solvation dynamics and photoinduced spin transition in aqueous $Fe(bipy)_3^{2+}$. J Phys Chem A 116:9878–9887
14. Thiel D et al (1993) Microsecond-resolved XAFS of the triplet excited state of $Pt_2(P_2O_5H_2)^{4-}{}_4$. Nature 362:40–43
15. Christensen M et al (2009) Time-resolved X-ray scattering of an electronically excited state in solution. Structure of the $^3A_{2u}$ state of Tetrakis-μ-pyrophosphitodiplatinate(II). J Am Chem Soc 131:502–508
16. Novozhilova I et al (2003) Theoretical analysis of the triplet excited state of the $[Pt_2(H_2P_2O_5)_4]^{4-}$ ion and comparison with time-resolved X-ray and spectroscopic results. J Am Chem Soc 125:1079–1087
17. Jun S et al (2010) Photochemistry of $HgBr_2$ in methanol investigated using time-resolved X-ray liquidography. Phys Chem Chem Phys 12:11536–11547
18. Smolentsev G et al (2007) Three-dimensional local structure refinement using a full-potential XANES analysis. Phys Rev B 75:144106-1-5

19. Harvey P (2001) Chemistry, properties and applications of the assembling 1,8-diisocyano-p-menthane, 2,5-dimethyl-2' ,5'-diisocyanohexane and 1,3-diisocyanopropane ligands and their coordination polynuclear complexes. Coord Chem Rev 219–221
20. Zipp A (1988) The behavior of the tetra-μ-pyrophosphito-di-platinum anion $[Pt_2(H_2P_2O_5)_4]^{4-}$ and related species. Coord Chem Rev 84:17–52
21. Kim C et al (2002) Excited-state structure by time-resolved X-ray diffraction. Acta Crystallogr A 58:133–137
22. Haldrup K et al (2009) Structural tracking of a bimolecular reaction in solution by time-resolved X-ray scattering. Angew Chem 48:4244–4248
23. Christensen M et al (2010) Structure of a short-lived excited state trinuclear Ag–Pt–Pt complex in aqueous solution by time resolved X-ray scattering. Phys Chem Chem Phys 12:6921–6923
24. Kong Q et al (2012) Theoretical study of the triplet excited state of PtPOP and the exciplexes M-PtPOP (M = Tl, Ag) in solution and comparison with ultrafast X-ray scattering results. Chem Phys 393:117–122
25. Haldrup K et al (2011) Bond shortening (1.4 Å) in the singlet and triplet excited states of $[Ir_2(dimen)_4]^{2+}$ in solution determined by time-resolved X-ray scattering. Inorg Chem 50:9329–9336
26. March A et al (2011) Development of high-repetition-rate laser pump/x-ray probe methodologies for synchrotron facilities. Rev Sci Instrum 82:073110
27. Lima F et al (2011) A high-repetition rate scheme for synchrotron-based picosecond laser pump/x-ray probe experiments on chemical and biological systems in solution. Rev Sci Instrum 82:063111
28. Ejdrup T et al (2009) Picosecond time-resolved laser pump/X-ray probe experiments using a gated single-photon-counting area detector. J Synchrotron Radiat 16:387–390
29. Gawelda W et al (2009) Structural analysis of ultrafast extended x-ray absorption fine structure with subpicometer spatial resolution: application to spin crossover complexes. J Chem Phys 130:124520 1–9

Chapter 8
Single-Crystal to Single-Crystal Solid-State Photochemistry

Menahem Kaftory

8.1 Introduction

Understanding the mechanism of a chemical reaction is fundamental for the understanding of chemistry. Determining the structure of reacting molecules during a reaction is a major factor in exploring the mechanism of a process. There are, however, two drawbacks in the study of the structures of the reacting species during a chemical reaction when it is carried out in solution. The first is the absence of a method for determining the structure to the level of atomic resolution, and the second is the short time a reaction may take to go to completion. The best method for structure determination is X-ray diffraction, however, the limitation is that it is being used for crystalline materials and the time for data collection is few orders of magnitude longer than the reaction time for the majority of chemical processes. In order to overcome these obstacles one is limited to the study of reactions that take place in the solid and to shorten the time of measurements. While many reactions are known to take place in the solid-state the time of measurements are still longer than is needed to follow structural changes during reaction processes. Much progress was achieved in recent years to shorten the time of data collection and I leave this subject to the experts in the field. I would like to describe examples of structurally monitoring photochemical solid-state reactions which hopefully will be reinvestigated in years to come when the goal of very fast data collection has been achieved.

Most of the known and best studied solid-state photochemical reactions are those involving crystals containing pure photoreactive molecules. However, almost all of the reactions taking place with a deterioration of the crystal. This fact limits the monitoring of the structural changes as a result of a reaction. The product has to be recrystallized and only then its crystal structure can be determined. As a result, the structural information from the actual reaction is lost, partially or completely.

M. Kaftory (✉)
Schulich Faculty of Chemistry, Technion – Israel Institute of Technology, Haifa, Israel
e-mail: kaftory@tx.technion.ac.il

J.A.K. Howard et al. (eds.), *The Future of Dynamic Structural Science*,
NATO Science for Peace and Security Series A: Chemistry and Biology,
DOI 10.1007/978-94-017-8550-1_8,

One has to speculate on the events during the reaction and one can miss some of them. In contrast, solid-state reactions that undergo single-crystal to single-crystal (SCtSC) transformations are rewarding. In the era of photochemical reactions special experimental criteria were proposed for SCtSC reactions to occur [1, 2]. Irradiation of the crystalline sample at a wavelength that is close to the tail of the absorption band enables a SCtSC reaction to proceed successfully. However, in most cases the single crystal did not survive intact and resulted polycrystalline material.

The main reason for the limited number of examples undergoing topotactic photochemical reaction is that any substantial structural change experienced by a reactive molecule in the crystal will affect the neighbouring molecules undergoing the same structural variation and result in a local disorder. We may compare it to a crowd of people walking on a sidewalk in one direction and suddenly a few pedestrians change direction, this immediately will cause a disorder in the crowd.

A general method by Enkelmann [1, 2] describes the induction of a homogeneous photochemical reaction (SCtSC) by irradiating the crystal with light at a wavelength corresponding to the chromophore's absorption tail. This method enables monitoring structural changes during photoreactions. It may also provide another means by which new polymorphs of the product can be obtained [2]. However, in most cases if the photochemical reaction is executed on the pure light-sensitive compound, the reaction does not proceed to full conversion before the crystal disintegrates. Therefore, monitoring of the reaction process is very limited.

It has been demonstrated recently [3–8] that in co-crystals composed of light-stable host molecules and light-sensitive guest molecules the photochemical reaction may proceed to completion with the preservation of the integrity of the single-crystal. If the host molecules provide topochemical conditions required for mono or bimolecular reactions and the guest molecules are photochemically active, regio- and stereo-selective reactions are anticipated [9–11]. If the host molecule is chiral, it induces chirality into the space where the guest molecules react and as a result enantio-selective reactions take place [12–15]. Therefore, inducing photochemical reactions in co-crystals proved to be a unique method for synthesizing a large variety of compounds [16]. The volume of space available for the guest molecules determines whether the reaction will be heterogeneous or homogeneous. In cases where the size of the reaction pocket is large enough to accommodate the substrate and the product (not simultaneously), the reaction is expected to be homogeneous. This volume is also called "reaction cavity," originally introduced and developed by Cohen to describe reactions in crystals [9]. This model has been further developed by Weiss et al. [10], for performing photochemistry in co-crystals systems.

The advantages of performing photochemical reactions in co-crystals compared with carrying the photochemical reaction in the neat solid reactive compound are the following:

1. the dilution and the separation between potentially photochemical active molecules enables structural variation with no interference from molecules of the same kind, and this will lead to homogeneous reaction.

2. by changing the light-stable host molecules it is possible to identify the system where the reaction will be topotactic, and this also gives the opportunity to study the effect of different neighbouring molecules on the reactivity of the light-sensitive guest molecule.
3. replacing the host molecules may affect the conformation and the geometry of the reactive molecular target and may provide valuable data for the understanding of the reaction mechanism.
4. it reduces the number of photons required to achieve a certain amount of conversion.
5. it provides the ability to engineer templates for carrying out special synthesis.

In the next sections we will show examples of SCtSC photochemical reactions that will demonstrate the advantages in using co-crystals as the chemical environment for the understanding of photochemical reactions.

8.2 Experimental Setting

In the examples given below the following experimental conditions were used:

Irradiation: A single crystal was irradiated either with UV LED in the range 350–390 nm with the maximum of 365 nm set to 95 mW or with the third (355 nm) harmonics of a circularly polarized pulsed Nd:VO4 laser. The laser was set to pulse width of 35 ns and 10 kHz. The average power was less than 1 mW for a spot with a diameter of ~1 mm. The crystal revolved while it was exposed to the light.

X-Ray Crystal Structure: X-Ray diffraction intensities were measured either with Kappa CCD Nonius Diffractometer or with a Bruker Smart APEX2 CCD diffractometer installed on a rotating anode source (Mo-K_{α}, $\lambda = 0.71073$ Å) and equipped with an Oxford Cryosystems nitrogen gas flow apparatus. Diffraction intensity data were measured before irradiation and after each cycle of exposure. The transformation from a single-crystal to single-crystal took place with no significant effect on the quality of the crystal or on the diffraction intensities and mosaicities.

The crystal structure at each stage was solved by direct methods (SHELX97). All non-hydrogen atoms were refined anisotropically. The positions of the hydrogen atoms were located in difference Fourier maps and refined by riding on their parent atoms.

The refinement of the crystal structure that includes the unreacted guest molecules and the partially produced product was carried in a similar procedure as used above, however the occupancy factor of the atoms of the unreacted molecules and the product molecules was refined. At lower and higher conversion, the atoms of the minor species were refined isotropically, at conversion closer to 50 % both unreacted molecules and the product molecules were refined anisotropically.

8.3 Chiral Discrimination [17]

1-Methyl-5,6-diphenylpyrazin-2-one (1) crystallizes in two modifications, one of which is light-stable and the other light-sensitive. The light-sensitive modification is known to undergo photodimerization in the solid state (Scheme 8.1). The product can be one of the four isomers shown, depending on the relative arrangement of the reacting pairs.

This polymorph crystallizes in the monoclinic space group $P2_1$ with two crystallographically independent molecules in the asymmetric unit. The molecules are packed in stacks running parallel to the unique *b* axis. The two independent molecules are arranged alternately along the stack. In principle, there are two different pairs of molecules within a stack that can undergo photodimerization, and each should form a different enantiomer (Scheme 8.2).

A large crystal was irradiated and a solution of the product was separated by HPLC. The optical purity of the (+)-enantiomeric sample was estimated to be greater than 90 %. This finding indicates that only one of the two pairs undergoes photoreaction. The structure of a single crystal of the pyrazinone was elucidated by X-ray diffractometry before and after irradiation with a laser at a wavelength of 488 nm and the conversion was found to be 19 %. The results of the crystal-structure determinations provide additional evidence that only one of the two pairs of molecules undergoes photodimerization although there are no significant differences in the distances between the reacting centres. Furthermore, the latter results suggest that weak hydrogen bonds are a dominant factor that determines which of the two pairs is dimerized upon irradiation (Fig. 8.1.). Indeed, the produced enantiomer is the one formed by dimerization of the pair of molecules held by closer hydrogen bonds shown in Fig. 8.1.

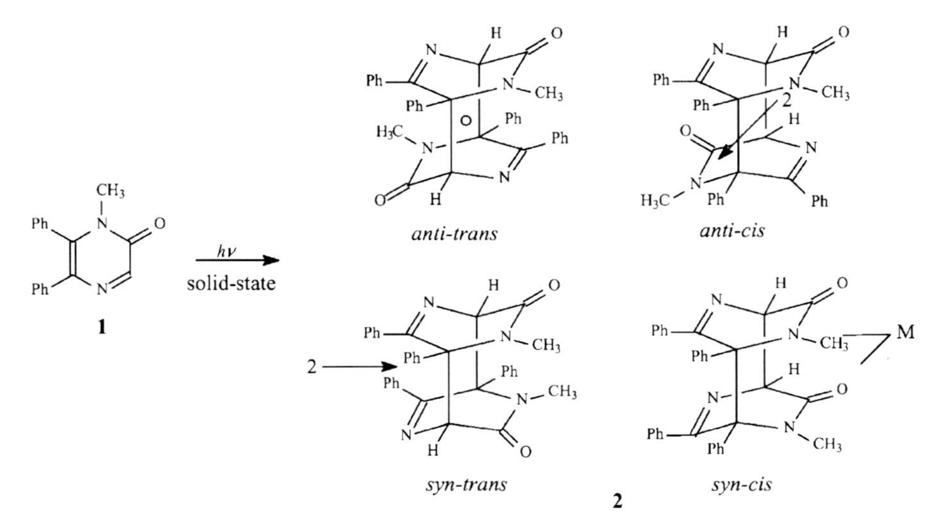

Scheme 8.1 Different isomers that might be obtained from the [4+4] photodimerization of 1-Methyl-5,6-diphenylpyrazin-2-one (**1**) in the solid state

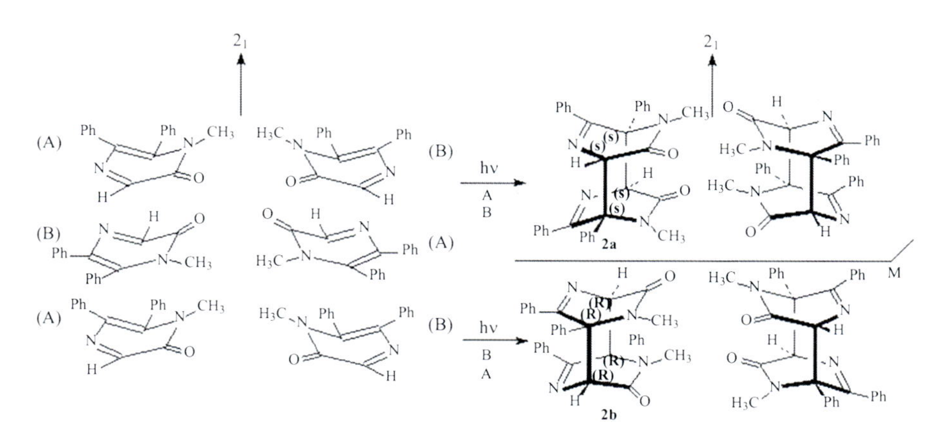

Scheme 8.2 Two possible photodimerization routes that might lead to different enantiomers

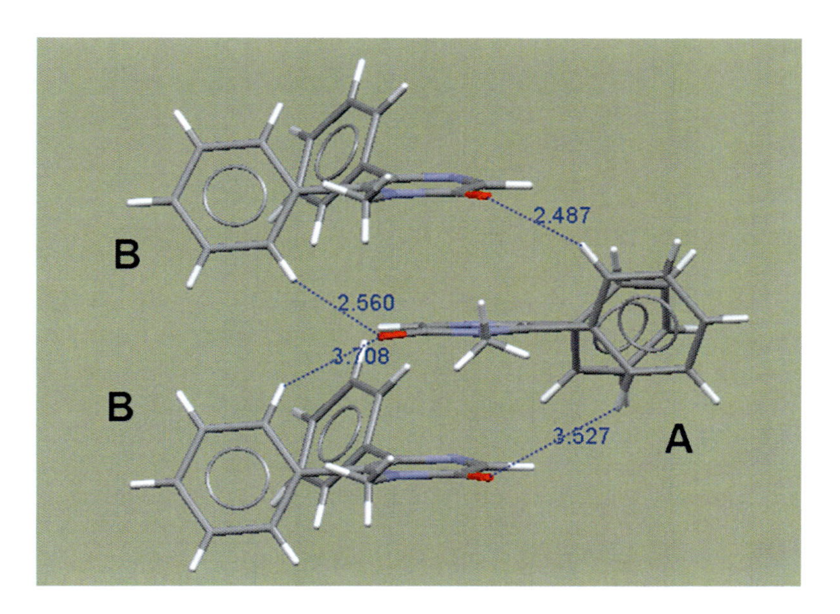

Fig. 8.1 Hydrogen bonds between molecules in a stack

8.4 Ring Closure [18]

Six inclusion compounds containing a photoreactive guest molecule, 4-oxo (phenylacetyl)morpholine or 1-(phenylglyoxylyl)piperidine, with different host molecules have been crystallized. The guest molecules underwent photochemical reaction upon irradiation. Examining their structures suggests that α-hydrogen abstraction by oxygen, the first step in cyclization of α-oxoamides, should be

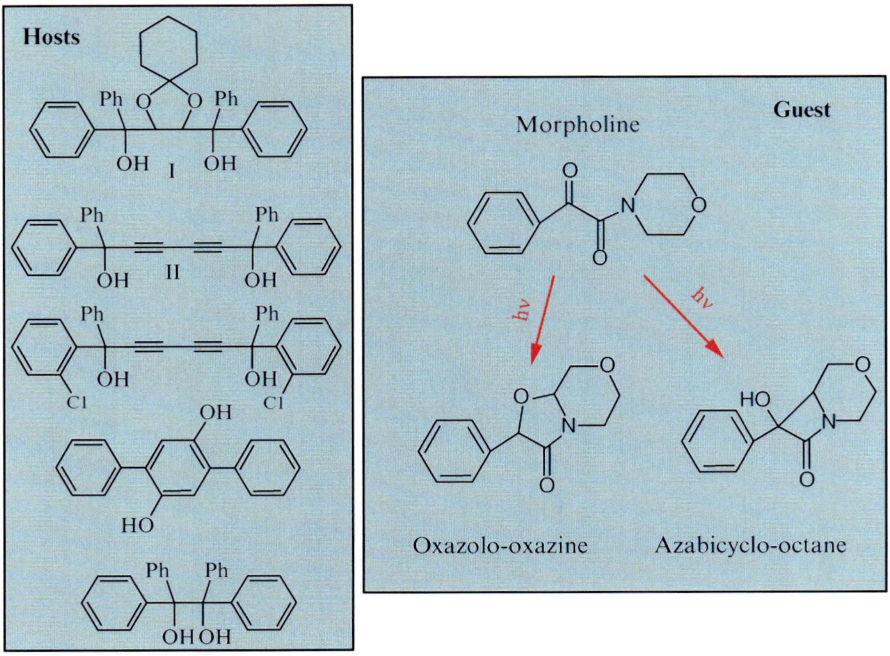

Hosts

Guest

Morpholine

Oxazolo-oxazine

Azabicyclo-octane

Scheme 8.3 Two different ways that the cyclisation of morpholine can occur when embedded in different host molecules

possible in all cases. In four cases crystallinity was maintained during and at the end of the conversion process *i.e.* the process was a single-crystal to single-crystal transformation. The crystal structure of the product crystals revealed that in two of the inclusion compounds the product obtained is a result of formation of a four-membered ring while in the other two the product obtained results from formation of a five-membered ring (example in Scheme 8.3).

It was found that the morpholine molecules adopt two different conformations which lead to the two different ring-sizes in the product. The photochemical reaction may take different pathways either as a result of the different conformation or as a result of the different shapes of the cavities provided by the host molecules. The differences between the conformations of the two guest molecules that lead to closure to either 4 or 5-membered ring is shown in Fig. 8.2.

8.5 Unexpected Events [19, 20]

The best demonstration of the advantage of using co-crystals composed of light-stable host and light-sensitive guest molecules for single-crystal to single-crystal photodimerization is given next. In the following examples we will show that water

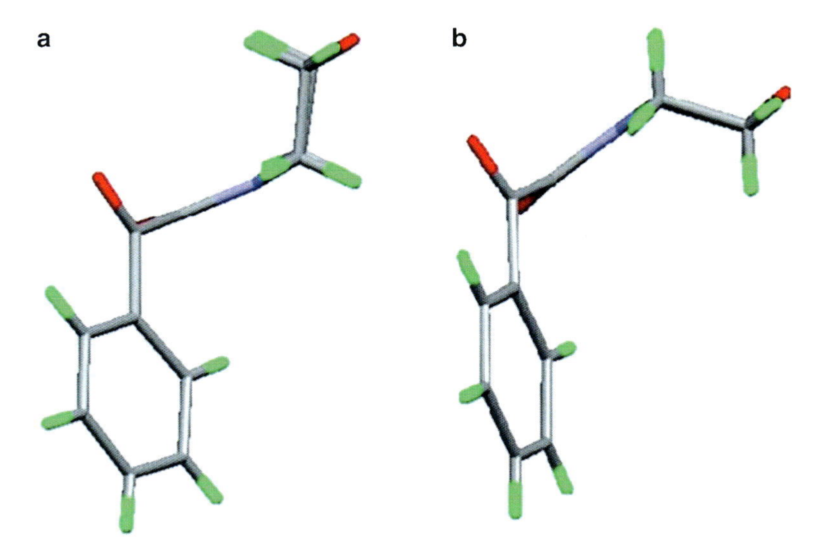

Fig. 8.2 Two different conformations of morpholine in different co-crystals

molecules can penetrate into the crystal in an ordered fashion without changing the crystal integrity. It will also be shown that molecules can flip in the crystal with preservation of the single-crystal entity. These observations could not have been detected if the transformation was not of the SCtSC type.

Three co-crystals of the light-stable compound 1,1,6,6-tetraphenyl-2,4-hexadiyne-1,6-diol (**I**) with light-sensitive molecules 2-methyl-2(1H)-pyridinone (**a**), 6-methyl-2(1H)-pyridinone (**b**) and 1,2-dimethyl-2(1H)-pyridinone (**c**) (Scheme 8.4) were exposed to UV light.

Due to the ability to monitor the structure after different periods of irradiation time, it was found that crystals of **I-a** underwent physical and chemical changes. During the photodimerization the molecules undergo rotation by *ca.* 180° (flip), moreover, after full conversion was reached an ordered pattern of water molecules have been detected which form hydrogen bonds and the unit-cell volume was tripled. After standing in the open air for few weeks, the water molecules leave the crystal and the crystal remains intact. The variation of the structure of the guest molecules is shown in Fig. 8.3. The light-sensitive molecules in crystals of **I-b** are disordered in 85/15 % ratio. The disordered molecules are related to each other by a pseudo 2-fold axis. The light-sensitive molecules are arranged is pairs that are related to each other by inversion centres. Of these pairs, the proximity of the reactive atoms that meet Schmidt rule [21] is detected only between molecules with the minor occupancy (15 %). However, the photodimerization goes to full conversion. The reason is that the pairs of molecules with major occupancy (85 %) undergo rotation by *ca.* 180° (flip) and then photodimerize. We showed that the flip takes place also upon heating the crystal. Measurements of the dependence of the occupation of the disordered molecules with temperature revealed an estimate for activation energy of the flip of 9.72 kJ/mol. The flip was rationalized as an equilibrium effect. After full conversion

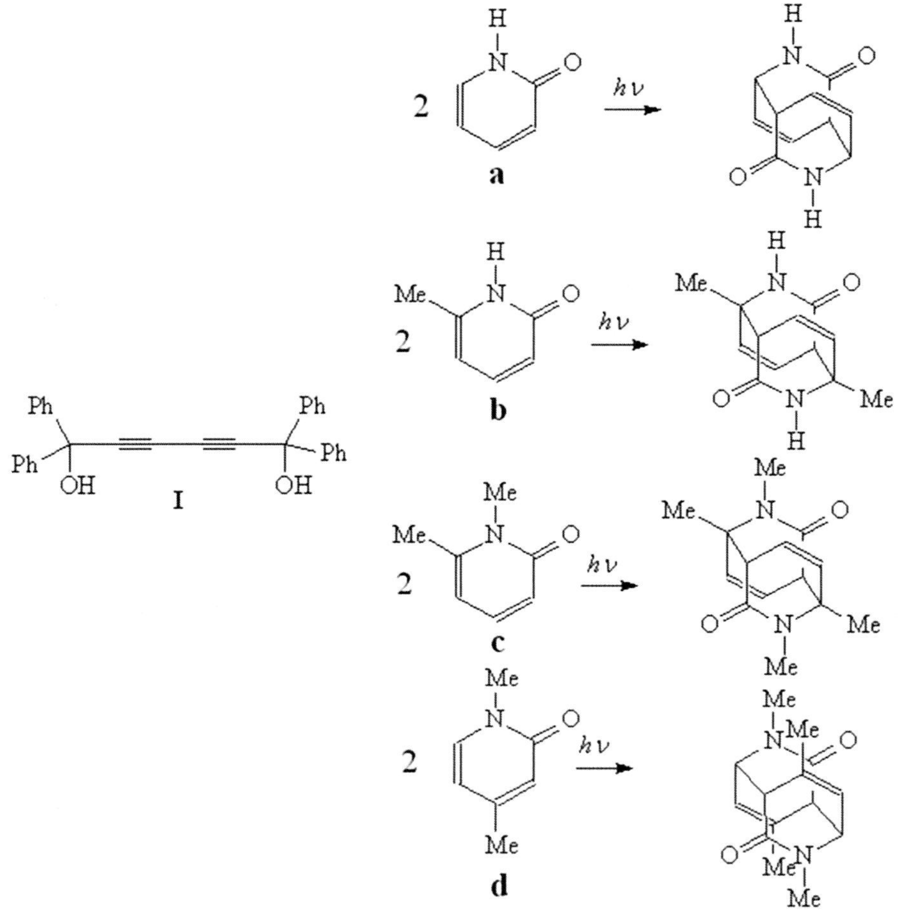

Scheme 8.4 The products of solid state photodimerization of four isomers of substituted pyridones in co-crystal with 1,1,6,6-tetraphenyl-2,4-hexadiyne-1,6-diol (**I**)

was reached, water molecules penetrated the crystal and are arranged in an ordered fashion forming hydrogen bonds causing doubling of the unit cell volume. After standing in the open air for few weeks, the water molecules leach out and the crystal remains intact. The variation of the structure of the guest molecules is shown in Fig. 8.3.

Crystals of **I-c** include disordered light-sensitive molecules in 50/50 % ratio. In a similar way as found in **I-b**, only one pair of molecules meet Schmidt's rule. However the photodimerization goes to completion. The meaning is that molecules of the "wrong" relative geometry undergo the flip. The variation of the structure of the guest molecules is shown in Fig. 8.4. Water molecules do not penetrate into the crystal. This can be explained by the absence of free hydrogen acceptors and donors in the crystal as a result of replacing N-H by N-Me in the guest molecule.

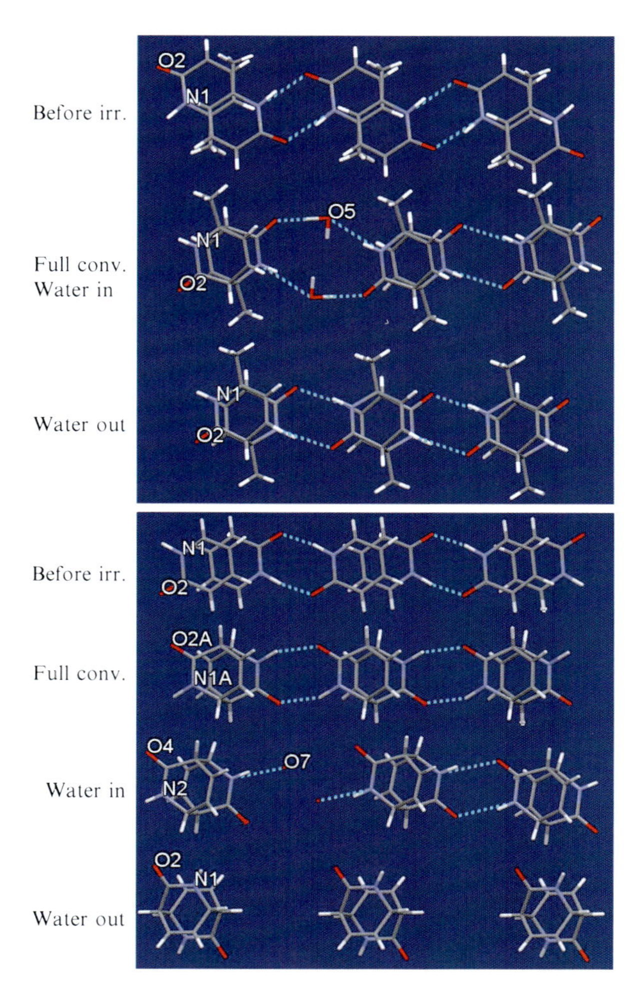

Fig. 8.3 Variation of the structure of the guest light-sensitive molecules at different stages of the photodimerization in **I-a** (*left*) and **I-b** (*right*)

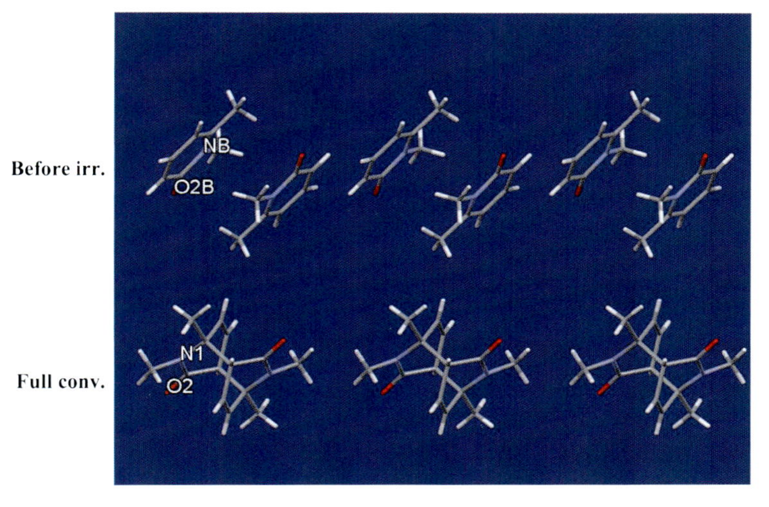

Fig. 8.4 Variation of the structure of the guest light-sensitive molecules at different stages of the photodimerization in **I-c**, before irradiation (*top*), at full conversion (*bottom*)

8.6 Kinetic Measurements [22]

Single-crystal to Single-crystal transformation allow kinetic data to be obtained. The crystal structure is monitoring after different periods of illumination. The occupancy factor of the product dimer at the different time periods is then obtained from the refinement of the crystal structures. These data can be analysed using the JMAK [23–26] equation:

conversion $= 1 - \exp(-kt^n)$ (where conversion $= 1 -$ occupancy factor) to describe the mechanism of crystal growth.

This equation was used as a model for the kinetics of phase transitions involving nucleation and growth mechanisms. In the equation, "conversion" is the fraction of the dimer, t is the exposure time, k is the constant of the growth rate, and n is the Avrami exponent that shows the dimensionality of the growth. When $n = 2$, 3, or 4 the dimensionality of the growth is 1, 2, and 3, respectively. When $n = 1$ it means that the reaction is homogeneous with equal probability to occur in any region of the sample.

In order to understand the mechanism of the reaction and growth of the product, a single crystal of **I-a** was illuminated and checked periodically to establish the progress of the conversion.

Figure 8.5 shows the fraction of the photodimer produced (expressed in term of conversion) by the irradiation of a single crystal of **I-a** with 355 nm UV-LED.

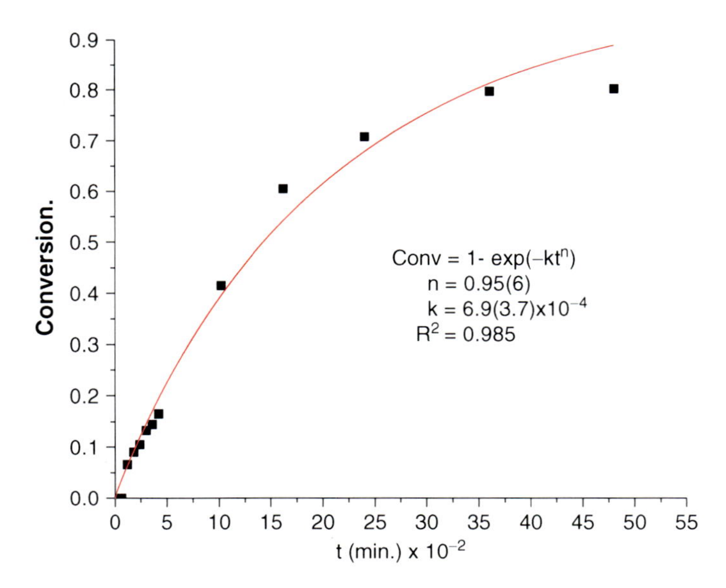

Fig. 8.5 Plot of the fraction of the light-sensitive molecules of **I-a** as a function of the irradiation time at room temperature. The data points (*black squares*) are the refined occupancy factors, and the line is a fit using the equation and the parameter shown. The time scale is based simply on fluence of the Xe-light and are not an absolute property of the sample

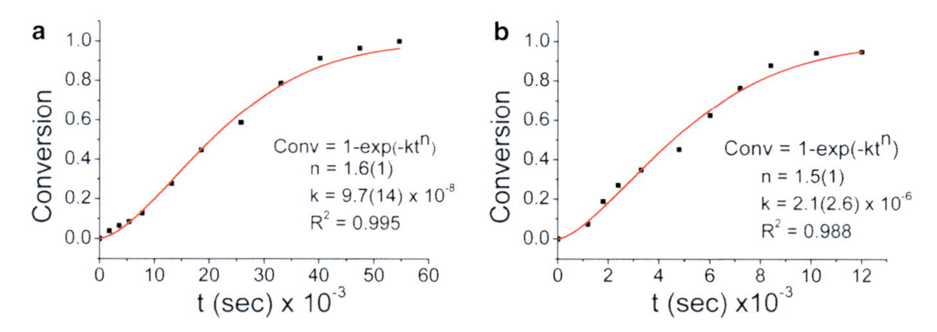

Fig. 8.6 Plot of the fraction of the guest molecule in **I-d** as a function of the irradiation time at 230 (**a**) and 280 K (**b**). The data points (*black squares*) are the refined occupancy factors, and the lines are a fit using the equation and the parameters shown. The time scale is based simply on fluence of the laser and are not an absolute property of the sample

We have found that $n = 0.95(4)$, indicating that the growth mechanism is homogeneous, similar to the kinetics of *o*-methoxy cinnamic acid [27].

This behaviour can be explained by the fact that the pairs of the light-sensitive molecules occupy isolated cavities formed by the host light-stable molecules.

We have carried similar illumination experiment and data collection for **I-d** (Fig. 8.6) at two different temperatures.

We have found that $n = 1.6(1)$ and 1.5(1) at 230 and 280 K, respectively, indicating that the growth mechanism is similar to the two other examples described in the literature [28, 29]. The rate constants at the two temperatures were used for estimating the activation energy for the photodimerization in the solid state to be 32.5 kJ/mol.

8.7 Manipulation of Packing Arrangement [30]

[4 + 4] Photodimerization of isoquinolin-3(2H)-one (isoquinolinone) (1) may yield 12 different isomers depending on the relative geometry of the two monomers prior to the reaction (Scheme 8.5).

Irradiation of the pure solid compound shows that of all possible dimers only the dimer between the two hetero-rings, having an inversion centre, is produced. On the other hand, exposure to UV light of solid molecular compounds composed of isoquinolinone as the guest molecule and different host molecules shows that other isomers of the dimer are produced. The use of different host molecules affects the packing of the molecules in the crystal. As a result, the relative geometries between two monomer molecules are varied enabling photochemical dimerization to yield different isomers. When 1,1,6,6-tetraphenyl-2,4-hexadiyne-1,6-diol (**I**) is used as the host molecule, two polymorphic forms of the molecular compound are obtained. Irradiation of single-crystal of the two polymorphs yields a mixture of isomers of

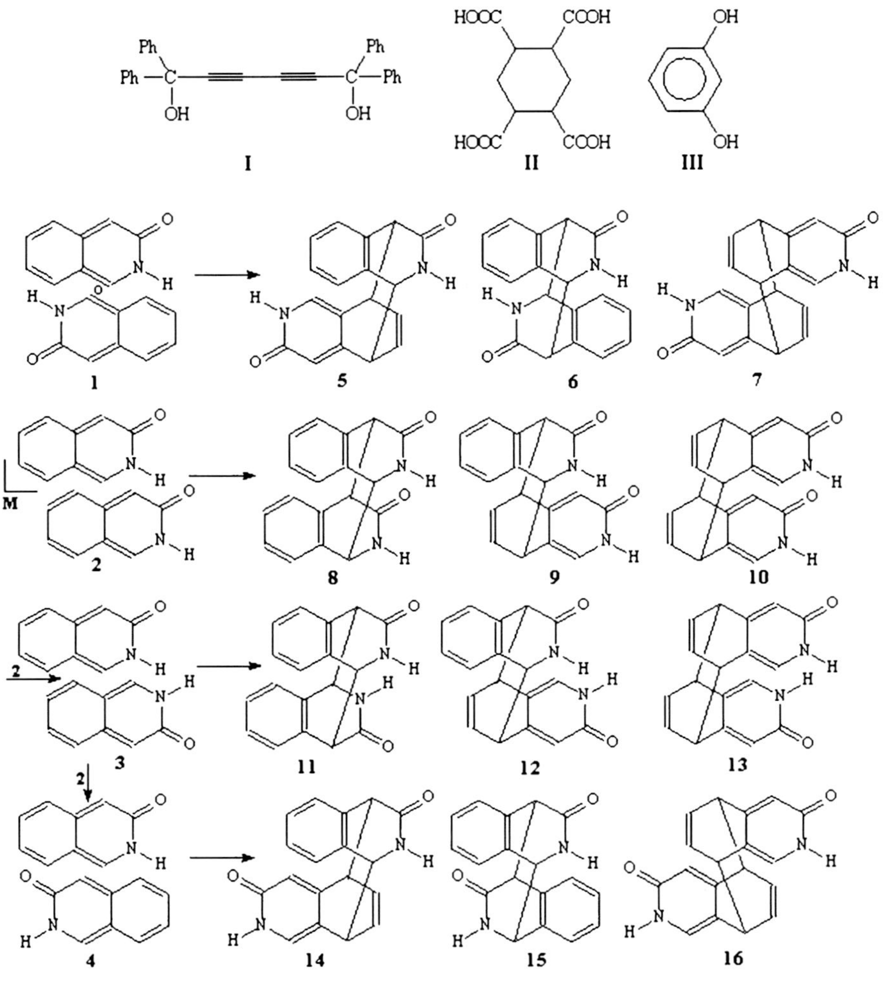

Scheme 8.5 Three different host molecules (**I–III**) used to examine the possibility of controlling the packing of isoquinolinone with 4 different geometric relationships (1–4) and their twelve (5–16) possible photodimers

the dimer. The monoclinic form produced isomer **5**, and the orthorhombic form produced a mixture of isomers **5** and **6**. When the host molecule is cyclohexanol-1,2,4,5-tetracarboxylic acid (**II**) three isomers are expected, and when the host molecule is 1,3-benzenediol (**III**) a single isomer is formed which is identical to that obtained from the neat isoquinolinone (6).

Although the ambitious manipulation was not completed, it demonstrate the principle of a new way of synthesising different pure isomers in the solid-state which overcomes the difficulties of isomers separation when the reaction takes place in solution.

8.8 Summary

It has been demonstrated that the use of co-crystals composed of light-stable hosts and light-sensitive guests is a rewarding system by providing opportunities for structural monitoring single-crystal to single-crystal photoreactions. It was also shown that there might be structural static and dynamic events that we might miss when systems showing single-crystal to single-crystal transformations are not being used.

References

1. Enkelmann V, Wegner G, Novak K, Vagener KB (1993) J Am Chem Soc 115:10390–10391
2. Novak K, Enkelmann V, Wegner G, Wagener KB (1993) Angew Chem Int Ed Engl 32:1614
3. Amirsakis DG, Elizarov AM, Garcia-Garibay MA, Glink PT, Stoddart JF, White AJ, William DJ (2003) Angew Chem Int Ed 42:1126–1132
4. Toda F, Bishop B (2004) Separation and reactions in organic supramolecular chemistry: perspectives in supramolecular chemistry, vol 8. Wiley, New York; Kaftory M (2011) In: Ramamurthy V, Yoshihisa Inoue (eds) Supramolecular photochemistry: controlling photochemical processes. Wiley, Hoboken, pp 229–266
5. Ananchenko GS, Udachin KA, Ripmeester JA, Perrier T, Coleman AW (2006) Chem Eur J 12:2441–2447
6. Halder G, Kepert C (2006) Aust J Chem 59:597–604
7. Zouev I, Lavy T, Kaftory M (2006) Eur J Org Chem 4164–4169
8. Coppens P, Zheng S-L, Gembicky M, Messerschmidt M, Dominiak PM (2006) CrystEngCommun 8:735–741
9. Cohen MD (1975) Angew Chem Int Ed Engl 14:386
10. Weiss RG, Ramamurthy V, Hammond G (1993) Acc Chem Res 26:530
11. Kaftory M, Tanaka K, Toda F (1985) J Org Chem 50:2154–2158
12. Kaftory M, Yagi M, Tanaka K, Toda F (1988) J Org Chem 53:4391–4393
13. Tanaka K, Mizutani H, Miyahara I, Hirotsu K, Toda F (1999) CrystEngCommun 3:8
14. Tanaka K, Toda F, Mochizuki E, Yasui N, Kai Y, Miyahara I, Hirotsu K (1999) Angew Chem Int Ed 38:3523
15. Ohba S, Hosomi H, Tanaka K, Miyamoto H, Toda F (2000) Bull Chem Soc Jpn 73:2075–2085
16. Tanaka K, Toda F (2002) In: Toda F (ed) Organic solid-state reactions. Kluwer Academic, Dordrecht, pp 109–158
17. Kaftory M, Shteiman V, Lavy T, Scheffer JR, Yang J, Enkelmann V (2005) Eur J Org Chem 2005:847–853
18. Lavy T, Sheynin Y, Sparkes HA, Howard JAK, Kaftory M (2008) CrystEngCommun 10:734–739
19. Lavy T, Kaftory M (2007) CrystEngCommun 9:123–127
20. Sreevidya TV, Cao D-K, Lavy T, Botoshansky M, Kaftory M (2013) Cryst Growth Des 13:936–941
21. Schmidt GMJ (1971) Pure Appl Chem 27:647–678
22. Cao D-K, Sreevidya TV, Botoshansky M, Golden G, Benedict JB, Kaftory M (2010) J Phys Chem A 114:7377–7381
23. Avrami M (1939) J Chem Phys 7:1103–1112
24. Avrami M (1940) J Chem Phys 8:212–224
25. Avrami M (1941) J Chem Phys 9:177–184

26. Christian JW (2002) The theory of transformations in metals and alloys, Part I, vol 1. Elsevier Science, Oxford
27. Fonseca I, Hayes SE, Bertmer M (2009) Phys Chem Chem Phys 11:10211–10218
28. Bertmer M, Nieuwendaal RC, Barnes AB, Hayes SEJ (2006) Phys Chem B 110:6270–6273
29. Benedict J, Coppens PJ (2009) Phys Chem A 113:3116–3120
30. Cao D-C, Sreevidya TV, Botoshansky M, Golden G, Benedict JB, Kaftory M (2011) CrystEngCommun 13:3181–3188

Chapter 9
Concepts of Structural Dynamics Investigations Chemical Research

Simone Techert

The following notes will mostly cover the mathematical and analytical content covered during the course and are divided into: Introduction, theoretical concepts of structural dynamics, impact on chemical research and concepts in chemical physics. In the literature chapter the experimental concepts of structural dynamics as well as additional scientific examples in the field of time-resolved synchrotron and free electron laser research are listed.

9.1 Introduction

In the past decade, sustained progress has been made in the field of time-resolved X-ray diffraction and photo-crystallography. Laser systems have been developed rapidly, and the combination of pulsed laser sources with pulsed X-ray sources, particularly by using synchrotron X-ray radiation (3rd generation synchrotrons) and X-rays generated by plasma sources, has made the application of pump-probe schemes routine. Free Electron Laser science is on the way towards that direction. Concerning the investigations of solid state structural dynamics, the techniques applied are time-resolved single crystal crystallography, powder diffraction and diffuse scattering. In the liquid phase and in solution, time-resolved wide angle and time-resolved small angle x-ray scattering have been developed.

S. Techert (✉)
FS-Structural Dynamics of (Bio)Chemical Systems,
Deutsches Elektronensynchrotron DESY, Hamburg, Germany

Structural Dynamics of (Bio)Chemical Systems, Institute for X-ray Physics,
Georg August University, Göttingen, Germany

Structural Dynamics of Biochemical Systems, Max Planck Institute
for Biophysical Chemistry, Göttingen, Germany
e-mail: simone.techert@desy.de

J.A.K. Howard et al. (eds.), *The Future of Dynamic Structural Science*,
NATO Science for Peace and Security Series A: Chemistry and Biology,
DOI 10.1007/978-94-017-8550-1_9,

In a kinetic and reaction mechanistic context, time-resolved diffraction techniques addresses some of the following questions: (i) the refinement of structural changes during the time course of a reaction (filming the "molecular movie"), (ii) possible relations between transient structural changes and the intermediates found by other time-resolved and dynamical techniques, as optical spectroscopy or NMR ("dynamical structure function relations"), and (iii) structural changes in states of matter far from equilibrium ("non-equilibrium structural dynamics").

In the following, we will give an overview about the different kinetic descriptions underlying the analysis of time-resolved X-ray diffraction experiments. We intent of applying them for rationalizing structural dynamics schemes starting from small organic molecular systems up to structural dynamics schemes of complex protein assemblies. Finally, a diagram called "periodic table of structural dynamics" will be derived which is puzzled together from about 100 reactions studied during the last one and half decade in the workgroup.

The notes collect the different kinetic descriptions which are applied in the lecture and tutorial.

9.2 Kinetic Descriptions in Structural Dynamics

Time dependences for first, second and third order reaction laws. For a general reaction

$$\alpha A + \beta B + \gamma C \rightarrow \text{products}, \tag{9.1}$$

where α, β and γ are the stoichiometric factors and A, B *and* C are the concentrations of the reactants, the time law is given by

$$-\frac{1}{\alpha}\frac{dA}{dt} = k_n A^p B^q C^r. \tag{9.2a}$$

p, q and r are the order of the reaction. Analogous, the time laws for B and C are defined as

$$-\frac{1}{\beta}\frac{dB}{dt} = k_n A^p B^q C^r \quad \text{and} \quad -\frac{1}{\gamma}\frac{dC}{dt} = k_n A^p B^q C^r. \tag{9.2b}$$

The total order of the reaction n is $n = p + q + r$.

For a reaction of first, second and third order, the time dependences are well known and described as

$$\text{first order}: \qquad k\,t_1 = \ln\frac{b}{b-x}, \tag{9.3}$$

$$\text{second order}: \qquad k\,t_2 = \frac{1}{b-x} - \frac{1}{b}, \tag{9.4}$$

$$\text{third order}: \qquad k\,t_3 = \frac{1}{2(b-x)^2} - \frac{1}{2b^2}, \tag{9.5}$$

where b and x are determined by the concentrations of the reactants and products (see the following description).

As a start, we assume that the changes in the Bragg diffraction pattern of a time-resolved experiment are only due to the concentration change of the compounds in a reaction mixture. For a unimolecular reaction of type

$$D \rightarrow E \tag{9.6}$$

the Bragg diffraction peaks will change in time because of the decreasing concentration in B and the increasing concentration in C. For this reaction, the intensities of a Bragg diffraction pattern are

$$I(\Theta) = \delta[D] + \varepsilon[E] + \alpha. \tag{9.7}$$

At the beginning of the reaction, $[D] = b$ and $[E] = 0$ so that the intensity at $t = 0$ is defined as

$$I(\Theta, t = 0) = \delta[D] + \alpha. \tag{9.8}$$

At the end of the reaction, it holds $[D] = 0$ and $[E] = b$ and

$$I(\Theta, t = \infty) = \varepsilon[E] + \alpha. \tag{9.9}$$

For the time point t, $[D] = \delta(b - x)$ and $[E] = x$ so that the intensity is

$$I(\Theta, t) = \delta(b - x) + \varepsilon x + \alpha \tag{9.10}$$

For the n-th Bragg diffraction peak, normalization leads to the correlation function

$$C_n(t) = \frac{I(\Theta, t) - I(\Theta, t = \infty)}{I(\Theta, t = 0) - I(\Theta, t = \infty)}. \tag{9.11}$$

Since $I(\Theta, t) - I(\Theta, t = \infty) = b - x$ and $I(\Theta, t = 0) - I(\Theta, t = \infty) = b$ one can derive the following time laws for the reaction of first, second and third order from Eqs. 9.3, 9.4, 9.5, and 9.6 by substitution of $C_n(t)$:

$$\text{first order}: k t_1 = \ln \frac{b}{b - x} = \ln \left(\frac{1}{C_n(t)} \right), \tag{9.12}$$

$$\text{second order}: k t_2 = \frac{1}{b - x} - \frac{1}{b} = \frac{1}{b} \left[\frac{1}{C_n(t)} - 1 \right], \tag{9.13}$$

$$\text{third order}: k t_3 = \frac{1}{2(b - x)^2} - \frac{1}{2b^2} = \frac{1}{2b^2} \left[\frac{1}{C_n(t)} - 1 \right]. \tag{9.14}$$

9.2.1 Time-Dependent Correlation Functions of the Bulk

In the previous paragraph, a general description of the time law of first, second and third order as a function of a correlation function has been derived. In this paragraph, we will concentrate on a description of the correlation function for the case that the intensity of a Bragg reflection is modulated by the change in composition of the reaction mixture during a proceeding reaction (i.e., their occupancies) and by a change in the structure factor caused by geometrical changes. Time-dependent changes in isotropic or anisotropic temperature factors or the Debye Waller factor are – to a first assumption and for clarity of the description – neglected. Furthermore, the description is approached within the framework of kinematic theory.

For the stationary case, the intensity of a single Bragg reflection is defined as

$$I(\Theta, t=\infty) = \sum_{hkl} p_{hkl}(2\Theta - 2\Theta_{hkl}) I_{hkl} \quad \text{and} \quad I_{hkl} = E_{hkl} A_{hkl} M_{hkl} L_{hkl} |F_{hkl}|^2, \tag{9.15a}$$

where $p_{hkl}(2\Theta - 2\Theta_{hkl})$ = profile function of the peak at position hkl, E_{hkl} = extinction correction, A_{hkl} = absorption correction, M_{hkl} = multiplicity of Bragg diffraction peak and L_{hkl} = Lorentz and polarization correction.

In the case of powder diffraction, the description is extended by the correction of the preferred orientation P_{hkl} to

$$I(\Theta, t=\infty) = \sum_{hkl} p_{hkl}(2\Theta - 2\Theta_{hkl}) I_{hkl} \quad \text{and} \quad I_{hkl} = E_{hkl} A_{hkl} P_{hkl} M_{hkl} L_{hkl} |F_{hkl}|^2, \tag{9.15b}$$

and the peak profile function $p_{hkl}(2\Theta - 2\Theta_{hkl})$ is developed as Voigt I and Voigt II profiles.

Let us now define an "apparatus constant" which does not change upon the chemical reaction proceeds. For single Bragg reflections, the constant is defined as

$$K_{hkl} = \sum_{hkl} p_{hkl}(2\Theta - 2\Theta_{hkl}) E_{hkl} A_{hkl} M_{hkl} L_{hkl}. \tag{9.15c}$$

For single powder diffraction peaks, the constant is defined as

$$K_{hkl} = \sum_{hkl} p_{hkl}(2\Theta - 2\Theta_{hkl}) E_{hkl} A_{hkl} P_{hkl} M_{hkl} L_{hkl}. \tag{9.15d}$$

Thus, it follows that $I(\Theta, t=\infty) = \sum_{hkl} K_{hkl} |F_{hkl}|^2$, generalized for single crystal crystallography Eq. 9.15c or powder diffraction Eq. 9.15d.

The correlation function can then be re-written as:

$$c_n(t) = \left[\left(\sum_{hkl} K^t_{hkl} N^t_{hkl} |F_{hkl}(t)|^2\right) - \left(\sum_{hkl} K^{t=\infty}_{hkl} N^{t=\infty}_{hkl} |F_{hkl}(t=\infty)|^2\right)\right] \cdot \left[\left(\sum_{hkl} K^{t=0}_{hkl} N^{t=0}_{hkl} |F_{hkl}(t=0)|^2\right) - \left(\sum_{hkl} K^{t=\infty}_{hkl} N^{t=\infty}_{hkl} |F_{hkl}(t=\infty)|^2\right)\right]^{-1}, \quad (9.16)$$

where N^t_{hkl} is the number of photo-excited molecules in a lattice at the time point t. Furthermore, the occupancy of the excited state structures are normalized as

$$N^t{}_{hkl} = {}^1N_{photo}(t), N^{t=0}_{hkl} = {}^1N_{photo}(t=0) \text{ and } N^{t=\infty}_{hkl} = {}^1N_{photo}(t=\infty). \quad (9.17a)$$

${}^1N_{photo}(t)$ is the occupancy of the excited state molecules at a time point t normalized against a unity volume, ${}^1N_{photo}(t = 0)$ is the occupancy of the excited state molecules for the start of the reaction $t = 0$ and ${}^1N_{photo}(t = \infty)$ is the excited state molecules occupancy for $t = \infty$. In the case of a time-independent instrumental function, the product of the single correction terms equals $K^t_{hkl} = K^{t=0}_{hkl} = K^{t=\infty}_{hkl} = K_{hkl}$.

The structure factor $F_{hkl}(t)$ is defined as $F_{hkl}(t) = \sum_j f_j \exp\left[2\pi i\left(hx_j + ky_j + lz_j\right)\right]$ where f_j is the atomic form factor of the atom j.

Putting K_{hkl} outside the brackets and canceling it down, Eq. 9.16 is simplified to

$$C_n(t) = \left[\left(\sum_{hkl} {}^1N_{photo}(t) |F_{hkl}(t)|^2\right) - \left(\sum_{hkl} {}^1N_{photo}(t=\infty) |F_{hkl}(t=\infty)|^2\right)\right] \cdot \left[\left(\sum_{hkl} {}^1N_{photo}(t=0) |F_{hkl}(t=0)|^2\right) - \left(\sum_{hkl} {}^1N_{photo}(t=\infty) |F_{hkl}(t=\infty)|^2\right)\right]^{-1}. \quad (9.18)$$

If a reaction is of first order, and if it is monitored by time-resolved x-ray diffraction the time dependence of Eq. 9.12 is given by

$$k t_1 = \ln\left(\frac{\left(\sum_{hkl} {}^1N_{photo}(t=0) |F_{hkl}(t=0)|^2\right) - \left(\sum_{hkl} {}^1N_{photo}(t=\infty) |F_{hkl}(t=\infty)|^2\right)}{\left(\sum_{hkl} {}^1N_{photo}(t) |F_{hkl}(t)|^2\right) - \left(\sum_{hkl} {}^1N_{photo}(t=\infty) |F_{hkl}(t=\infty)|^2\right)}\right). \quad (9.19)$$

Similarly, a second order time dependence is defined as

$$A k t_2 = \frac{\left(\sum_{hkl} {}^1N_{photo}(t=0)|F_{hkl}(t=0)|^2\right) - \left(\sum_{hkl} {}^1N_{photo}(t=\infty)|F_{hkl}(t=\infty)|^2\right)}{\left(\sum_{hkl} {}^1N_{photo}(t)|F_{hkl}(t)|^2\right) - \left(\sum_{hkl} {}^1N_{photo}(t=\infty)|F_{hkl}(t=\infty)|^2\right)} - 1, \tag{9.20}$$

and the third order time dependence

$$2A^2 k t_3 = \frac{\left(\sum_{hkl} {}^1N_{photo}(t=0)|F_{hkl}(t=0)|^2\right) - \left(\sum_{hkl} {}^1N_{photo}(t=\infty)|F_{hkl}(t=\infty)|^2\right)}{\left(\sum_{hkl} {}^1N_{photo}(t)|F_{hkl}(t)|^2\right) - \left(\sum_{hkl} {}^1N_{photo}(t=\infty)|F_{hkl}(t=\infty)|^2\right)} - 1 \tag{9.21}$$

with $A = \left(\sum_{hkl} {}^1N_{photo}(t=0)|F_{hkl}(t=0)|^2\right) - \left(\sum_{hkl} {}^1N_{photo}(t=\infty)|F_{hkl}(t=\infty)|^2\right)$.

As in common kinetic experiments, the order of reaction can now be determined by plotting the correlation function $C_n(t)$ (Eq. 9.18) against t. The order of reaction can be distinguished by a logarithmic, linear ($A\,k$) or quadratic dependence (A^2k) of the Bragg intensity correlation function from t.

The current description assumes that within the pump/probe experiment the photo-excitation and photo-transformations schemes are so efficient that the crystalline bulk which is transformed forms entirely new domains within the crystal or crystallites of investigation. Then the ground state structures and excited state structures can be treated as two separated phases.

9.2.2 *Reversible Bulk Reactions*

For a complete back-relaxing system, where the geometry at the beginning of the reaction is equal to the geometry at the end of the reaction, $|F_{hkl}(t=0)|^2 = |F_{hkl}(t=\infty)|^2$. For the population, ${}^1N_{photo}(t=\infty) = 0$, since the product state retransforms to the reactant state, simplifying the correlation function Eq. (9.18) to

$$C_n(t) = \left[\sum_{hkl} {}^1N_{photo}(t)|F_{hkl}(t)|^2\right] \cdot \left[\sum_{hkl} {}^1N_{photo}(t=0)|F_{hkl}(t=0)|^2\right]^{-1} \tag{9.22}$$

and $A = \left(\sum_{hkl} {}^1N_{photo}(t=0)|F_{hkl}(t=0)|^2\right)$ in Eqs. 9.19, 9.20, and 9.21.

When investigating ultrafast processes with an apparatus of low time resolution, for ${}^1N_{photo}(t = 0)$ a maximum value is reached since the early rising time of the defect occupancy from zero to maximum population cannot be resolved. For normalised units it means ${}^1N_{photo}(t = 0) = 1$.

9.2.3 Monoexponential Time Law of a Reversible Reaction

Taking the correlation function Eq. 9.12 and Eq. 9.22 into account, the monoexponential time law for a first order reaction is

$$\left[\sum_{hkl} {}^1N_{photo}(t)|F_{hkl}(t)|^2\right] = \left[\sum_{hkl} {}^1N_{photo}(t=0)|F_{hkl}(t=0)|^2\right] \cdot \exp(-kt_1). \tag{9.23}$$

The time-dependence of the population of the defect structure and the structure factor itself are defined as

$${}^1N_{photo}(t) = {}^1N_{photo}(t=0) \cdot \exp\left(-k_{N_{photo}} t\right). \tag{9.24}$$

and

$$|F_{hkl}(t)|^2 = |F_{hkl}(t=0)|^2 \cdot \exp(-k_{F_{hkl}} t). \tag{9.25}$$

Substitution in Eq. 9.23 and reforming yields in

$$C_n(t) = \sum_{hkl} \exp\left(-k_{N_{photo}} t\right) \exp(-k_{F_{hkl}} t) = \sum_{hkl} \overline{U}_{N_{photo}} \overline{U}_{F_{hkl}} \tag{9.26}$$

where $\overline{U}_{N_{photo}} = \exp\left(-k_{N_{photo}} t\right)$ and $\overline{U}_{F_{hkl}} = \exp(-k_{F_{hkl}} t)$.

In an ultrafast structural distortion, there is a coincidental occupancy increase of the defect structure to its maximum. This is, for example, the case when after a Franck Condon transition an electronically excited state relaxes to a local minimum configuration, which does not correspond to the electronic ground state configuration but has a lifetime longer than all other geometrical relaxation processes. In this case $|F_{hkl}(t = 0)|^2$ changes with a step function at $t = 0$; $\overline{U}_{F_{hkl}}$ of the time correlation function then acts as a constant offset or as a Heavyside function and $C_n(t)$ is modulated dominantly through $\overline{U}_{N_{photo}}$ characterizing the decay of the defect structure population. This special case of Eq. 9.26 can be found for the description of a local minimum structure of an electronically excited state after all geometrical relaxation processes have taken place.

For structural distortions occurring on the picosecond to milliseconds time-scale, the two time scales have to be considered in turn, i.e. $|F_{hkl}(t)|^2$ changes faster than $N_{photo}(t)$. Consequently, $\overline{U}_{N_{photo}}$ now acts as Heavyside function in $C_n(t)$, i.e. for each formed geometry the population is the same. This case can be found when the correlation function Eq. 9.26 describes geometrical relaxation processes from a non-equilibrium point of the potential energy hypersurface (PES) to a local or global minimum on the PES.

In order to distinguish both these processes – the description of structural changes *during* structural relaxation and *after* structural relaxation via Eq. 9.26 – the time scales on which these processes occur have to be separated, so that $\overline{U}_{N_{photo}} =$ constant for processes involving lifetimes of vibrational states (femto- to picosecond time scale) and $\overline{U}_{F_{hkl}} =$ constant for processes where the redistribution of the energy over the vibrational modes is already over meaning processes involving electronic lifetimes (pico- to millisecond time scale).

9.2.4 Time-Dependent Correlation Functions of Photo-Disorder

In the case of homogeneous but small amounts of photo-transformation the photo-excited molecules do not form entirely crystallite domains or phases anymore. In this case, assuming a statistically random excitation distribution of the chromophore centers in the crystal lattice, each excited chromophore can be described as a defect structure within the periodic lattice. Crystallographically, these defect structures are treated within the framework of an averaged crystallochemical formula. Similar to the description of the bulk transformation, the occupancy of the defect structure is then normalized as in Eq. 9.17a to

$$N^t{}_{hkl} = {}^1N_{defect}(t), N_{hkl}^{t=0} = {}^1N_{defect}(t), N_{hkl}^{t=0} = {}^1N_{defect}(t=0) \text{ and } N_{hkl}^{t=\infty} = {}^1N_{defect}(t=\infty). \tag{9.17b}$$

${}^1N_{defect}(t)$ is the occupancy of the defect molecules at a time point t normalized against a unity volume, ${}^1N_{defect}(t=0)$ is the occupancy of the defect molecules (= excited chromophores) for $t = 0$ and ${}^1N_{defect}(t=\infty)$ is the defect structure occupancy for $t = \infty$. In the case of a time-independent instrumental function, the product of the single correction terms equals $K_{hkl}^t = K_{hkl}^{t=0} = K_{hkl}^{t=\infty} = K_{hkl}$.

The correlation function is then defined as

$$C_n(t) = \begin{array}{c} \left[\left(\sum_{hkl} |F_{hkl}(t)|^2\right) - \left(\sum_{hkl} |F_{hkl}(t=\infty)|^2\right)\right] \\ \cdot\left[\left(\sum_{hkl} |F_{hkl}(t=0)|^2\right) - \left(\sum_{hkl} |F_{hkl}(t=\infty)|^2\right)\right]^{-1} \cdot \end{array} \tag{9.27}$$

with the structure factor defined as $F_{hkl}(t) = \sum_j {}^1N_{defect,j} f_j \exp\left[2\pi i\left(hx_j + ky_j + lz_j\right)\right]$ (the occupancy weights here the structure factor (!)) and the occupancy ${}^1N_{defect}$ of the photo-excited molecules consisting of atoms j. Similarly, their time dependence is an exponential function of the form

$$^{1}N_{defect}(t) = {}^{1}N_{defect}(t=0) \cdot \exp\left(-k_{N_{defect}} t\right). \tag{9.28}$$

Note that the big difference between bulk kinetic description and defect distributed kinetic description is given by the concentration respectively occupancy description of the photo-excited molecules outside and inside the structure factor, leading to a quadratic and linear dependence within the given correlation functions.

So far, the current chapter has mostly described the theoretical concepts of structural dynamics investigations. Without going into detail, experimental examples will be listed in the following paragraph, in "Selected Literature".

Selected Literature – Concepts of Structural Dynamics in Physical Chemistry

Methods – Selected Examples of Synchrotron and Free Electron Laser Research

1. Techert S, Schotte F, Wulff M (2001) Picosecond X-ray diffraction probed transient structural changes in organic solids. Phys Rev Lett 86:2030–2034
2. Busse G, Frederichs B, Petrov NK, Techert S (2004) XRD and spectroscopic studies of thiacyanine dye J-aggregates in thin films: comparison between spectroscopy and wide angle X-ray scattering. Phys Chem Chem Phys 6:3309–3314
3. Davaasambuu J, Durand P, Techert S (2004) Experimental conditions for light-induced reactions in powders investigated by time-resolved X-ray diffraction. J Synchrotron Rad 11:483–489
4. Cavalieri AL, Fritz DM, Lee SH, Bucksbaum PH, Reis DA, Rudati J, Mills DM, Fuoss PH, Stephenson GB, Kao CC, Siddons DP, Lowney DP, MacPhee AG, Weinstein D, Falcone RW, Pahl R, Als-Nielsen J, Blome C, Dsterer S, Ischebeck R, Schlarb H, Schulte-Schrepping H, Tschentscher T, Schneider J, Hignette O, Sette F, Sokolowski-Tinten K, Chapman HN, Lee RW, Hansen TN, Synnergren O, Larsson J, Techert S, Sheppard J, Wark JS, Bergh M, Caleman C, Huldt G, van der Spoel D, Timneanu N, Hajdu J, Akre RA, Bong E, Emma P, Krejcik P, Arthur J, Brennan S, Gaffney KJ, Lindenberg AM, Luening K, Hastings JB (2005) Clocking femtosecond X-rays. Phys Rev Lett 94:114801–114804
5. Lindenberg AM, Gaffney KJ, Arthur J, Brennan S, Luening K, Hastings JB, Larsson J, Synnergren O, Hansen TN, Sokolowski-Tinten K, Blome C, Duesterer S, Ischebeck R, Schlarb H, Schulte-Schrepping H, Tschentscher T, Schneider J, Sheppard J, Wark JS, Caleman C, Bergh M, Huldt G, Van Der Spoel D, Timneanu N, Hajdu J, MacPhee AG, Weinstein D, Lowney DP, Allison TK, Matthews T, Falcone RW, Cavalieri AL, Fritz DM, Lee SH, Bucksbaum PH, Reis DA, Rudati J, Fuoss PH, Kao CC, Siddons DP, Pahl R, Als-Nielsen J, Von Der Linde D, Hignette O, Sette F, Chapman HN, Lee RW, Techert S, Akre RA, Bong E, Krejcik P (2005) Atomic-scale visualization of inertial dynamics. Science 308:392–395
6. Blome C, Tschentscher T, Davaasambuu J, Durand P, Techert S (2005) Femtosecond time-resolved powder diffraction experiments using hard X-ray free-electron lasers. J Synchrotron Radiat 12:812–819

7. Strüder L, Epp S, Rolles D, Hartmann R, Holl P, Lutz G, Soltau H, Eckart R, Reich C, Heinzinger K, Thamm C, Rudenko A, Krasniqi F, Kühnel K-U, Bauer C, Schröter C-D, Moshammer R, Techert S, Miessner D, Porro M, Hälker O, Meidinger N, Kimmel N, Andritschke R, Schopper F, Wei-denspointner G, Ziegler A, Pietschner D, Herrmann S, Pietsch U, Walenta A, Leitenberger W, Bostedt C, Möller T, Rupp D, Adolph M, Graafsma H, Hirsemann H, Gärtner K, Richter R, Foucar L, Shoeman RL, Schlichting I, Ullrich J (2010) Large-format, high-speed, X-ray pnCCDs combined with electron and ion imaging spectrometers in a multipurpose chamber for experiments at the 4th generation light sources. Nucl Instrum Methods A 614:483–496
8. Rajkovic I, Busse G, Hallmann J, More R, Petri M, Quevedo W, Krasniqi F, Rudenko A, Tschentscher T, Stojanovic N, Düsterer S, Treusch R, Tolkiehn M, Techert S (2010) Diffraction properties of periodic lattices under free electron laser radiation. Phys Rev Lett 104:125503–6
9. Rajkovic I, Hallmann J, Grübel S, More R, Quevedo W, Petri M, Techert S (2010) Development of a multipurpose vacuum chamber for serial optical and diffraction experiments with free electron laser radiation. Rev Sci Instrum 81:045105-1-6
10. Grossmann P, Rajkovic I, Moré R, Norpoth J, Techert S, Jooss C, Mann K (2012) Time-resolved NEXAFS-spectroscopy on photoinduced phase changes using a table-top XUV spectrometer. Rev Sci Instrum 83:053110-1–053110-4
11. Jain R, Burg T, Petri M, Kirschbaum S, Feindt H, Steltenkamp S, Sonnenkalb S, Becker S, Griesinger C, Menzel A, Techert S (2013) Efficient microchannel device for scattering experiments with high flux X-ray sources and their perspective towards application in neutron scattering science. Eur. J. Phys. E 36:109–118

Ab initio Descriptions of Structural Dynamics – Diffraction

12. Debnarova A, Techert S, Schmatz S (2006) Ab initio treatment of time-resolved X-ray scattering: application to the photoisomerization of stilbene. J Chem Phys 125:224101-1–224101-9
13. Debnarova A, Techert S, Schmatz S (2010) Ab-initio studies of ultrafast X-ray scattering of the photodissociation of iodine. J Chem Phys 133(12):124309-1–124309-8
14. Debnarova A, Techert S, Schmatz S (2011) Computational studies of the X-ray scattering properties of laser aligned stilbene. J Chem Phys 134:054302-1–054302-7
15. Debnarova A, Techert S, Schmatz S (2012) Limitations of high-intensity soft X-ray laser fields for the characterisation of water chemistry: coulomb explosion of the octamer water cluster. Phys Chem Chem Phys 14(27):9606–9614

Statistical Descriptions of Structural Dynamics – Kinetics – Examples of Synchrotron and Free Electron Laser Research

16. Techert S (2004) First, second and third order correlation function in time-resolved X-ray diffraction experiments. J Appl Cryst 37:445–450
17. Busse G, Tschentscher T, Plech A, Wulff M, Frederichs B, Techert S (2003) First investigations on the kinetics of the topochemical reaction of p-formyl-trans-cinnamic acid by time-resolved X-ray diffraction. Faraday Discuss 122:105–117

18. Techert S, Zachariasse KA (2004) Structure determination of the intramolecular charge transfer state in crystalline 4-(diisopropylamino)benzonitrile from picosecond X-ray diffraction. J Am Chem Soc 126:5593–5600
19. Ramos A, Techert S (2005) Influence of the water structure on the acetylcholinesterase efficiency. Biophys J 89:1990–2003
20. Quevedo W, Petri M, Busse G, Techert S (2008) On the mechanism of photo-induced phase transitions in ternary liquid crystal systems near thermal equilibrium. J Chem Phys 129:024502-1–024502-10
21. Hallmann J, Morgenroth W, Paulmann C, Davaasambuu J, Kong Q-Y, Wulff M, Techert S (2009) Time-resolved X-ray diffraction of the photochromic a-styrylpyrylium (TFMS) crystal films reveals ultrafast structural switching. J Am Chem Soc 131:15018–15025
22. Moré R, Busse G, Hallmann J, Paulmann C, Scholz M, Techert S (2010) The photodimerisation of crystalline 9-anthracenecarboxylic acid – the case of a Non-topotactic autocatalytic transformation. J Phys Chem C 114:4142–4148
23. Quevedo W, Peth C, Busse G, Mann K, Techert S (2010) Nanosecond dynamics of photo-excited lyotropic liquid crystal structures. J Phys Chem B 114:8593–8599
24. Hallmann J, Grübel S, Rajkovic I, Quevedo W, Busse G, Moré R, Petri M, Techert S (2010) First steps towards probing chemical systems and dynamics with free electron laser radiation - case studies at the FLASH facility (special issue five years of FLASH). J Phys B At Mol Opt Phys 43:194009–194016
25. Petri M, Menzel A, Bunk O, Busse G, Techert S (2011) Concentration effects on the dynamics of liquid crystalline self-assembly: time-resolved X-ray scattering studies. J Phys Chem A 115:2176–2183
26. Sørensen HO, Schmidt S, Wright JP, Vaughan GBM, Techert S, Garman EF, Oddershede J, Davaasambu J, Paithankar KS, Gundlach C, Poulsen HF (2012) Multigrain crystallography. Z Kristallogr 227:63–78
27. Petri M, Frey S, Menzel A, Görlich D, Techert S (2012) Structural characterization of nanoscale meshworks within a nucleoporin FG hydrogel. Biomacromolecules 13(6):1882–1889
28. Hallmann J, More R, Morgenroth W, Paulmann C, Kong Q, Wulff M, Techert S (2012) Evidence for point transformations in photoactive molecular crystals by the photoinduced creation of diffuse diffraction patterns. J Phys Chem B 116(36):10996–11003
29. Quevedo W, Busse G, Hallmann J, Moré R, Petri M, Krasniqi F, Rudenko A, Tschentscher T, Stojanovic N, Düsterer S, Treusch R, Tolkiehn M, Techert S, Rajkovic I (2012) Ultrafast time dynamics studies of periodic lattices under free electron laser radiation. J Appl Phys 112:093519–5

Chapter 10
Structure and Dynamics of Light-Excited States

Eric Collet

The optical control of materials and related physical properties (electronic state, magnetic, optical...) by laser irradiation has gained tremendous interest within the emerging field of photoinduced phase transitions. Light-induced changes of molecular systems involve subtle coupling between the electronic and structural degrees of freedom, which are essential to stabilize the photo-excited state, different in nature from the stable state. Therefore the new experimental field of photocrystallography plays a key role. Its outreach goes far beyond simple structural analysis under laser excitation. By playing on different physical parameters and developing the techniques and analysis, one can investigate new out of equilibrium physics through light-driven cooperative dynamics and transformations in materials, or follow a chemical reaction in real time. These notes review different aspects of the use of photo-crystallography to investigate the nature, the mechanisms and the dynamics of photoinduced phase transitions.

With the advent of control science the current challenge is not only to observe matter on ever smaller scale but also to direct its functionality at the relevant length, time and energy scales. Control with light is one of these fascinating perspectives. This emerging field deals with transformation of molecules or materials upon absorption of light. Thus, the aim of these driven transformations, either under continuous light flux or a brief light pulse is to force the matter towards a new state far from thermal equilibrium. Such manipulation may trigger spectacular switching of the macroscopic physical properties, for example in a material from non magnetic to magnetic, from paraelectric to ferroelectric, or from insulator to conductor.

E. Collet (✉)
Institut de Physique de Rennes, UMR 6251 UR1-CNRS, University Rennes 1,
BAT 11A Campus de Beaulieu, 35042 Rennes Cedex, France
e-mail: eric.collet@univ-rennes1.fr

J.A.K. Howard et al. (eds.), *The Future of Dynamic Structural Science*,
NATO Science for Peace and Security Series A: Chemistry and Biology,
DOI 10.1007/978-94-017-8550-1_10,

10.1 Introduction

Significant photo-stationary change of state during chemical or material transformation is observed if the excited state is stabilized over long time, at least at low temperature [1]. This occurs when the intramolecular structural reorganization which follows the electronic photo-excitation gives rise to an energy barrier between the trapped excited molecular state and the ground state. In some situations the molecular processes are independent, but in many others intermolecular interactions lead to cooperative photoinduced phenomena. There are certain materials, such as spin crossover crystals [2], exhibiting light driven bistability under continuous photo-irradiation. The steady flow of energy from photon flux competes with the dissipation in the thermal bath. The cooperative interactions lead to a feedback mechanism and so to nonlinear kinetic law for the thermally activated relaxation process (the medium is no more passive but active). A universal framework relying on the physics of nonlinear dynamics and bifurcations provides a general description for these bistable processes. Let us notice that when the lifetime is very short time-resolved techniques have to be used [3].

The physics of ultrafast transformation induced by an ultra-short femtosecond laser pulse is different. Understanding and tracking how matter works during elementary dynamic processes is faced with several new and challenging basic questions. For instance, ultrafast information processing based on the control of light-driven switching of the physical properties of materials requires the answer to key questions such as how to direct the system through a complex pathway from atomic to material scales and what fundamentally limits the transformation speed. This feat can be accomplished thanks to the increase of sophisticated instrumentation in ultra-fast science. A nice example is that of femtochemistry, a field awarded a Nobel Prize in 1999 [4]. In the photo-excited electronic state the interatomic potentials change and determine the motion of atomic wavepackets prepared by a femtosecond light pulse. This may induce drastic effects such as breaking of bonds and subsequent chemical transformation from transient states.

Recent advances in pulsed electron and X-ray sources make it now possible to probe structural processes at a much finer level than the time average studies would allow before [3, 5]. Thus, as the pulse length can today be as short as 100 femtoseconds (1 fs = 10^{-15} s), an optical pump coupled to X-ray or electron probe experiments allow tracking the atomic or molecular motions on their intrinsic elementary time-scale, *i.e.* the periods of electronic vibrations. Over the recent years we have witnessed stunning achievements in structural dynamics, providing atomic level views of transient phenomena in inorganic, molecular and even biomolecular systems. Those directly contribute to the understanding of material transformation through non-equilibrium, non-linear, cooperative photoinduced processes. These notes provide a review of techniques and scientific cases illustrating these new possibilities.

10.2 Time-Resolved X-Ray Scattering

Tracking structural information in the time domain provides insights for understanding dynamical processes in matter and its transformation [3]. Depending on time resolution of the experiment, different processes can be investigated, spanning from intrinsic atomic or molecular dynamics to macroscopic propagation or diffusion effects. In some cases which do not require fast measurement, conventional diffraction will suffice for tracking slow evolution of the sample. But when the phenomena to investigate are fast, or when the excited state is transient, more sophisticated techniques are required. These are based on pump-probe method, in which a laser flash drives a photo-switching process and a probe pulse (X-ray or electrons) visualize the excited structure at a given delay *dt* after laser excitation. The stable and the excited states are compared by analysing the evolution of the diffraction pattern (Fig. 10.1).

Detailed structural analyses of photo-excited structures were performed with microsecond or 100 ps time resolutions in systems as diverse as molecules in solution [6–11], molecular crystals [12–14] or biological crystals [15–17]. However, because the typical period of atomic motions is on the order of 100 fs, these methods are still unable to resolve structural changes prior to a system settling in a quasi-thermodynamic equilibrium. Femtosecond pump-probe X-ray diffraction makes it possible to study such nonequilibrium structures on time scales approaching the fastest atomic vibrational periods [18–22]. The time resolution relies on the duration of the cross-correlation between the optical pump and probe pulses, as well as on the precision of the synchronization between them. The development of new sources of short-pulse hard X-rays with fine synchronization to a laser source open the way toward femtosecond crystallography. Successful experiments have used the X-rays emitted by Thomson scattering from relativistic electrons [23],

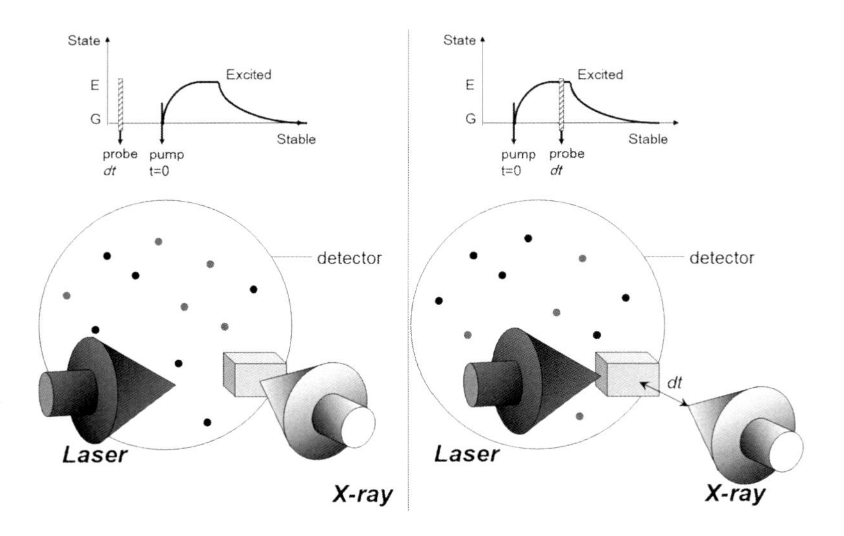

Fig. 10.1 Time-resolved diffraction pump-probe set-up

laser-generated plasmas [18, 19], and synchrotron-based "slicing" of the stored electron beam [22, 24]. Other important issues limiting the so far developed use of femtosecond X-ray pulses are the divergence of the beam, the X-ray flux and the stability. Much more intense pulses have been available at a test facility of a linear accelerator [25], and now hard X-ray beamlines at X-ray free electron lasers offer both much greater per-pulse intensity and shorter pulses [26].

Electron diffraction also represents an interesting alternative. Electrons are strongly scattered by matter and it is now possible to generate electron bunches approaching 100 fs duration. Several scientific cases have by now been successfully studied with this technique [27, 28].

Besides diffraction, other types of structural information can be obtained, for example through X-ray absorption spectroscopy (XAS) [29–31]. Time-resolved extension uses the same pump-probe principle described above for diffraction. Only the nature of the signal measured and its analysis are different. The XAS techniques are of particular interest for probing subtle changes of molecules in solution.

A slicing method capable of producing femtosecond pulses of synchrotron radiation was proposed for the Advanced Light Source [32]. Slicing technique uses the very high electric field achievable by femtosecond laser to modulate the energy of a "slice" (100 fs) of an electron bunch (100 ps) in the storage ring. This slice has a larger energy spread than the rest of electrons in the bunch, thereby its emission deviates when passing through a radiator. This is because the electrons entering the insertion devise with different energies will follow different paths.

Among the variety of schemes for generating femtosecond X-ray pulses, laser-driven plasma sources are well suited for the time scale of small atomic displacements [3, 33, 34]. The irradiation of metal targets by femtosecond laser pulses, with peak intensity higher than 10^{16} W/cm^2, results in generation of a plasma. The acceleration of free electrons into the target by the very high electric field of the pulses and the generation of characteristic radiation and Bremsstrahlung by interactions with target atoms emits radiation in the hard X-ray range. An interesting intrinsic property of this technique is that the time $t = 0$ is very well defined since it is the same optical laser that excites the crystal to be studied and generates the plasma X-rays. However, the X-ray pulse is emitted through a large solid angle and the source is highly divergent. Only a small part of the typically 10^9 emitted X-ray photons per pulse can be collected for keeping the time resolution. A recent review by T. Elsaesser described both the technique and some applications [35]. For a target thickness of the order of 10–20 μm, the characteristic emission has duration of the order of 100 fs.

X-ray Free Electron Lasers (X-FEL) are the next generation X-ray sources. The principle is similar to that of synchrotrons. The major differences are the following: X-FEL's are not electron storage devices and so the electron bunches are used only once; the magnetic structure used to generate X-rays are linear sections hundreds of meters long. Such long interaction volume forces all the electrons in the bunch to interact constructively via the generated X-ray field, which shortens X-ray pulse duration tremendously (100 fs), enhances coherence, flux (10^{12} photons per pulse in 0.1 % bandwidth), and emittance. The time duration of the X-ray pulses can be decreased down to a few fs. The first experiments have just started in the new LCLS machine operational at Stanford [36], and several other machines are in construction in Europe [37] and Japan [38].

10.3 Dynamical Structural Science

10.3.1 Coherent Atomic Motion

A powerful aspect of diffraction techniques is that they are able to track the organization of matter at the atomic scale. In the frame of dynamical structural science, experiments with time resolution comparable to typical atomic dynamics time-scales, it is now possible to track atomic motions and follow their dynamics in real time with 100 fs resolution. The most popular example of such studies is the coherent optical phonons generated in bismuth by light excitation [19, 21, 22, 39]. An ultra-short laser pulse can control the interatomic potential in terms of equilibrium distance between the two Bi atoms in the unit cell, as well as curvature of the potential. This ultrafast modification of the potential by the ultra-short light pulse drives the generation of on optical phonon at the centre of the Brillouin zone. As the structure factor is coherently modified within the different unit cell of the crystal (responding in phase), the structure factor is time dependent. There are two atoms of Bi in the unit cell along the diagonal of the cell. Then, the structure factor reaches:

$$F_{hkl} = 2f_{Bi}cos\left[\pi(h+k+l)x\right],$$

where x is the reduced Bi-Bi distance along the diagonal (x $\approx$ 0.47 at equilibrium). The A_{1g} phonon involved corresponds to a Bi-Bi stretching along the diagonal, which means that x depends on time, and so does the structure factor:

$$F_{hkl}(t) = 2f_{Bi}cos\left[\pi(h+k+l)x(t)\right].$$

As indicated in Fig. 10.2, the variation x, which is just smaller than 0.5, affect in different ways the intensity of the (111) peak, which decreases when x approaches 0.5, and the one of the (222) peak, which increases when x approaches 0.5, since:

$$I_{111}(t) = 4f_{Bi}^2 cos^2[3\pi x(t)]$$

$$and$$

$$I_{222}(t) = 4f_{Bi}^2 cos^2[6\pi x(t)],$$

K. Sokolowski-Titen et al. [19] directly observed from the changes in the Bragg peaks intensity the large-amplitude coherent atomic vibrations, with clear oscillations of the (111) and (222) reflections. The atomic displacements affect in an opposite way the two reflections mentioned above. As x oscillates I(111) is minimum and

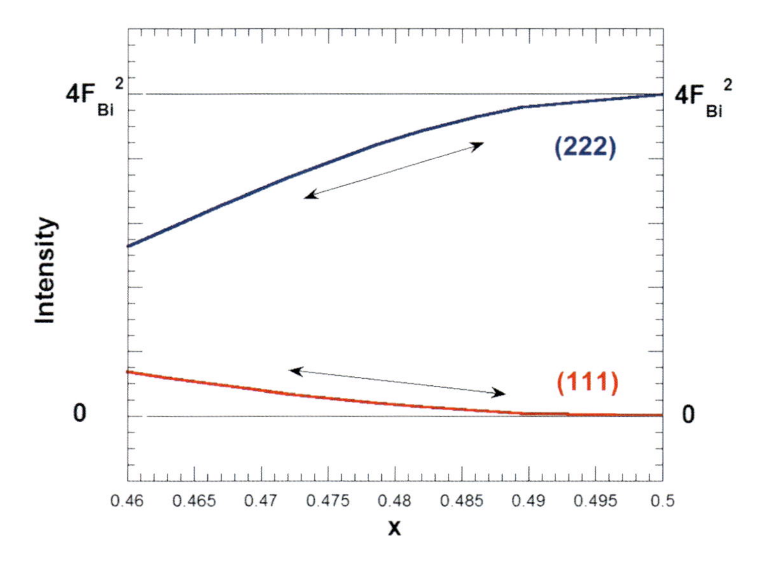

Fig. 10.2 Dependence of the intensities of the (111) and (222) Bragg peaks with the reduced Bi-Bi distance x

I(222) is maximum when x is maximum, and vice versa. From the measurements, it is possible to extract the phonon frequency, the phonon amplitude generated by light excitation (approx. 0.1–0.2 Å) and the variation of the average Bi-Bi distance in the perturbed equilibrium position.

More recent works using slicing technique or high-brightness linear electron accelerator–based X-ray source deepen the study of this mechanism and the appearance of softening as well as squeezed phonon states. Coherent atomic dynamics were also studied in more complex systems such as tellurium [40], superlattice structures [41] or polariton dynamics [42].

10.3.2 Dynamical Processes of Spin-Crossover Materials

Spin-crossover materials can be switched either in solids or in solution by a laser pulse between LS and HS states within 300 fs, as demonstrated by time-resolved optical spectroscopy. In solution, only XAS techniques can provide structural information on the intra-molecular reorganization. In solid, diffraction makes it possible to have detailed view of the structural reorganization in the time domain while different processes unfold at molecular, unit cell, as well as the macroscopic crystal scales.

One of the key questions related to bi-stable systems like spin-crossover materials is the nature and the time-scale of the intrinsic switching dynamics. Recently,

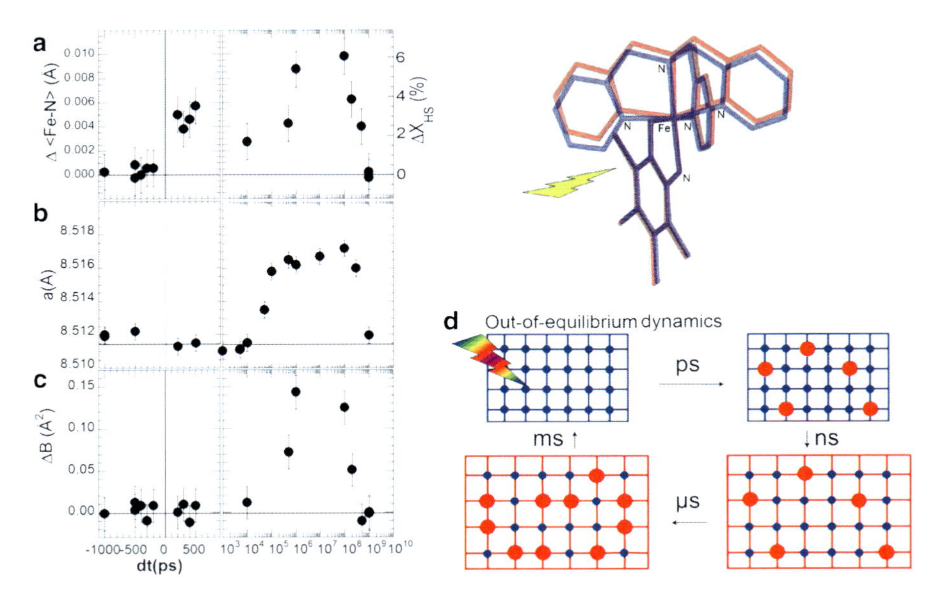

Fig. 10.3 Structural changes after femtosecond laser excitation: (**a**) ΔX_{HS} and Δ <*Fe–N*>, (**b**) lattice parameter *a* and (**c**) isotropic temperature factor variation ΔB. (**d**) Schematic drawing of the dynamics: HS molecules (*red circles*) generated within 1 ps by laser pulse in the cold (*blue*) lattice with mainly LS molecules (*blue circles*), warm lattice (*red*) expansion on ≈50 ns, thermal stabilization of HS population within μs. The HS (*red*) and LS (*blue*) structures are also represented at the top right (Figure from Ref. [13])

Bressler et al. and Chergui [29, 31] used the slicing technique to carry out an optical pump/X-ray probe experiment to study the ultrafast dynamics of $[Fe^{(II)}(bpy)_3]^{2+}$.

Recent reports combining time-resolved optical and X-ray diffraction techniques, demonstrated that the switching of the spin state in a macroscopic crystal constituted of bi-stable molecules involves different mechanisms in time and space. The studies were performed in a Fe(III) solid compound, triggered by a femtosecond laser flash [13, 14]. The ensuing dynamics span from sub-picosecond non-thermal molecular switching to microsecond diffusive heating processes through the lattice. The experiment was performed by using the optical pump/X-ray probe technique developed at the ESRF synchrotron (ID09B beamline).

The existence of different steps, summarized in Fig. 10.3, reflects a sequence of physical processes, hidden in the time domain, leaving different fingerprints for molecular transformation, cell deformation and macroscopic crystal switching:

- Step 1: local LMCT to HS relaxation cascade. It is characterized by a rapid elongation of the <Fe–N> bond length, a well-known fingerprint of increased spin multiplicity from electron transfer to antibonding orbitals.
- Step 2: unit cell expansion observed here through the evolution of the lattice parameter *a* within 100 ns.

- Step 3: thermal switching characterized by an additional increase of the <Fe–N> bond length (μs).
- Step 4: recovery to the thermal equilibrium state, limited by heat exchange with sample environment.

These results shed new light on the complex switching pathway from the molecular to material length and time scales. They pave the way for studying by diffraction techniques the out-of-equilibrium dynamics following laser pulse excitation, what is of fundamental interest in a large variety of materials, since it is important for the design of materials with enhanced functionality.

10.3.3 Photoinduced Ferroelectric Order

In the field of photoinduced phase transition, one of the main goals is to switch the macroscopic physical properties of a material. The example of light-induced ferroelectric order in tetrathiafulvalene- p chloranil (TTF-CA) is a nice example of what can be learned from diffraction regarding intra-and inter-molecular reorganizations. In TTF-CA, the electron donor (D) and electron acceptor (A) molecules alternate along the same stack. There are two electronic states: one is mainly neutral with regular alternation of molecules located on inversion symmetry sites: N state ... $D^\circ A^\circ D^\circ A^\circ D^\circ A^\circ$.... As the charge-transfer takes place, there are two degenerate, polar and dimerized I states, ...$(D^+A^-)(D^+A^-)(D^+A^-)$...or ...$(A^-D^+)(A^-D^+)(A^-D^+)$... In addition to the possibility to induce this ferroelectric transition by changing temperature or pressure, the fascinating option to switch between paraelectric and ferroelectric states by pulsed laser excitation was evidenced in TTF-CA by time-resolved X-ray diffraction [12, 43].

Direct evidence of a laser-induced ferroelectric order [12] and the opposite photo-induced transition [43] was obtained by diffraction. The Neutral-Ionic (N-I) phase transition involves coupled changes of molecular electronic states and structural reorganization. The symmetry-breaking associated with the generation of the ferroelectric phase with dimerized stacks is characterized by the change of space group. The space group of the paraelectric N phase is $P2_1/n$ manifested in the diffraction pattern by two types of systematic absences, $(0k0)$: $k = 2n + 1$ due to the presence of a two-fold axis and $(h0l) : h + l = 2n + 1$ due to the presence of a glide plane. The space group of the ferroelectric I phase is Pn and therefore the symmetry breaking associated with the ferroelectric order is detected by the appearance of $(0k0)$: $k = 2n + 1$ Bragg reflections (space group Pn).

The nature of the photo-induced phase has been determined from complete data collections at −2 and +1 ns, i.e. before and after the laser excitation. Appearance of *(030)* reflection (Fig. 10.4) shows the macroscopic ferroelectric nature of the

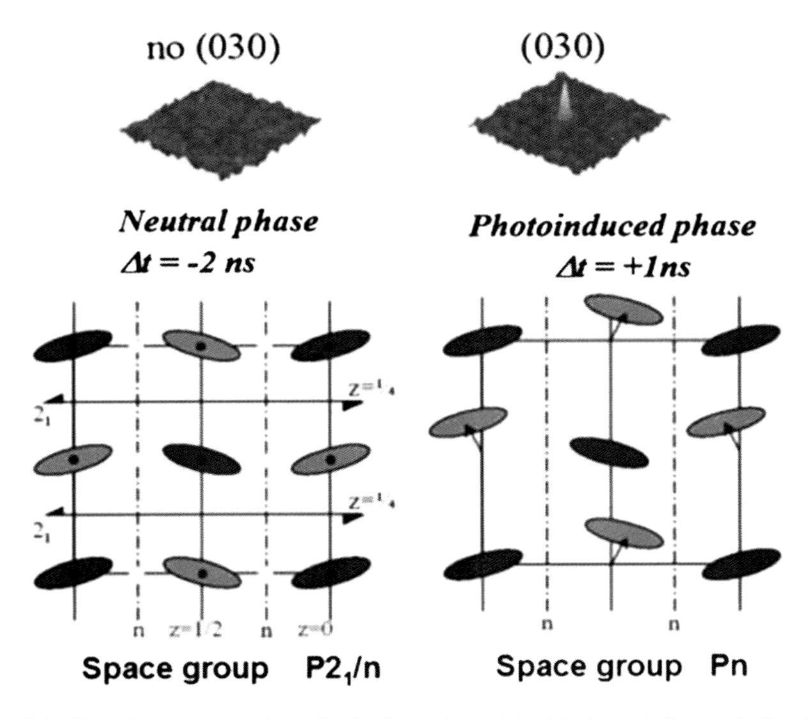

Fig. 10.4 (*Top*) Reconstructed intensity in the reciprocal (a, b) planes before and after laser irradiation (Δt = −2 ns and +1 ns respectively). Appearance of (030) reflection signs the ferroelectric nature of the 3D photo-induced phase. (*Bottom*) Schematic drawing of the space groups in both thermal equilibrium ($P2_1/n$) and photo-induced (*Pn*) states (Figure from Ref. [12])

photo-induced state (space group *Pn*, i.e. the one of the I phase) and the occurrence of a dimerization process.

The structures were refined at −2-ns (Fig. 10.5, left) and +1ns. The structure at −2 ns [12] is in perfect agreement with the neutral structure obtained at thermal equilibrium, both for the intra- and inter-molecular geometries, which means that in the stroboscopic mode the crystal recovers the thermal equilibrium state before a new laser excitation arrives. At +1-ns, because of the symmetry breaking, the structure was refined in the *Pn* space group, assuming a full transformation of the crystal.

The results obtained from the refinement clearly agree with a dimerization process in the photoinduced state at +1-ns mainly along the stacking axis **a** (Fig. 10.5, right), related to the loss of the inversion centre previously located on the molecules. The observed structural changes, both for intra- or inter-molecular geometries, are again close to the ones observed in the low temperature I phase, but the dimerization amplitude is lower (1/3). The reason for this is probably that the conversion is incomplete with neutral and transformed ionic ferroelectric phases coexisting, as explained in Ref. [12].

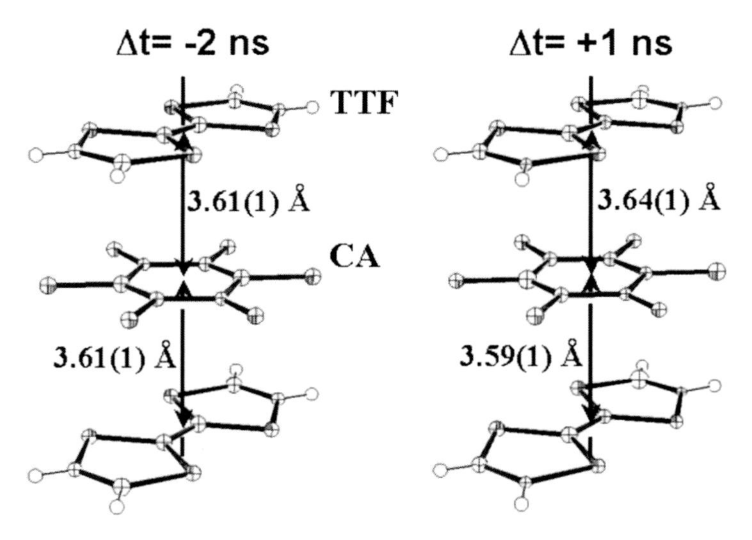

Fig. 10.5 The structure of TTF-chloranil before (*left*) and after (*right*) application of the laser pump pulse inducing electron transfer and pair formation. From Ref. [12]

10.4 Diffuse Scattering Probing Local Nonequilibrium Phenomena

The direct observation of precursor phenomena is an essential key for understanding the mechanism driving the photo-transformation of materials, with challenging issues for both fundamental and applied aspects in materials science. Precursor phenomena for photoinduced phase transition may be described in two limit cases: first a collective mechanism (soft mode) and second the generation of localized excitations (precursor clusters). The first one was investigated by 100-femtosecond X-ray diffraction where a coherent phonon triggers the time-dependence of the Bragg reflections. The next section presents investigations of the second phenomenon.

Recently we used time-resolved X-ray diffuse scattering to capture the second mechanism in the charge-transfer molecular compound TTF-CA. The photo-induced phase transitions undergone by this material are discussed in the literature as resulting from a new class of collective excitations, the so-called 1D lattice-relaxed charge-transfer exciton-strings. These nano-scale objects, represented by $\ldots D^0A^0(D^+A^-)(D^+A^-)(D^+A^-)D^0A^0\ldots$ are made of a train of dimerized I molecules extending along the crystalline stacking axis *a*. The direct experimental evidence of the thermally-induced 1D exciton-strings has been possible for TTF-CA. Time-resolved X-ray diffuse scattering experiments performed by Guérin et al. [44] at the Photon-Factory Advanced Ring synchrotron shows the time dependence of the diffuse plane, increasing just after the laser excitation (Fig. 10.6). This is a direct signature of the photo-generation of the local precursor

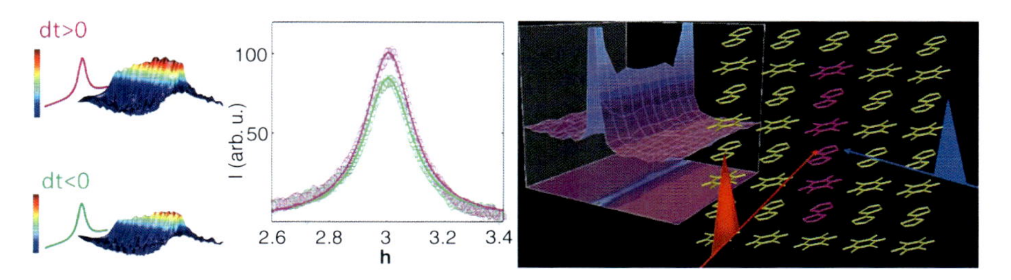

Fig. 10.6 Evolution of the diffuse scattering before (−100 ps) and just after (50 ps) laser excitation. The increase of the diffuse plane is directly related to the photo-excitation of 1D cluster along the stack, as schematically indicated on the *right*

clusters with short-range structural order, appearing in the first steps of the photo-induced transformation. In other words, the laser excitation drives a cooperative 1D transformation along the stacking axis a. The observed rising time [44] of the diffuse scattering between −50 and 50 ps is limited by the convolution of the signal with the 50 ps time resolution used, and so takes place on an ultrafast time-scale.

The present results directly evidence the photo-generation of short-range precursor clusters of the photoinduced phase transition. Regarding the dynamics, two fascinating questions have to be discussed for future investigations. First, what is the size of the photoinduced exciton-string and what is their dynamics of formation? This should strongly depend on the pumping photon energy. Second, on which time scale the interstack ordering between the exciton-strings appears and how does it proceed?

10.5 Symmetry Breaking

X-ray diffraction is a perfect technique for probing ordering phenomena between molecules and change of symmetry. This is especially true when symmetry breaking is associated with the appearance of new Bragg reflections. It is well known from Landau theory of phase transitions, that every density of probability (electronic, atomic position, spin,...) describing the crystalline structure in the two phases can be expressed as

$$\rho = \rho_0 + \eta \Delta\rho$$

where the first term ρ_0 is totally symmetric with respect to the high-symmetry phase and the second term $\Delta\rho$ describes the symmetry lowering arising in the low-symmetry phase. This last term is non-zero only in this low-symmetry phase and is proportional to an order parameter η associated with the amplitude of the symmetry breaking. In general structural phase transitions are associated with a change of space group manifesting symmetry breaking. When symmetry breaking manifests itself by the appearance of new Bragg peaks, their intensities are

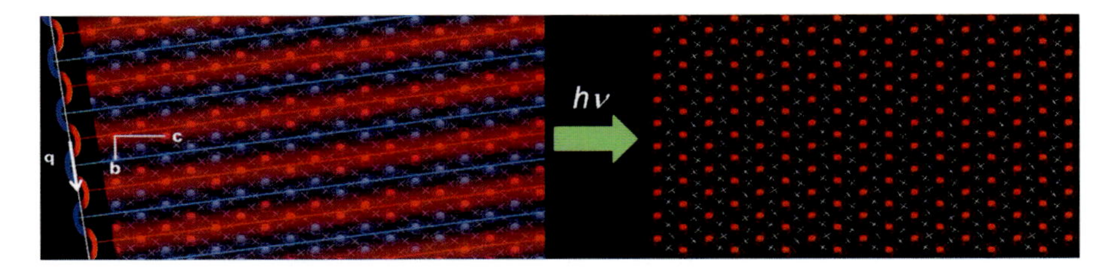

Fig. 10.7 Spatial modulation of the molecular spin state in 3D space: *blue* (LS states) and *red* (HS states) colours schematically reflect the spatial occupation modulation (*left*). Light excitation selectively switching the molecules to the HS state erases the wave

proportional to the square of the related order parameter η. Then its evolution can be easily determined from diffraction measurements.

In SC materials for example, change of magnetic molecular state is often isostructural (without change of symmetry) and the photoinduced HS state at low temperature is often similar to the HS state stable at high temperature. But some materials show symmetry breaking of structural nature. We demonstrated in the SC materials $[Fe^{II}H_2L^{2\text{-}Me}][PF_6]_2$ and $[Fe^{II}H_2L^{2\text{-}Me}][AsF_6]_2$ that light can drive symmetry breaking and that the photoinduced HS state generated at low temperature is different from the HS state existing at room temperature [45, 46]. The photocrystallography experiments performed at 15 K, after *cw* photo-excitation at 532 nm and generating the metastable photo-induced HS state, revealed different translation symmetry of the photoinduced HS state (*a*,*b*,3*c*) compared to the HS (*a*, *b*,*c*) state at room temperature. This is associated with an ordering of distorted HS molecules losing 2 fold axis corresponding to $\Delta\rho$. This symmetry breaking which occurs upon generating the photo-induced HS phase (PIHS) demonstrate that different competing false ground states exist and that some of them can only be reached under non equilibrium condition by light irradiation.

The relationship between molecular bistability in the solid state and aperiodicity has not been considered so far. A novel type of ordering phenomenon associated with the appearance of an incommensurate molecular spin state modulation of a bi-stable spin-crossover compound was recently evidenced [47]. It results from a concerted interplay between the symmetry breaking associated with the aperiodic ordering and the thermal balance between the two functional states. Our detailed study of the $[Fe^{II}H_2L^{2\text{-}Me}][SbF_6]_2$ material shows that the incomplete conversion of bi-stable molecules is associated with an incommensurate spatial modulation of the molecular spin-state and a description of the structure is obtained in (3 + 1)-dimensional superspace (Fig. 10.7). The aperiodic structure can be interpreted as resulting from a spatial modulation of the HS fraction, i.e. of the spin-state of the molecules, simply written as:

$$\gamma_{HS}(\mathbf{r}) = \gamma_{HS} + \eta \times \cos(\mathbf{q}.\mathbf{r}),$$

where γ_{HS} is the average HS fraction, $\mathbf{q}$ is the wave vector and η is the amplitude of the wave. In several SC systems, superstructure ordering of γ_{HS} over crystalline sites were reported [45] and the spin-state ordering occurs around an average value of $\gamma_{HS} = 0.5$ corresponding to a cell doubling along **c** axis, since $\mathbf{q} = \mathbf{c}^*/2$. This description can be extended to the incommensurate order, for which the wavelength of the modulation of $\gamma_{HS}(\mathbf{r})$ around $\gamma_{HS} = 0.5$ is incommensurate with the average 3D lattice (Fig. 10.7). The sequence of HS and LS sites therefore never repeats periodically in the 3D space. The particularity of the present material is that, unlike conventional modulated crystals with occupational, positional or conformational modulation, the primary modulated parameter is the spin-state of bi-stable molecules. As such, this aperiodic structure is demonstrated to be erased by selective laser population of a single molecular state (HS). By crossing the borders between aperiodicity, multi-stability and photoinduced phenomena we have shown that new perspectives appear for switching between different regimes of periodicity and for exploring new phases of matter related to function, such as photochromism and photomagnetism. Laser excitation, selectively populating a single molecular state, can erase the incommensurate phase and switch the crystal structure from 4D to 3D periodic.

10.6 Conclusion

In a recent past, tremendous advances for providing structural information of excited states were performed. The reviewed results presented here give an illustration of what can be done and open exciting perspectives, especially for studies in the time domain as more and more powerful X-ray or electron sources are developed. Among the large contributions that this new advanced method may provide, an important goal is to understand and control phase transitions driven by light. A deep understanding of such phenomena at different scales requires the combined use of X-ray or electron diffraction over different time scales (fast and ultra-fast), but also of temporal X-ray absorption and optical spectroscopies. This is the key for elucidating this new kind of coherent manipulation of matter by light and for controlling ultra-fast macroscopic switching of materials. This broad spectrum of recent advances in this field set the stage for imminent breakthroughs in the understanding of our dynamical world.

References

1. Woike T, Schaniel D (2008) Photocrystallography. Z Krsyt 223(4–5):235–369
2. Gütlich P, Goodwin HA (eds) (2004) Topics in current chemistry, spin-crossover in transition metal compounds I, II, III, vols 233, 234, 235. Springer, New York
3. Collet E, Schwarzenbach D (2010) Dynamical structural science. Acta Cryst A 66:133–280
4. Zewail A, Angew H (2000) Femtochemistry: atomic-scale dynamics of the chemical bond using ultrafast lasers (Nobel lecture). Chem Int 39:2586–2631

5. Neutze R, Wouts R, Techert S, Davidson J, Kocsis M, Kirrander A, Schotte F, Wulff M (2001) Visualizing photochemical dynamics in solution through picosecond X-ray scattering. Phys Rev Lett 87:195508
6. Plech A, Wulff M, Bratos S, Mirloup F, Vuilleumier R, Schotte F, Anfinrud PA (2004) Visualizing chemical reactions in solution by picosecond X-ray diffraction. Phys Rev Lett 92:125505
7. Ihee H, Lorenc M, Kim TK, Kong QY, Cammarata M, Lee JH, Bratos S, Wulff M (2005) Ultrafast X-ray diffraction of transient molecular structures in solution. Science 309:1223–1227
8. Wulff M, Bratos S, Plech A, Vuilleumier R, Mirloup F, Lorenc M, Kong Q, Ihee H (2006) Recombination of photodissociated iodine: a time-resolved X-ray-diffraction study. J Chem Phys 124:034501
9. Cammarata M, Levantino M, Schotte F, Anfinrud PA, Ewald F, Choi J, Cupane A, Wulff M, Ihee H (2008) Tracking the structural dynamics of proteins in solution using time-resolved wide-angle X-ray scattering. Nat Methods 5:881–886
10. Christensen M, Haldrup K, Bechgaard K, Feidenhans'l R, Kong Q, Cammarata M, Russo ML, Wulff M, Harrit N, Nielsen MM (2009) Time-resolved X-ray scattering of an electronically excited state in solution. Structure of the 3A2u state of tetrakis-μ-pyrophosphitodiplatinate(II). J Am Chem Soc 131:502–508
11. Haldrup K, Christensen M, Nielsen MM (2010) Analysis of time-resolved X-ray scattering data from solution-state systems. Acta Cryst A 66:261–269
12. Collet E, Lemée-Cailleau MH, Buron-Le Cointe M, Cailleau H, Wulff M, Luty T, Koshihara S, Meyer M, Toupet L, Rabiller P, Techert S (2003) Laser-induced ferroelectric structural order in an organic charge-transfer crystal. Science 300:612–615
13. Lorenc M, Hébert J, Moisan N, Trzop E, Servol M, Buron-Le Cointe M, Cailleau H, Boillot ML, Pontecorvo E, Wulff M, Koshihara S, Collet E (2009) Successive dynamical steps of photoinduced switching of a molecular Fe(III) spin-crossover material by time-resolved X-ray diffraction. Phys Rev Lett 103:028301
14. Cailleau H, Lorenc M, Guérin L, Servol M, Collet E, Buron-Le Cointe M (2010) Structural dynamics of photoinduced molecular switching in the solid state. Acta Cryst A 66:189–197
15. Srajer V, Teng T, Ursby T, Pradervand C, Ren Z, Adachi S, Schildkamp W, Bourgeois D, Wulff M, Moffat K (1996) Photolysis of the carbon monoxide complex of myoglobin: nanosecond time-resolved crystallography. Science 274:1726–1729
16. Schotte F, Lim M, Jackson TA, Smirnov AV, Soman J, Olson JS, Phillips GN Jr, Wulff M, Anfinrud PA (2003) Watching a protein as it functions with 150-ps time-resolved X-ray crystallography. Science 300:1944–1947
17. Tomita A, Sato T, Nozawa S, Koshihara S, Adachi S (2010) Tracking ligand-migration pathways of carbonmonoxy myoglobin in crystals at cryogenic temperatures. Acta Cryst A 66:220–228
18. Rousse A, Rischel C, Fourmaux S, Uschmann I, Sebban S, Grillon G, Balcou P, Förster E, Geindre JP, Audebert P, Gauthier JC, Hulin D (2001) Non-thermal melting in semiconductors measured at femtosecond resolution. Nature 410:65–68
19. Sokolowski-Tinten K, Blome C, Blums J, Cavalleri A, Dietrich C, Tarasevitch A, Uschmann I, Förster E, Kammler M, Hornvon-Hoegen M, von der Linde D (2003) Femtosecond X-ray measurement of coherent lattice vibrations near the Lindemann stability limit. Nature 422:287–289
20. Lindenberg AM et al (2005) Atomic-scale visualization of inertial dynamics. Science 308:392–395
21. Fritz DM, Reis DA, Adams B, Akre RA, Arthur J, Blome C, Bucksbaum PH, Cavalieri AL, Engemann S, Fahy S, Falcone RW, Fuoss PH, Gaffney KJ, George MJ, Hajdu J, Hertlein MP, Hillyard PB, Horn-von Hoegen M, Kammler M, Kaspar J, Kienberger R, Krejcik P, Lee SH, Lindenberg AM, McFarland B, Meyer D, Montagne T, Murray ÉD, Nelson AJ, Nicoul M, Pahl R, Rudati J, Schlarb H, Siddons DP, Sokolowski-Tinten K, Tschentscher T, von der Linde D,

Hastings JB (2007) Ultrafast bond softening in bismuth: mapping a solid's interatomic potential with X-rays. Science 315:633–636
22. Beaud P, Johnson SL, Streun A, Abela R, Abramsohn D, Grolimund D, Krasniqi F, Schmidt T, Schlott V, Ingold G (2007) Spatiotemporal stability of a femtosecond hard–X-ray undulator source studied by control of coherent optical phonons. Phys Rev Lett 99:174801
23. Schoenlein R, Leemans W, Chin A (1996) Femtosecond X-ray pulses at 0.4 Å generated by 90° Thomson scattering: a tool for probing the structural dynamics of materials. Science 274:236–238
24. Schoenlein RW, Chattopadhyay S, Chong HHW, Glover TE, Heimann PA, Shank CV, Zholents AA, Zolotorev MS (2000) Generation of femtosecond pulses of synchrotron radiation. Science 287:2237–2239
25. Cavalieri AL et al (2005) Clocking femtosecond X rays. Phys Rev Lett 94:114801
26. McNeil B (2009) Free electron lasers: first light from hard X-ray laser. Nat Photon 3:375–377
27. Baum P, Yang DS, Zewail AH (2007) 4D visualization of transitional structures in phase transformations by electron diffraction. Science 318:788–792
28. Gedik N, Yang D, Logvenov G, Bozovic I, Zewail AH (2007) Nonequilibrium phase transitions in cuprates observed by ultrafast electron crystallography. Science 316:425–429
29. Bressler C, Milne C, Pham V-T, ElNahhhas A, Van der Veen RM, Gawelda W, Johnson S, Beaud P, Grolimund D, Kaiser M, Borca CN, Ingold G, Abela R, Chergui M (2009) Femtosecond XANES study of the light-induced spin crossover dynamics in an Iron(II) complex. Science 323:489–492
30. Chen LX, Zhang X, Lockard JV, Stickrath AB, Attenkofer K, Jennings G, Liu DJ (2010) Excited-state molecular structures captured by X-ray transient absorption spectroscopy: a decade and beyond. Acta Cryst A 66:240–251
31. Chergui M (2010) Picosecond and femtosecond X-ray absorption spectroscopy of molecular systems. Acta Cryst A 66:229–239
32. Zholents AA, Zolotorev MS (1996) Femtosecond X-ray pulses of synchrotron radiation. Phys Rev Lett 76:912
33. Bargheer M, Zhavoronkov N, Woerner M, Elsaesser T (2006) Recent progress in ultrafast X-ray diffraction. ChemPhysChem 7:783–792
34. Siders CW, Cavalleri A, Sokolowski-Tinten K, Toth C, Guo T, Kammler M, von Hoegen MH, Wilson KR, von der Linde D, Barty CPJ (1999) Detection of nonthermal melting by ultrafast X-ray diffraction. Science 286:1340–1342
35. Elsaesser T, Woerner M (2010) Photoinduced structural dynamics of polar solids studied by femtosecond X-ray diffraction. Acta Cryst A 66:168–178
36. For more information: http://lcls.slac.stanford.edu/
37. For more information: http://xfel.desy.de/
38. For more information: http://www.spring8.or.jp/
39. Johnson SL, Beaud P, Vorobeva E, Milne CJ, Murray ED, Fahy S, Ingold G (2010) Nonequilibrium phonon dynamics studied by grazing-incidence femtosecond X-ray crystallography. Acta Cryst A 66:157–167
40. Johnson SL, Vorobeva E, Beaud P, Milne CJ, Ingold G (2009) Full reconstruction of a crystal unit cell structure during coherent femtosecond motion. Phys Rev Lett 103:205501
41. Bargheer M, Zhavoronkov N, Gritsai Y, Woo JC, Kim DS, Woerner M, Elsaesser T (2004) Coherent atomic motions in a nanostructure studied by femtosecond X-ray diffraction. Science 306:1771–1773
42. Cavalleri A, Wall S, Simpson C, Statz E, Ward DW, Nelson KA, Rini M, Schoenlein RW (2006) Tracking the motion of charges in a terahertz light field by femtosecond X-ray diffraction. Nature 442:664–666
43. Guérin L, Collet E, Lemée-Cailleau MH, Buron M, Cailleau H, Plech A, Wulff M, Koshihara S, Luty T (2004) Probing photoinduced phase transition in a charge-transfer molecular crystal by 100 picosecond X-ray diffraction. Chem Phys 299:163

44. Guérin L, Hébert J, Buron-Le Cointe M, Adachi S, Koshihara S, Cailleau H, Collet E (2010) Capturing one-dimensional precursors of a photoinduced transformation in a material. Phys Rev Lett 105:246101
45. Bréfuel N, Watanabe H, Toupet L, Come J, Kojima M, Matsumoto N, Collet E, Tanaka K, Tuchagues JP (2009) Concerted spin crossover and symmetry breaking yield three thermally and one light-induced crystallographic phases of a molecular material. Angew Chem Int Ed 48:9304–9307
46. Bréfuel N, Collet E, Watanabe H, Kojima M, Matsumoto N, Toupet L, Come J, Tanaka K, Tuchagues JP (2010) Nanoscale self-hosting of molecular spin-states in the intermediate phase of a spin-crossover material. Chem Eur J 47:14060–14068
47. Collet E, Watanabe H, Bréfuel N, Palatinus L, Roudaut F, Toupet L, Tanaka K, Tuchagues JP, Fertey P, Ravy S, Toudic B, Cailleau H (2012) Aperiodic spin state ordering of bistable molecules and its photoinduced erasing. Phys Rev Lett 109:257206

Chapter 11
Direct Observation of Various Reaction Pathways in Crystalline State Photoreactions

Yuji Ohashi

11.1 Introduction

We found that the chiral cyanoethyl group bonded to the cobalt atom in a cobaloxime complex crystal was racemized without crystal decomposition on exposure of the crystal to X-rays or visible light as shown in Fig. 11.1 [1]. Similar reactions were also found for the crystals with different axial base ligands. Such reactions are very attractive since the reaction process is directly observable by structure analysis at any intermediate stage. Therefore, we defined these processes as crystalline state reactions to discriminate them from usual solid state reactions, in which the crystals are easily broken during the reactions. The racemization rate of the chiral group was significantly decreased when the temperature of the crystal was decreased and the crystal showed no change at temperatures below 173 K. This indicated that the reaction rate was strongly influenced by the environment around the reactive group in the crystal structure. We defined the reaction cavity for the reactive group [2]. When the axial base ligand was replaced with a variety of amines and phosphines, the crystal structures and the reaction cavities for the chiral cyanoethyl group became different. The racemization rate was found to be linearly dependent upon the size of the reaction cavity [3]. It has been made clear that the size of the reaction cavity controls the reactivity or reaction rate and that the shape of the reaction cavity control the reaction pathway [4].

In this lecture, two recent typical examples of crystalline state reactions will be shown and their reaction mechanisms will be well explained by using the concept of the reaction cavity.

Y. Ohashi (✉)
Ibaraki Quantum Beam research Center (IQBRC), 162-1 Shirakata,
Tokai, Ibaraki 319-1106, Japan
e-mail: yujiohashi@jcom.home.ne.jp

J.A.K. Howard et al. (eds.), *The Future of Dynamic Structural Science*,
NATO Science for Peace and Security Series A: Chemistry and Biology,
DOI 10.1007/978-94-017-8550-1_11, © Springer Science+Business Media Dordrecht 2014

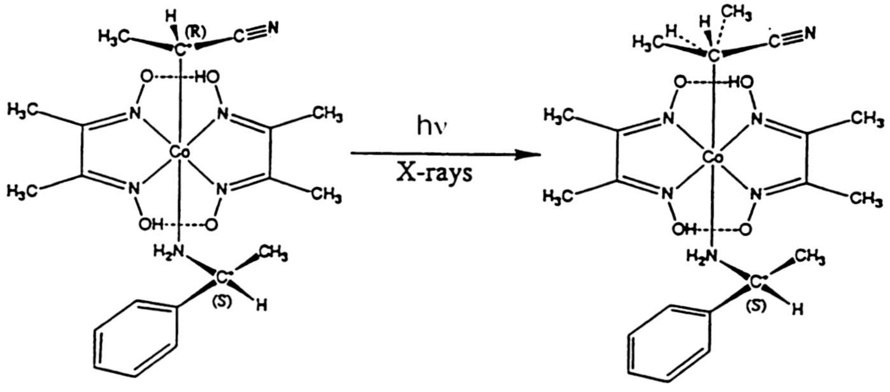

Fig. 11.1 Crystalline state racemization of cobaloxime complexes by X-rays or visible light

11.2 Chirality Inversion Only by Photoirradiation

The first example shows that the chirality of the chiral group is inverted only by photoirradiation. The alkyl group was replaced with a bulkier group, 1-(ethoxycarbonyl)ethyl (1-ece) instead of 1-cyanoethyl group. The complex crystal with the 1-ece group showed an unusual phenomenon, that is, the chirality inversion in a crystal only by photo-irradiation [5]. The crystal structure of [(*S*)-1-cyclohexylethylamine]bis(dimethylglyoximato)[*S*-1-(ethoxycarbonyl)ethyl]- cobalt(III), (*S*-cha) (*S*-1-ece)cobaloxime, is shown in Fig. 11.2. The space group is $P2_12_12_1$. There is one molecule in an asymmetric unit. The chiral *S*-1-ece groups have contacts with each other as a ribbon along the 2_1 axis parallel to the *a* axis. Figure 11.3 shows the molecular structure. The methyl group of 1-ece group, C5, takes the *syn* conformation to the carbonyl group, O2–C3–C4–C5, and the ethyl of ece group has a *trans* conformation around the O1–C2 bond. When the crystal was irradiated with a halogen lamp with a long-path filter (R64), the unit-cell dimensions were gradually changed; the *a* and *b* axes slightly contracted, whereas the *c* axis and the unit cell volume *V* significantly expanded. The crystal structure, however, is nearly the same as before, except that the 1-ece group has a disordered structure due to the partial inversion. Figure 11.4 shows the structural change of the 1-ece group viewed along the normal to the cobaloxime plane. The configuration of the *S*-1-ece group changed from *S* to *R* with retention of the single crystal form.

The 1-ece groups of most molecules were inverted from *S* to *R*. The *S*:*R* ratio changed from 100:0 to 18:82 after 24 h exposure. The ratio of 18:82 was not changed after prolonged irradiation. It is clear that the chirality was inverted only by photoirradiation. This is the first observation that the chirality of most molecules in a crystal can be inverted to the opposite configuration only by photoirradiation.

It is very interesting to examine whether or not the chirality of the 1-ece group would be inverted when the crystal of the diastereoisomer, (*S*-cha)(*R*-1-ece) cobaloxime, was exposed to visible light. The crystal was prepared and the structure

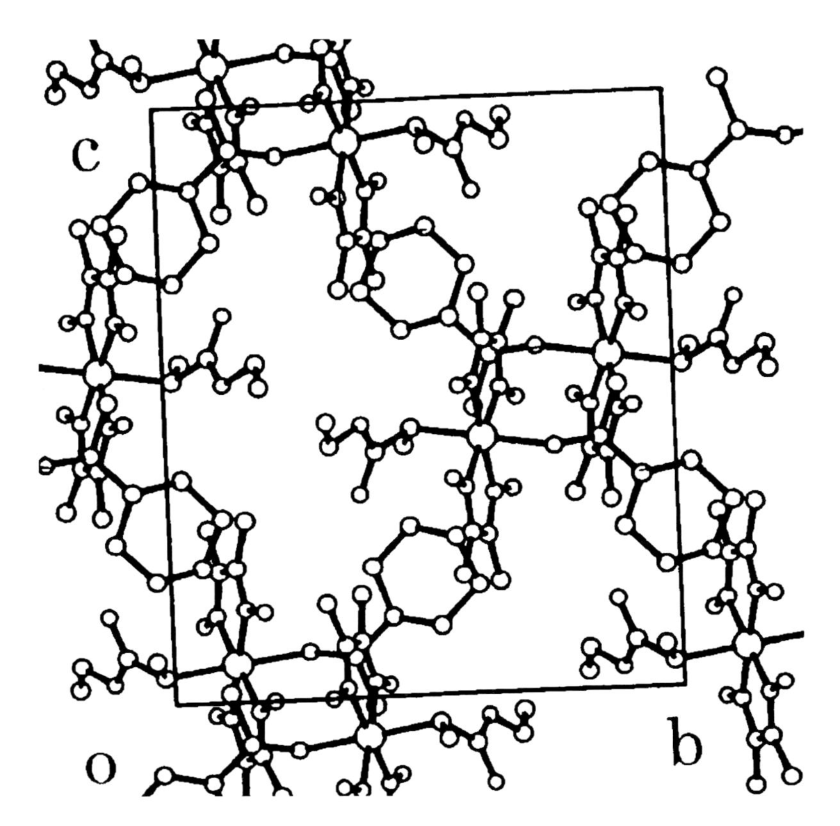

Fig. 11.2 Crystal structure of (*S*-1-ece)(*S*-cha)cobaloxime viewed along the a axis

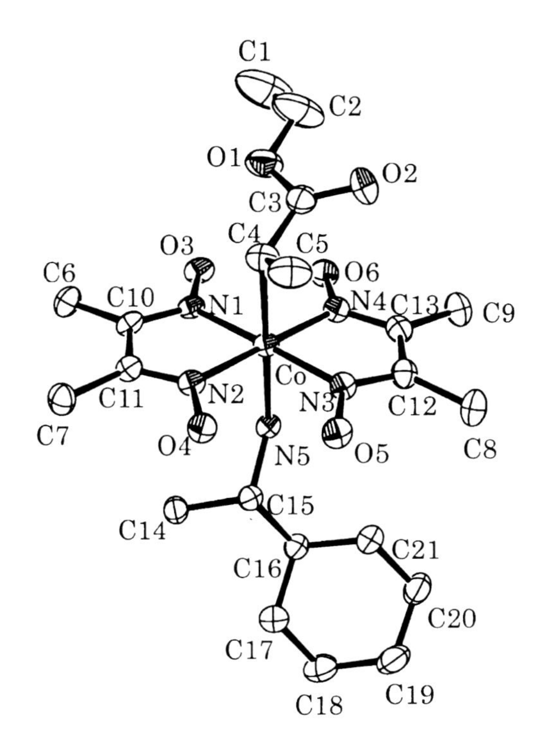

Fig. 11.3 Molecular structure of (*S*-1-ece)(*S*-cha)cobaloxime

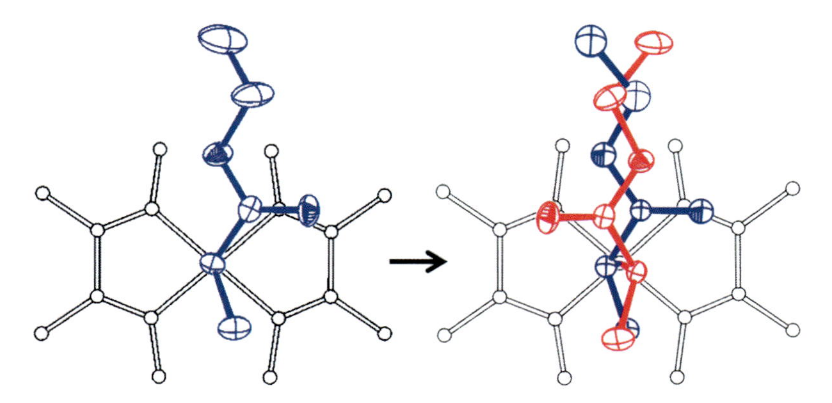

Fig. 11.4 Structural change of 1-ece group after irradiation. *Blue colored* 1-ece group has *S* configuration whereas *red colored* has *R*

before irradiation was analyzed by X-rays. To our surprise, the crystal structure is isomorphous to the previous one with the *S*-1-ece group. The structure of the *R*-1-ece group is the mirror image of that of the *S*-ece group as shown in Fig. 11.3 and the same as the corresponding one to the photo-produced disordered ece group as shown in Fig. 11.4. When the crystal was irradiated with the halogen lamp, the cell dimensions gradually changed. They converged to the values observed in the photoirradiated crystal with the *S*-1-ece group. The structure analyzed by X-rays after 24 h exposure revealed that it is essentially the same structure as the photo-product shown in Fig. 11.4. The *S:R* ratio converged to 18:82, which is the same as that observed in the photoirradiated crystal with the *S*-isomer.

In order to check the ratio of 18:82, the mixed crystal with equal amounts of *S* and *R* isomers, that is, the *racemic*-isomer, was prepared. The crystal structure analyzed by X-rays showed that it is isomorphous to the crystals with *S*- or *R*-1-ece groups and that has a disordered structure of the ece group with the *S:R* ratio of 50:50. When the *racemic* crystal was irradiated with the halogen lamp, the cell dimensions gradually changed and converged to the values observed in the photoirradiated crystals of *S* and *R* isomers. The *S:R* ratio in the photoirradiated crystal is 18:82. It is clear that the *S:R* ratio of 18:82 is the most stable when the crystal with any composition of the S and R isomers is exposed to the halogen lamp.

The question why the ratio of 18:82 is the most stable should be answered. Figure 11.5a–c show the reaction cavities for the *S*-1-ece, *(racemic)*-1-ece and *R*-1-ece groups, respectively, before photoirradiation. Each cavity is divided into two parts by the plane, which passes through the Co atom and is perpendicular the cobaloxime plane and parallel to the C-C bonds of the five-membered rings of cobaloxime. The volume ratio of the left to right parts in each cavity is 38:61, 46:54 or 47:53 for *S*-1-ece, *(racemic)*-1-ece or *R*-1-ece complex crystal, respectively. It must be emphasized that the volume ratio of *(racemic)*-1-ece crystal is not 50:50 but is 46:54. This is probably due to the asymmetric environment around the *(racemic)*-1-ece group in the chiral $P2_12_12_1$ cell. After the photoirradiation the 1-ece

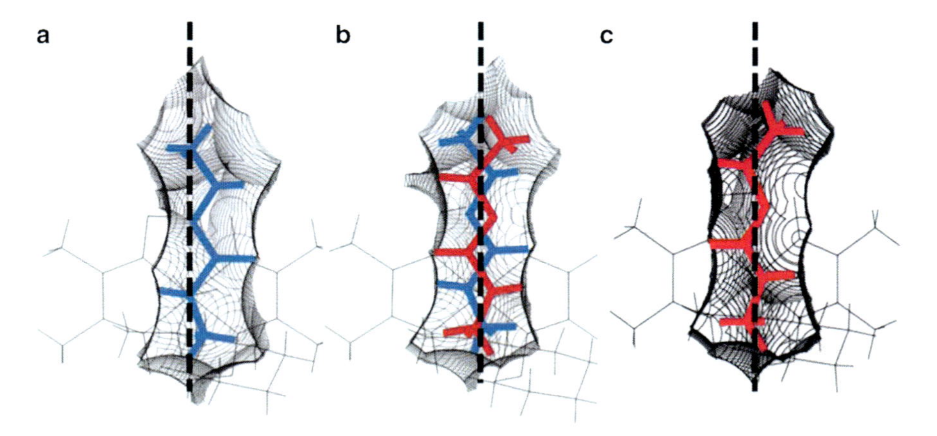

Fig. 11.5 Reaction cavities of 1-ece groups in the crystals of (**a**) (*S*-1-ece)(*S*-cha)cobaloxime, (**b**) (*rac*-1-ece)(*S*-cha)cobaloxime and (**c**) (*R*-1-ece)(*S*-cha)cobaloxime

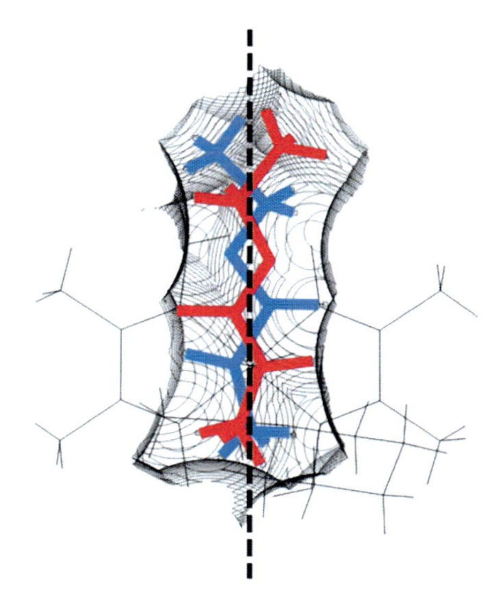

Fig. 11.6 Reaction cavity after irradiation. The ratio of the *left* and *right* parts of the cavity becomes ca. 50:50 on either of the *R*- and *S*-1-ece groups after photoirradiation. The enthalpy term of the ece group plays an important role in the chirality inversion process

groups of the three crystals became essentially the same disordered structure with the *S*:*R* ratio of 18:82. The reaction cavity for the disordered 1-ece group is shown in Fig. 11.6. The ratio of the left to right sides in the cavity for the photoproduced 1-ece group of *S*-1-ece, *(racemic)*-1-ece and *R*-1-ece crystals became 51:49, 51:49 and 51:49, which are almost the same as 50:50. It is clear that the steric repulsion from the surrounding atoms should have equal effect.

In the crystallization of (*S*-1-ece)(*S*-cha)cobaloxime, pseudo polymorphic crystal was obtained, which has one water molecule as a solvent. On exposure to

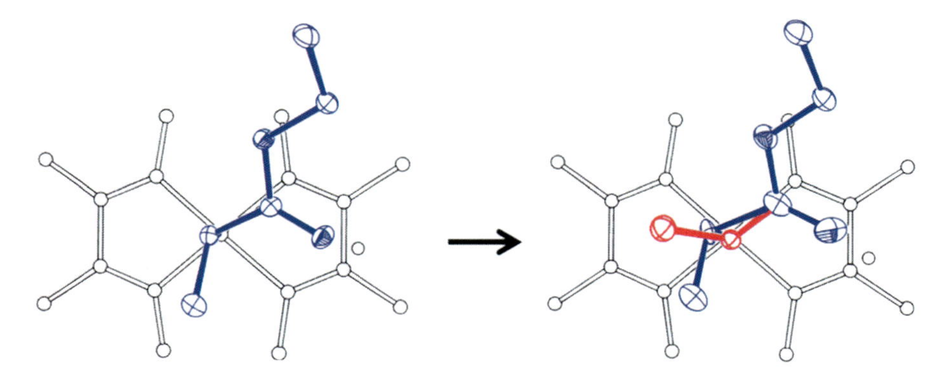

Fig. 11.7 Structural change of *S*-1-ece group in the monohydrate crystal during the photo-reaction. A half of the group is changed to *R* configuration

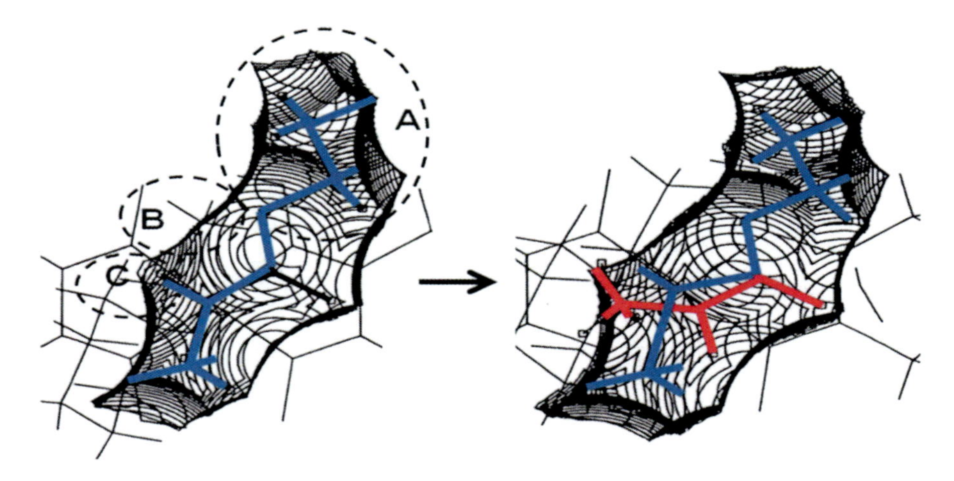

Fig. 11.8 Reaction cavities before and after the photo-reaction

the halogen lamp, the cell dimensions gradually changed with retention of the single crystal form. The structural change of the *S*-1-ece group before and after photoirradiation is shown in Fig. 11.7. Only the methyl group is inverted to the opposite configuration. As shown in Fig. 11.8, the reaction cavity in the monohydrate crystal is smaller and narrower than that in the non-hydrate crystal. The parts of cavity indicated as A and B, which should accommodate the inverted 1-ece group, are too small. Only the volume of C is available for the racemization. After 24 h exposure the *S*:*R* ratio became 50:50. This type of photoracemization has been observed in many cobaloxime crystals. The reason why two crystal forms show different reaction pathways is easily explained with the size and shape of reaction cavities for *S*-1-ece groups in the two crystals.

11.3 Intermediate Structure in the Process of Photoisomerization

The cobaloxime complex with the 4-cyanobutyl (4-cb) group was found to be isomerized only to the 3-cyanobutyl group (3-cb) group with retention of the single crystal form. The NMR spectroscopic data indicated that about 6 % of 4-cb group was transformed to the 2-cyanobutyl (2-cb) group [6]. Since many attempts to observe the photoisomerization from 4-cb to 1-cb groups in the crystalline state were unsuccessful, the diphenylboron bridge was introduced to the equatorial ligands to observe the photo-isomerization from 4-cb to 1-cb, as shown in Fig. 11.9. Of more than 10 complexes with different axial ligands studied, the complexes with pyridine (py), 4-ethylpyridine (4epy), and 3,4-lutidine (lut), showed the photo-isomerization with retention of the single crystal form. The crystals with py, 4epy and lut before photo-irradiation are called **I**, **II** and **III**, respectively [7].

The molecular structure of **I** is shown in Fig. 11.10. One phenyl group of diphenylboron bridge is almost perpendicular to the cobaloxime plane whereas another one is almost parallel to the plane. The conformation of 4-cb group is all *trans* and is approximately parallel to the perpendicular phenyl group. The 4-cb groups and perpendicular phenyl groups of two molecules make a pair around an inversion center of the $P2_1/c$ unit-cell. When a crystal of **I** was exposed to the halogen lamp for 24 h, the cell volumes were increased by 69.6 $Å^3$. The molecular structure after photo-irradiation is shown in Fig. 11.11. The colored 3-cb group was produced. The occupancy factor of 3-cb group is 0.24(1). Only one enantiomer of the chiral carbon C4B was obtained at one site around an inversion center. Since the crystallinity was retained after 24 h exposure, the photo-irradiation was continued for further 48 h. The cell volume was decreased by 54.8 $Å^3$ after a total of 72 h exposure. The molecular structure after 72 h exposure is shown in Fig. 11.12. It is

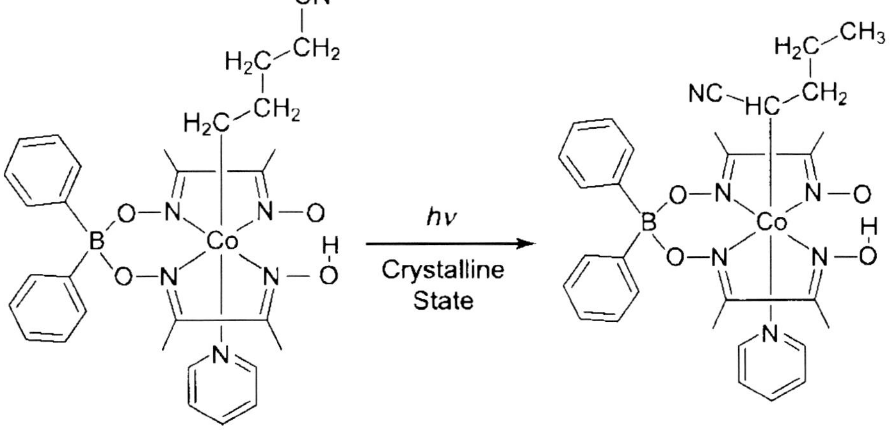

CN
H_2C CH_2
H_2C CH_2
O—N N—O
B Co H
O—N N—O
N
hv
Crystalline State
H_2C CH_3
NC HC CH_2
O—N N—O
B Co H
O—N N—O
N

Fig. 11.9 Photo-isomerization from 4-cb to 1-cb in the diphenylboron bridge cobaloxime

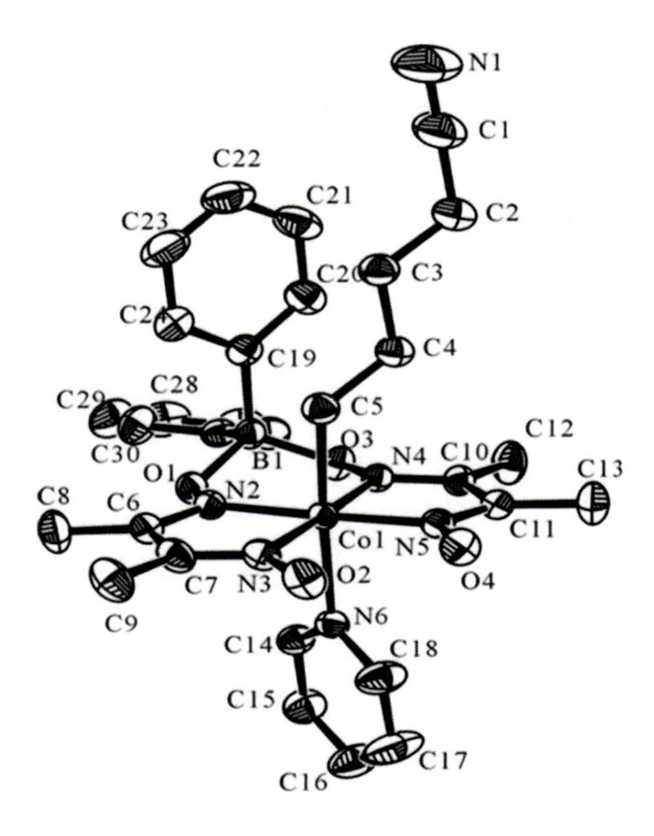

Fig. 11.10 Molecular structure of I before photoirradiation

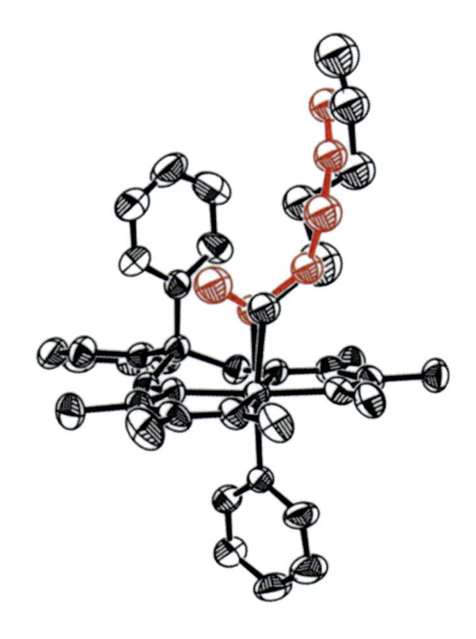

Fig. 11.11 Molecular structure after 24 h photoirradiation. The newly appeared group is 3-cb

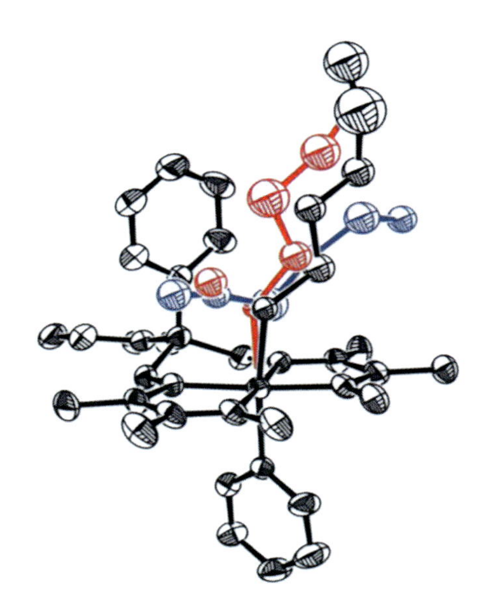

Fig. 11.12 Molecular structure after 72 h exposure

clear that the another colored(blue) 1-cb group was produced after 72 h exposure in addition to the 3-cb group. Only one enantiomer of the chiral carbon, C2C, was observed at one site of an inversion center. The occupancy factors of 4-, 3- and 1-cb groups became 0.69(1), 0.15(1) and 0.16(1), respectively. Although the change from 4-cb to 3-cb is small, about a half of 3-cb was transformed to 1-cb. The decrease in cell volume is caused by the transformation from 3-cb to 1-cb. In order to examine the conformational change of the cyanobutyl group in the photo-isomerization, the structures of the original and photo-produced cyanobutyl groups viewed along the normal to the cobaloxime plane are shown in Fig. 11.13. Considerably large movement of the terminal CN group is necessary to occupy the opposite side to the cobalt atom at the final stage.

The crystal of **II** was exposed to a halogen lamp for 24 h. Only the 3-cb group appeared. Although the occupancy factor of 3-cb group became 0.71(1), there was no indication of 2-cyanobutyl (2-cb) or 1-cb group. The cell volume was decreased by 95 Å^3 in the photoreaction. Further exposure was impossible due to the degradation of the crystallinity.

The molecular structure of **III** is shown in Fig. 11.14a. The conformation of diphenylboron bridge is almost the same as those of **I** and **II**. Such a conformation has been found in the related complex crystals. The conformation of 4-cb group is *trans*, *gauche* and *gauche* around the C1–C2, C2–C3, and C3–C4, respectively. When the crystal was exposed to a halogen lamp, the cell dimensions gradually changed and the change was insignificantly small after 24 h exposure. The molecular structure after 24 h exposure is shown in Fig. 11.14b. The occupancy factor of photo-produced 1-cb group is 0.472(6), which means that about half of 4-cb was isomerized to 1-cb. Although the CN group moves to a great extent, the methylene chain of C1–C2–C3–C4 takes nearly the same position during the photoisomerization.

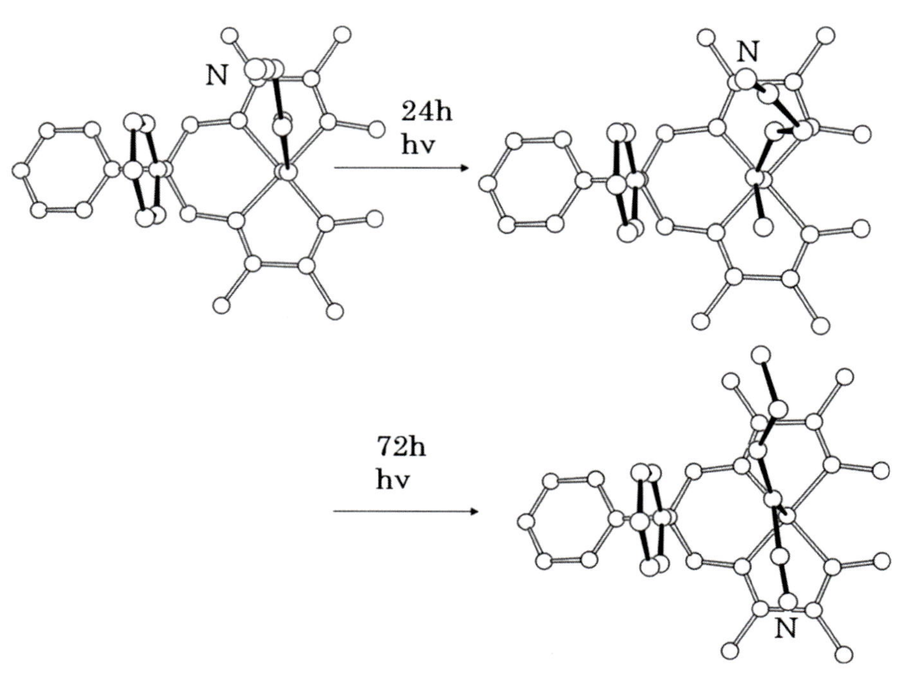

Fig. 11.13 Conformational change from 4-cb to 1-cb groups

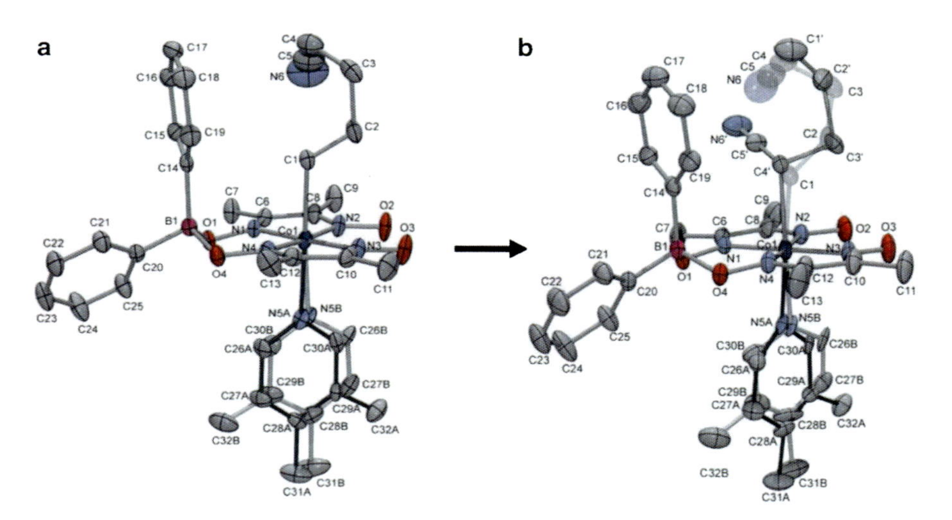

Fig. 11.14 Structural change (**a**) before and (**b**) after 24 h photoirradiation. Only 1-cb group appeared after the irradiation

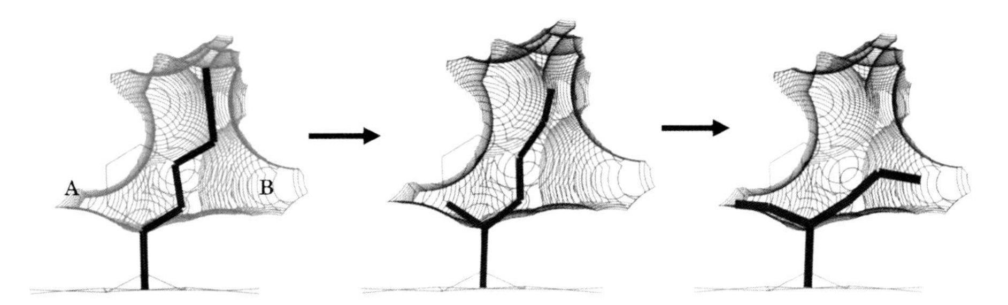

Fig. 11.15 Structural change of 4-cb group in the reaction cavity

The reaction cavities for the 4-cb group were drawn and the volumes were calculated for the crystals of **I**, **II** and **III** before photo-irradiation. Several KBr pellets for each crystal were irradiated with light and successive FT-IR measurements of the pellets were carried out one by one. The isomerization rate constant at early stages (within 40 min) was estimated from the intensity change due to stretching vibration mode of the C-N triple bond of the 4-cb group. Assuming first-order kinetics, the reaction rate constants for **I**, **II** and **III** are 6.2, 9.5, and $19.0 \times 10^{-3}\ s^{-1}$, respectively. The reaction cavity volumes of **I**, **II** and **III** are 38.1, 49.2 and 32.7 $Å^3$, respectively. Since the reaction cavity of **I** is smaller than that of **II**, the rate constant of **I** is smaller than that of **II**. This indicates that the reaction mechanism of **I** from 4-cb to 3-cb at early stages should be the same as that of **II**. However, in spite of the smallest reaction cavity, the rate constant of **III** is the largest and the 4-cb group is easily isomerized to 1-cb group. This clearly suggests that the reaction mechanism of **III** is different from those of **I** and **II**.

In order to make clear the reason why the 4-cb group of **I** is finally isomerized to 1-cb group, the reaction cavities for 4-cb group before photoirradiation are drawn in Fig. 11.15. There are void regions in the neighbourhood of the chiral carbon atom bonded to the cobalt atom, A and B. When the 4-cb group is changed to 3-cb group, the methyl and cyanopropyl groups can occupy the regions A and B, respectively. After 72 h exposure, the cyano and propyl groups are accommodated in A and B, respectively. This also indicates that the space A is too small for the ethyl group when the 2-cb group is produced. It seems adequate to consider that the 2-cb group was not observed in the intermediate stages because it is very unstable due to heavy steric repulsion.

Although it was clarified that the 4-cb group of **I** is isomerized to 1-cb through 3-cb and 2-cb, the 4-cb group of **III** was directly isomerized to 1-cb and the intermediate structures of 3-cb and 2-cb were not observed. In order to elucidate the reaction mechanism of **III**, deuterium atoms were introduced in the 4-cb group and the structure after photo-irradiation was analyzed by neutron diffraction, since the neutron diffraction can discriminate between deuterium and hydrogen atoms [8]. It may be possible to clarify which hydrogen atoms are replaced with deuterium atoms during the reaction.

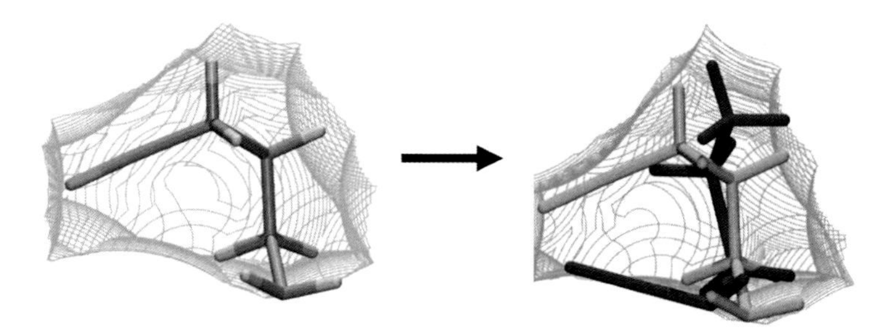

Fig. 11.16 Structural change from 4-cb to 1-cb in the reaction cavity for the initial 4-cb group

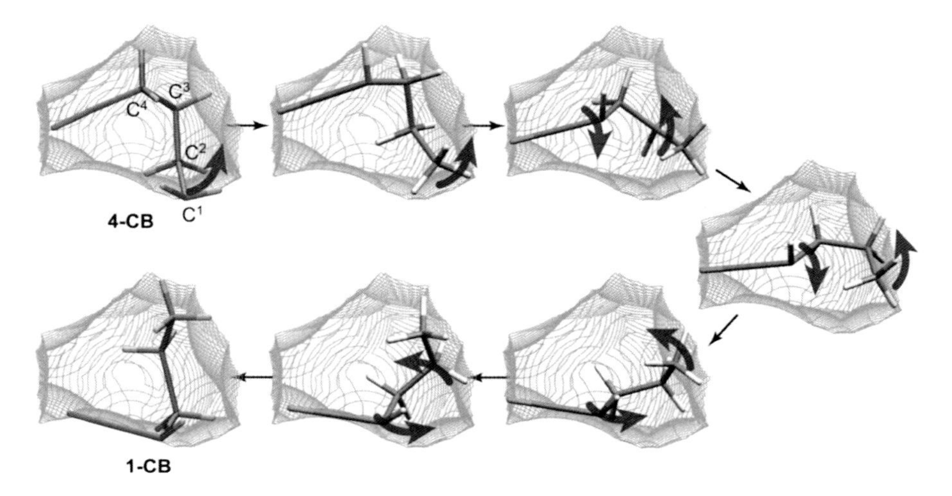

Fig. 11.17 Movement of the photo-produced radical in the cavity for the 4-cb group

The neutron diffraction made it clear the process is 'alkyl turn' isomerization. The C^1 atom of the photo-produced 4-cb radical abstracts one of the deuterium atoms bound to C^4, then the 4-cb radical is turned upside down. Finally, the bond between the cobalt and C^1 atom is formed. The distance between the C^1 and the deuterium atoms before the photoreaction is 2.66 Å, which is shorter than the sum of the van der Waals radii.

The 'alkyl turn' isomerization is very favorable in the crystalline state reaction. To explain the change of the cb group, the reaction cavity for 4-cb in the crystal structure before irradiation and that after irradiation, which includes both of 4-cb and 1-cb, is shown in Fig. 11.16. Since the two cavities are very similar, it is reasonable to assume that the reaction cavity does not change to a great extent during the photo-isomerization. This is the reason why the isomerization rate is the fastest in the three crystals.

The movement of the produced radical is estimated as shown in Fig. 11.17. Since the cavity is large in the lateral direction, the long $-CH_2CN$ part of photo-produced radical can use this space so as to rotate around the C^4–C^3 bond like the pedal motion of a bicycle. Therefore, this reaction can proceed without destruction of the single-crystal form. This is a reason why the 'alkyl turn' isomerization occurs in the crystal of **III**. It is clear that the different pathways from 4-cb to 1-cb in the crystals of **I** and **III** are caused by the different reaction cavities for 4-cb groups in the two crystal structures.

The crystalline-state reactions, in which the chemical reactions proceed with retention of the single crystal form, are very helpful when considering the reaction mechanism. The merit of crystalline-state reactions is that we can directly 'observe' the structural change that occurs in a crystal by X-ray or neutron diffraction, even if it is a multi-step reaction. Moreover, the reaction cavity is a good tool for analysing the reaction mechanism in the crystalline-state reactions quantitatively, although the concept is rather simple. The relation between the reaction rate and cavity size holds well in almost all crystalline-state reactions. The shape of the reaction cavity explains the asymmetric induction during the photoreaction. Recently a variety of the crystalline-state reactions of cobaloxime complexes have been reviewed [9].

References

1. Ohashi Y, Sasada Y (1977) X-ray analysis of Co-C bond cleavage in the crystalline state. Nature 267:142–144
2. Ohashi Y, Yanagi K, Kurihara T, Sasada Y, Ohgo Y (1981) Crystalline state reaction of cobaloxime complexes by X-ray exposure. 1. A direct observation of Co-C bond cleavage in [(*R*)-1-cyanoethyl][(*S*)-α-methylbenzylamine]bis(dimethylglyoximato)-cobalt(III). J Am Chem Soc 103:5805–5812
3. Ohashi Y (1988) Dynamical structure analysis of crystalline-state racemization. Acc Chem Res 21:268–274
4. Ohashi Y (1998) Real-time in situ observation of chemical reactions. Acta Cryst A54:842–849
5. Nitami T, Sekine A, Uekusa H, Ohashi Y (2011) Chirality inversion in a crystal only by photoirradiation. Bull Chem Soc Jpn 84:1066–1074
6. Vithana C, Uekusa H, Sekine A, Ohashi Y (2001) Successive crystalline-state photoisomerization of the 4-cyanobutyl group in a cobaloxime complex. Bull Chem Soc Jpn 74:287–292
7. Hirano A, Sekine A, Uekusa H, Ohashi Y (2011) Direct observation of different reaction pathways by X-rays in the photoisomerization of cobaloxime complexes with dephenylboron bridging. Bull Chem Soc Jpn 84:496–504
8. Hosoya T, Uekusa H, Ohashi Y, Ohhara T, Kuroki R (2006) A new photoisomerization process of the 4-cyanobutyl group in a cobaloxime complex crystal observed by neutron diffraction. Bull Chem Soc Jpn 79:692–701
9. Ohashi Y (2013) Dynamic motion and various reaction paths of cobaloxime complexes in crystalline-state photoreaction. Cryst Rev 19(1):2–146

Chapter 12
Time-Resolved XAS Spectroscopy Probing Dynamic Species in Homogeneous Catalysis – Towards Faster Methods Providing More Information

Moniek Tromp

12.1 Introduction

A catalyst [1] is a material, which increases the rate of a chemical reaction without being consumed itself and can direct the reaction to specifically form desired products. Catalytic processes are widely used in daily life: A catalyst in a car converts the toxic exhaust gases from the engine into more environmentally friendly gases. In the pharmaceutical and flavour and fragrance industry, catalysts are used to produce the required medicines or perfumes. Understanding the catalytic process itself, i.e. how the catalyst works and what factors affect its performance will help us to optimise many industrially important processes, making them more efficient and environmentally sustainable for example by producing less waste or requiring milder process conditions. A way to study these catalytic systems in detail is using spectroscopy, different types of light to take pictures or movies of the catalysts while they are working.

Detailed information on the structural and electronic properties of the catalyst and how they change during reaction is required to understand their performance and thus their reaction mechanism. Like X-rays in a hospital can be used to visualise the bone structure of a skeleton in a body, X-rays can be used to visualise the structure and electronics of molecules and materials. High energy X-rays expose the detailed structure providing information on the type of atoms, their position and geometry. This so-called X-ray absorption spectroscopy (XAS) can be applied while the catalysts are working, providing insights into how reactants bind to the active centre.

X-ray Absorption spectroscopy (XAS) is a technique for detailed structural and electronic characterisation of materials [2]. Its strengths include the ability to perform experiments *in situ/operando* (variable temperature and pressure) and

M. Tromp (✉)
Catalyst Characterisation, Chemistry, Catalysis Research Center, Technische Universität München, Lichtenbergstrasse 4, 85748 Garching b. München, Germany
e-mail: Moniek.Tromp@tum.de

J.A.K. Howard et al. (eds.), *The Future of Dynamic Structural Science*,
NATO Science for Peace and Security Series A: Chemistry and Biology,
DOI 10.1007/978-94-017-8550-1_12,

time-resolved [3]. The technique is element specific and it does not require long-range order so amorphous systems and solutions can be studied, which makes it a powerful technique in catalysis. New technique developments are centered on increased time-resolution as well as increased energy resolution and thereby increased information content. In heterogeneous catalysis, the technique is well-known and applied extensively to study its structural and electronic properties ex situ as well as in situ and operando [4]. In homogeneous catalysis, the use of XAS and related techniques is still very limited, despite it's high potential [5].

Fast moving monochromators and energy dispersive acquisition methodologies have enabled the study of catalytic systems down to the millisecond time scales and reaction mechanisms have been obtained. For example, time-resolved XAFS in combination with UV-Vis has allowed detailed characterization of homogenous reaction intermediates *in situ/operando* and time-resolved [6], making it a powerful tool in revealing reaction mechanisms. For many catalytic systems, this time resolution is sufficient for the information required. However, an even higher time resolution is required when one wants to investigate photochemical or photocatalytic systems, as well as formation of some catalytic intermediates. The activated or intermediate states in general have lifetimes of picoseconds to nanoseconds. New XAS data acquisition methods as well as detector developments now allow XAS spectra down to a time resolution of nano- and picoseconds [7].

So far, XAS has mainly been used to obtain a structural picture, whereas the electronic structure is often poorly understood. Moreover, the major disadvantage of XAS is that it determines an average of all the different structures present, as such complicating the analysis and interpretation. New developments in XAS using new instrumentation and data acquisition methods, while selecting specific X-ray energies provide more detailed electronic information [8]. For example, charge transfer between reactants and metal particles has been visualised and identified to specific adsorption sides on the metal particles. Also, interaction between different organometallic centers during catalysis has been proven and the approach direction of reactants identified, which could not be done by any other technique before. Selectively probing certain X-ray energies will also allow one to obtain information on atoms with specific properties only; for example the structure of atoms with only one specific charge (oxidation state).

These novel XAS techniques, (fast) time-resolved and high energy resolution are currently being developed further and applied to catalytic systems. The properties and catalytic performance of catalysts are studied during their synthesis and performance, with a combination of complementary (spectroscopic) techniques. The influences of for example different ligands, supports, reactants and reaction conditions on the structural and electronic properties of the active site are investigated, as such obtaining detailed insights in reaction mechanisms and catalyst 'restructuring'. In this chapter, the different X-ray techniques will be described and a few different examples in homogeneous catalysis given, with an outlook to future time-resolved and high energy resolution experiments.

12.2 X-ray Absorption Spectroscopy

X-ray absorption fine structure (XAFS) spectroscopy experiments rely on, as the name already suggests, high-energy X-rays, normally produced by synchrotrons. Laboratory equipment is available, however, that provides a much lower X-ray intensity and produces only relatively low-energy X-rays [9].

In an X-ray absorption experiment, the sample is exposed to high energy X-rays. Part of the photons is absorbed by the sample, depending on its absorption characteristics. If the energy of the X-ray photons is high enough to excite a core electron to an empty energy state or to the continuum, a strong increase in the X-ray absorption, the so-called absorption edge, is observed, see Fig. 12.1. The excitation energy of the core electron is correlated to the binding energy of the core electron, making XAS element-specific. The core electron absorbs the X-ray energy, is excited and is transformed in a photo-electron. At higher and higher energies, the outgoing photoelectron is given more and more energy, with which it can 'travel through' the molecule or material structure.

The first part of the XAFS spectrum up to ~50 eV above the absorption edge is called the X-ray absorption near-edge spectrum (XANES). XANES probes empty energy states (empty density of states (DOS)) in the atom or molecule, providing detailed information on the electronic properties. The absorption edge is part of the XANES and is a spectroscopically allowed dipole transition. The exact energy position of the absorption edge gives information on the oxidation state, or the charge distribution within the molecule, of the element under investigation. Pre-edge peaks or features are observed below the absorption edge and originate

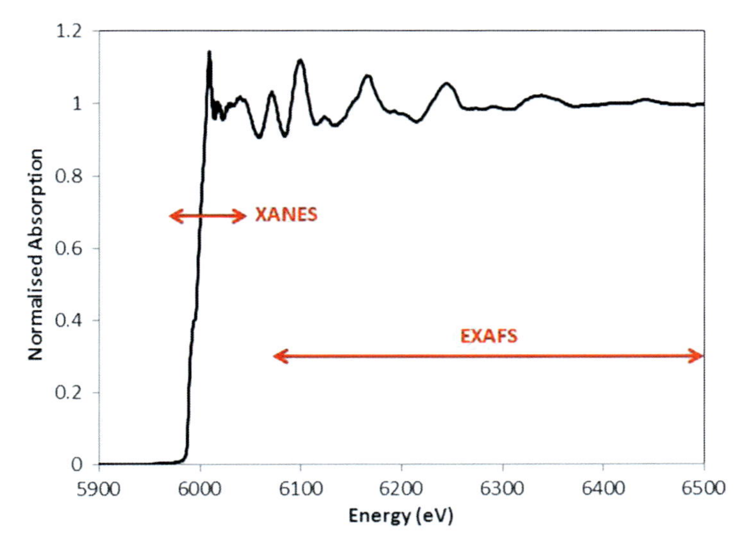

Fig. 12.1 Normalised Cr K-edge X-ray Absorption Spectrum, including XANES and EXAFS regions

from direct quadrupole transitions and/or transitions to dipole-allowed molecular orbitals that have hybridised with other molecular orbitals, which are in itself dipole-forbidden. This is possible in certain geometries of the molecule, and makes the transitions partially allowed and thus visible. Multiple scattering (i.e. scattering of the photoelectron from more than one neighbouring atom before returning to the central atom) features are present at energies above the absorption edge and are very sensitive to and indicative of the geometry around the absorbing atom.

Detailed XANES studies on transition-metal complexes have demonstrated the sensitivity of X-ray absorption K edges and their (pre-)edge features to their chemical environment [10], such as oxidation state and charge distributions, site symmetry, and crystal field splitting.

The second part of the spectrum, the higher-energy extended X-ray absorption fine structure (EXAFS) part, originates from scattering of the outgoing photoelectron against the electron density of neighboring atoms, creating a backscattered wave that interferes with the outgoing photoelectron wave [2, 11], either constructively or destructively. This interference pattern, observed as oscillations in the spectrum, gives information on the local structure around absorber atoms, i.e. the number and type of neighbours, the distance to those neighbours and their static and thermal disorder.

The advantage of XAFS is that it does not require long-range ordering of the system: amorphous materials and solutions can be studied and measurements can be performed *in situ/operando* and time-resolved. Detailed electronic and structural information about the catalysts in their dynamic chemically active environment can thus be obtained, and structure–performance relationships and reaction mechanisms derived. Moreover, a combination of spectroscopic techniques gives complementary information about the system under investigation, which can help to unravel the EXAFS data of multiple structures in the system.

12.2.1 Time-Resolved XAS: Quick-EXAFS and EDE

Quick-EXAFS (QEXAFS) can be performed in two different manners: (i) a continuous movement of the double crystal monochromator system to scan the energy range as in a conventional XAS experiment (Fig. 12.2a) [2, 3], or (ii) changing the Bragg angle of a cam-driven channel cut crystal in an oscillatory manner giving time resolutions of ~ 50 ms for wide energy ranges, also called piezo-XAS [12, 13]. Measurements can be performed in transmission, fluorescence, and total-yield detection modes and thus allow the characterisation of concentrated down to very dilutes systems. The QEXAFS techniques are limited by the speed of movement of the crystals, the integration time to obtain acceptable statistics, and the reproducibility of the crystal movements [3].

Another approach to acquire time resolved XAS data is based on an energy-dispersive EXAFS (EDE) methodology in which a curved crystal polychromator is

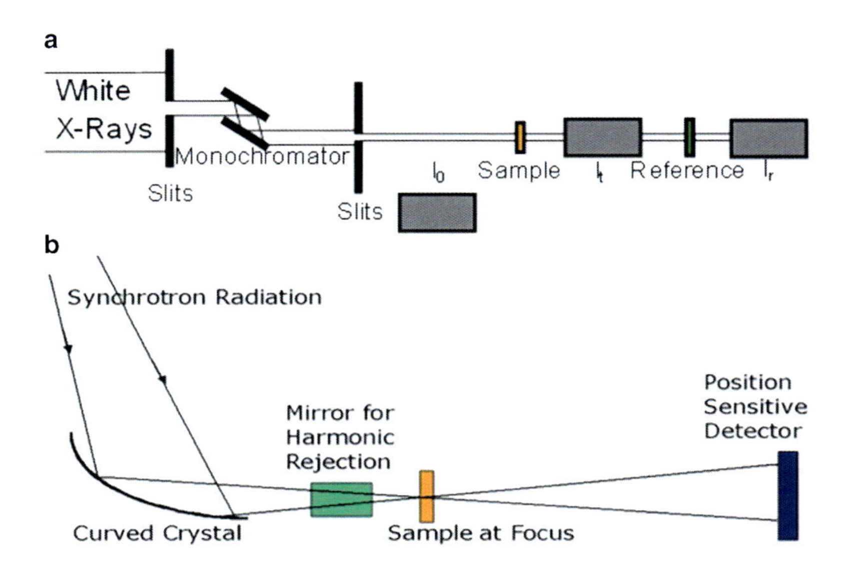

Fig. 12.2 Schematic representation of (*top*) scanning X-ray absorption fine structure (XAFS) data acquisition set-up and (*bottom*) an energy-dispersive mode. The figures are simplified, and many of the optical components are left out for clarity. In the normal scanning data acquisition mode (*top*), a double-crystal monochromator (as displayed) is used to select and scan through the required energy range. In the piezo-XAS data acquisition scheme, a channel cut crystal is employed. The absorption before (I0) and after (It) the sample (and reference Ir) are measured as a function of energy in the so-called transmission mode. The energy-dispersive data-acquisition mode (*bottom*) is based on a spherically bent crystal to focus a band of energies at the sample. All energies are dispersed onto a position-sensitive detector and detected at once

employed [14]. The crystal curvature provides a band spread of X-ray energies to span one (or more) XAFS spectra. The range of energies is focused on the sample and the spectrum measured in one shot using a position sensitive detector, generally resulting in a higher time resolution than for QEXAFS. The EDE technique has several advantages and disadvantages compared to the QEXAFS approach [3]: A small EDE spot size reduces the sample volume necessary, but thereby also requires a sample that is very homogeneous in composition. Since all spectral points are measured simultaneously, the acquisition time can be reduced to milliseconds or below. At the same time, the acquisition method relies on a very stable beam with high demands on optical components. Since the sample is exposed to the entire band of energies, radiation damage of samples and material components can be an issue and the demands on detector sensitivity, radiation hardness, linearity and readout time are high. Finally, the EDE data acquisition method is limited to transmission, which limits the number and type of applications (e.g. very dilute systems or low energy elements cannot be studied using EDE).

Each of these approaches to time-resolved studies can be used in combination with complementary techniques such as X-ray powder diffraction (XRD) and optical spectroscopies such as UV–visible, Infra-Red and/or Raman.

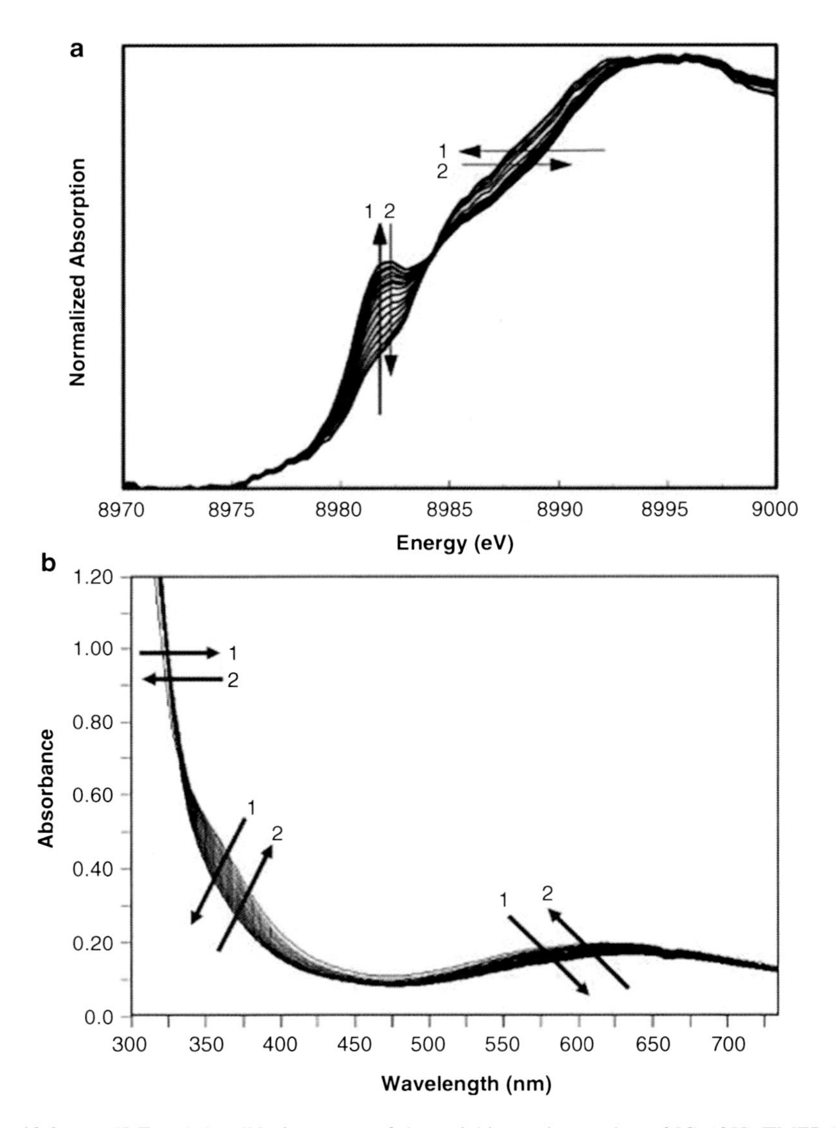

Fig. 12.3 (**a**) EDE and (**b**) UV-vis spectra of the stoichiometric reaction of [Cu(OH)(TMEDA)] 2Cl2 + imidazole + PhB(OH)2. Maximum after ~30 s (*arrow 1*), after which the features decrease again (*arrow 2*) [6]

The time-resolved approaches have been used to characterise catalytic reactions intermediates in many heterogeneous systems [4] for example providing insights in changing metal particle nuclearity as well as metal oxidation state and identify reactive surface species. A few homogeneous systems have been studied using these approaches too [5]. Combined EDE and UV-Vis (Fig. 12.3) have identified the reaction mechanism of a Cu-catalysed CN coupling.

Alternative methods to probe reaction intermediates at known reaction times, but still allowing longer XAS data acquisition to improve the signal to noise ratio are presented too [15].

12.2.2 Fast Time-Resolved XAS Approaches

The millisecond time-resolution as can be obtained by Quick-EXAFS and EDE approaches is enough for most catalytic systems (where transport phenomena are often in these time regimes). However, an even higher time resolution is required when one wants to investigate photochemical or photocatalytic systems, as well as formation of some catalytic intermediates, where activated or intermediate states only have lifetimes of picoseconds to nanoseconds.

Different approaches have been followed at different synchrotrons to increase the time resolution of XAS data acquisition (XANES and EXAFS). The technique is of great current interest, with major current developments being carried out at the APS [16] and SLS [17]. Each of these facilities utilizes a scanning monochromator maintained at fixed energy with a time series built up by repeat acquisitions at each point in time and energy. So the EXAFS spectrum is measured energy point by energy point, doing multiple pump-probe experiments per point, to get a good S/N ratio. Experimental acquisition times for this approach are up to 40 h [18] and recently down to 200 s per energy point at the APS [19] or ~4.5 h at the SLS [20] for transient XAFS spectra of organometallic complexes in solution.

An alternative and new data acquisition method for fast time-resolved experiments is the ED approach as was explained above [21]. The specific advantages of this approach for the fast time resolved experiments are: (i) the X-ray beam is focused and can thus be positioned within the laser irradiated volume only, (ii) the entire XAFS spectrum is obtained in one shot, and so has the potential to significantly reduce the total acquisition time (the pump-probe experiment still has to be done multiple times, to obtain a spectrum with high enough S/N). To obtain the high time-resolution, the detector was time-windowed around the electron bunch of the synchrotron, i.e. the machine clock signal was used to trigger the detector and laser. The time-resolution which can as such be obtained is then governed by the bunch length of the X-ray source, being ~70 ps at the ESRF, ~40 ps at the Diamond Light source, in standard filling modes, and ~5 ps at Diamond in a low α-mode [22].

Important to mention here again is the relatively high concentration of the species under investigation required for the ED approach, which can significantly influence the type and properties of the excited state species formed and thus has to be taken into account. This methodology was proven for pump-probe experiments on a well-studied Cu(phenanthroline) system [21]. For example the decay of the excited state species of [Cu(dbtmp)2] + (dbtmp =2,9-di-n-butyl-3,4,7,8-tetramethyl-1,10-phenanthroline) was monitored for up to 100 ns, at 10 ns time intervals (Fig. 12.4). A data acquisition time per spectrum of ~3.5 h could be achieved, which was mainly limited by the laser frequency being 10 Hz. In the future, using a kHz laser, a reduction to only a few minutes total data acquisition should be feasible.

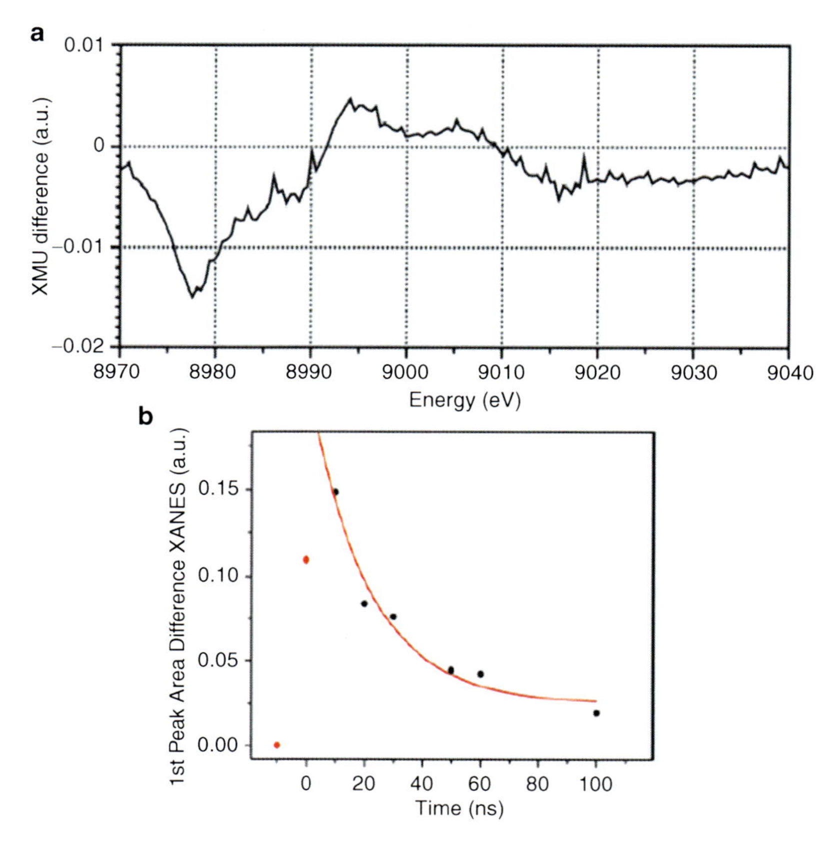

Fig. 12.4 (**a**) Cu K edge XANES differential spectrum (excited state – ground state XANES) for [Cu(dbtmp)2]Cl in CH3CN; (**b**) differential peak areas as a function in time, after excitation [21]

12.3 High Energy Resolution XAS/XES [23]

12.3.1 X-ray Absorption Near Edge Structure

The XANES region of the XAS spectrum is, in contrast to the EXAFS part, less understood and more difficult to analyse and interpret. In many studies in the literature, the XANES region is used as a fingerprint and samples of interest are compared to known references. However, much more detailed information is present.

The local electronic configuration of the metal ion is reflected in the pre-edge region of the XANES spectrum, which arises from resonant excitations into the lowest unoccupied orbitals. The K edge probes the dipole allowed transition of a 1s electron to the metal np states. K pre-edge features originate from direct quadrupole transitions and/or dipole transitions to p-states that have hybridised with the d (or s) states, depending on geometry. Detailed X-ray absorption near edge structure (XANES) studies on transition metal complexes have demonstrated the sensitivity of X-ray absorption K edges and their (pre)edge features to their chemical environment, like oxidation state, site symmetry and crystal field splitting [24]. By carefully observing and analysing the XANES spectra (pre-edge and main edge) in combination with theoretical calculations methods, detailed electronic and structural information can be obtained, for example the amount of charge redistribution and thus covalent interaction between catalyst and substrate [24].

Unfortunately, the intensity of the K pre-edges is normally small compared to that of the main edge and the features are therefore difficult to analyse. As a result the K pre-edge features can often only be analysed qualitatively or in a more global way, relating the centre of gravity and integrated intensities to the materials properties like oxidation state and charge [24]. The relatively low resolution of K edge XANES data is due broadening of 1s lifetime. For 3d transition metals the 1s lifetime broadening is in the range of 0.8–1.6 eV, whereas their L-edge lifetime broadening is significantly shorter, being ~0.1–0.5 eV [25], and thus potentially providing more detailed information.

The partially filled d states in transition metal systems are only probed indirectly with K-edge XAS, and the pre-edge features are only visible in certain favourable geometries. However, the XAS L-edge does directly probe the 2p to nd transitions. The K edges of 3d and 4d transition metals are found in the hard X-ray regime between 5 and 26 keV and can be measured under virtually any condition, for example at elevated temperatures and pressures as required in catalytic reaction conditions. However, the L-edges of 3d transition metals are found in the soft X-ray regime between 400 and 1,000 eV, which requires measurements to be performed under (near) vacuum conditions (upper limit in millibar range) [26]. XAS L-edge for 4d transition metals are in the range of 2–4 keV, still in a low energy regime, complicating *in situ/operando* experiments.

12.3.2 X-ray Emission Spectroscopy

The electronic states that give rise to the edge of an absorption spectrum are resonantly excited states that subsequently decay; i.e. excitation of a core electron, followed by emission of a higher orbital electron to fill the core hole. Many emission lines are possible. Depending on the core level probed, emission to different orbitals can occur with differing probabilities e.g. the K fluorescence lines resulting from 2p to 1s transitions are eight times stronger than the 3p–1s transitions and 50–100 times stronger than the transitions from higher orbitals to the 1s [27].

With a fixed incident energy at or above the energy of the absorption edge, the X-ray emission spectrum (XES) of a sample can be measured by scanning the energy of the fluorescence signal with a so-called secondary spectrometer (crystal analyzer) (with an energy bandwidth in the order of the lifetime broadening).

In comparison to X-ray photoelectron spectroscopy (XPS), XES can be performed *in situ*/operando on a wider range of samples, since the sampling depth is set by the X-ray attenuation length rather than the electron escape depth. X-ray emission spectroscopy has already been shown to be a valuable technique in 3d transition metal K pre-edge to separate the pre-edge features from the K main edge and resolve previously unobserved spectral features [28]. Similarly, high-resolution XES spectra obtained for the 4d transition metal molybdenum compounds are shown to be very sensitive to the chemical and electronic environment of the metal [29].

Non-resonant XES or valence-to-core XES with emission energies close to the Fermi level (or HOMO/LUMO) map the density of occupied states in the valence orbitals (d orbitals for transition metal systems) [23]. The spectra are very sensitive to the type of ligand enabling distinction between, e.g., H_2O and OH^-, while keeping the advantage of element selectivity that holds for all inner-shell spectroscopies. Non-resonant XES thus amends one of the major shortcomings of EXAFS, i.e. insensitivity to neighbouring atoms with similar atomic number. Examples include the discrimination of ligands with similar atomic number like O, N, and C in the first coordination sphere of organometallic complexes [30], as well as characterisation of solid materials [31].

Alternatively, the fluorescence of a single emission line can be determined as a function of incident energy [32, 33]. The core hole broadening observed in a 'normal' transmission or total fluorescence spectrum is then substituted by the broadening of the single emission line (i.e. final state lifetime), resulting in an experimental line width smaller than that due to the core hole lifetime. The multiple pre-edge features and shoulders present in the XANES region will thus be better resolved and easier to separate and analyse. For example, by measuring the fluorescence of a single L emission line for 5d transition metals, high energy resolution fluorescence detection (HERFD) XANES data have been obtained, providing insights in reactant adsorption sites and support vs. adsorbate effects [33].

12.3.3 Resonant Inelastic X-ray Scattering Spectroscopy (RIXS)

In resonant inelastic X-ray scattering (RIXS) spectroscopy, the X-ray intensity is measured as a function of incident and emitted energy, using a secondary spectrometer.

For example, in the case of a 3d transition metal ion, the radiative decay with the highest probability after 1s core hole creation is a 2p to 1s transition (Kα emission). The spectroscopy denoted as 1s2p RIXS (Fig. 12.5) describes the core hole in the intermediate (1s) and final (2p) state. The process can be viewed as an inelastic scattering of the incident photon at the 3d transition metal atom [34], Similarly, the 3p $\rightarrow$ 1s decay can be probed by monitoring Kα emission lines, i.e. 1s3p RIXS.

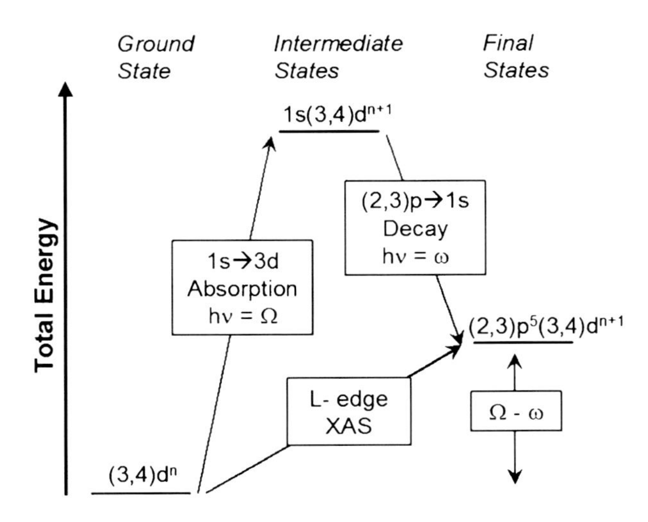

Fig. 12.5 RIXS energy scheme for the 1s2(3)p RIXS in a transition metal ion. For simplicity, atomic configurations are used and only 1s-3d excitations are shown (Amended from Ref. [8])

The RIXS final states, i.e. core hole in either the 2p or 3p orbital with an additional electron in the 3d orbital, are formally identical to respectively L- and M-edge. The final states in the XAS and RIXS process however are reached in a different way, and due to different selection rules with different line intensities, so the spectra obtained will not be identical [35]. At the same time, the energy splittings due to crystal field, multiplet and spin-orbit interactions are the same, thus providing similar information with the two techniques.

The incident energy Ω as well as the emitted energy ω are varied in a RIXS experiment. The RIXS data are generally presented in a 2D contour plot, with the final state energy transfer ($\Omega-\omega$) against the incident energy. This is done to be able to assign the total energy of an electronic state to the axes of the contour plots. The energy transfer axis of the plot then relates to the excitation energy in L or M edge XAS (Fig. 12.6). Different cross sections in the RIXS plane can then be taken relating to possible 1D experiments: (1) vertical – resonant X-ray emission spectra (constant incident energy), (2) horizontal – probing the same final state for all incident energies (constant energy transfer), (3) diagonal – HERFD XAS or lifetime removed spectra diagonal (constant emission energy) [32], as also explained above. Using this data acquisition set-up, the edge absorption pre-edge features can be separated out and interpreted/understood for e.g. materials electronic properties, incl. spin states, and the energy of the molecular orbitals assigned. A great example of increased information content is the application of 2p3d Mo RIXS to a series of different Mo inorganic materials. A rich structure with considerable more spectral information than in conventional XANES is observed providing detailed geometrical and electronic information, like crystal field splitting [37].

The energies of the fluorescence emission transitions are very sensitive to the chemical environment. When different emissions can be resolved and assigned to specific chemical environments, selective (HERFD) XAFS at a selected emission line can be performed to probe those specific environments. This allows the

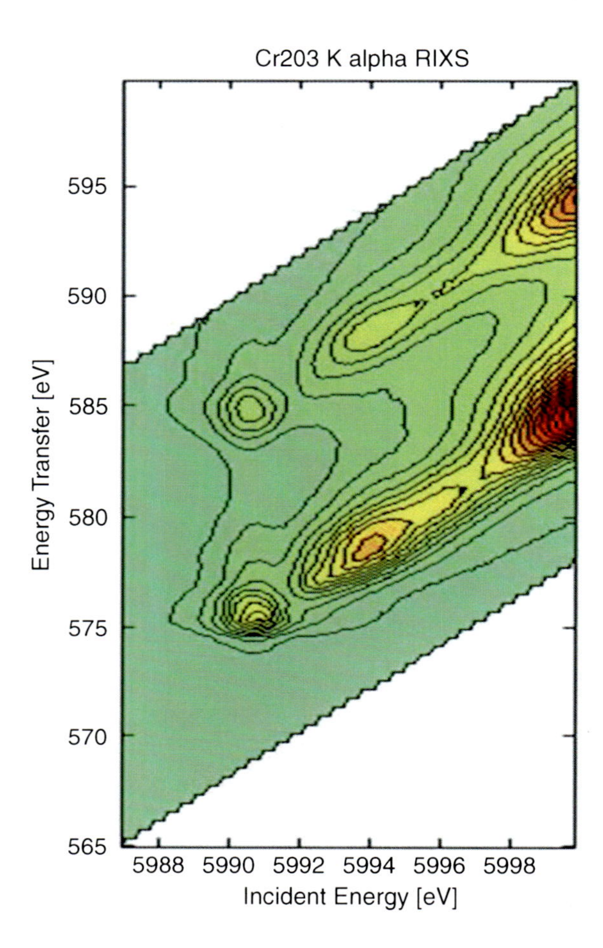

Fig. 12.6 1s2p Cr RIXS plane of Cr_2O_3 [36]

characterisation of individual components in chemical mixtures. A number of applications have been demonstrated [23], spin-selective K edges, oxidation state selective XAFS, and range extended XAFS using detection channels for a certain edge, valence and/or spin-state. All is done using hard X-rays and thus allowing *in situ*/*operando* experimentation.

We have recently applied these methods to Cr catalysed epoxidation reactions, thereby studying the changing structure and electronics of the Cr(salen) catalysts. Although there is not clear effect visible in the 1s2p RIXS of different salen-substituents applied, the RIXS was able to identify and unravel pre-edge features, not visible in the normal XANES, related to Cr dimer formation [38].

When the emission energy is chosen close to the Fermi level (like for valence-to-core XES), the resonant XES or valence band RIXS planes can be obtained. The resonant XES offers unique and direct access to the d-d transitions of metal ions in molecules, or Fermi level and electronic band structure in extended solids. While the XANES is sensitive to the unoccupied density of states, the valence to core XES is selective to the occupied density of states. The final state resembles the optical excited states as observed in UV-Vis spectroscopy (with the RXES being a second order process with two dipole transitions). This method has been used successfully to characterise the Fermi level in supported metal nanoparticles [39], also upon adsorption of molecules. In organometallic complexes, for example the self-assembly of square-planer platinum complexes has been studies [40], the preferred coordination of adsorbate molecules on a Ti site in TS1 [41].

12.3.4 Theory Considerations

Analysis of inner-shell spectra for transition metal systems is a complex problem. Depending on the electronic transition observed experimentally, different theoretical programs are more or less suited and successful. K-edge XAFS spectra above the pre-edge can be well-reproduced using the FEFF code [42]. The FEFF8 and FEFF9 codes allow one to perform ab initio self-consistent field, real-space full multiple scattering calculations utilising a muffin-tin potential. The calculations include a complex-valued exchange-correlation potential (or self-energy) and the effect of the screened core hole and therefore approximate effectively the continuum orbitals, and because of the full multiple scattering, also approximate the strong antibonding resonances present in this energy region. FEFF implements self-consistent field potentials for the determination of the Fermi-level and the charge transfer. This code, however, does not reproduce charge transfer effects adequately and does not fully account for intra-atomic electron-electron interactions that lead to the multiplet structure of the electronic energy levels.

In all cases where a core hole other than a 1s hole is present in the initial or final state, multiplet effects are important [43]. They determine the spectral shapes and influence the L_3 to L_2 branching ratio. To account for the ligand environment around the transition metal, the spherical symmetry of the wavefunctions in the atom is reduced to the point group symmetry of the metal ion in the compound. Group theoretical considerations then allow one to determine the orbital splitting, wavefunction mixing and electron transition selection rules. The full structure information as well as the complete description of the chemical bonding in coordination complexes, e.g. influence of different ligands etc. are not at all described in the multiplet approach.

Density functional theory calculations include full structural models and are able to reproduce the bonding orbitals of the system very well, but do not approximate the anti-bonding and continuum orbitals 10–50 eV above the Fermi level adequately because of basis set limitations and the use of the ground-state exchange-correlation potential [e.g. 44, 45]. So far, these ab initio calculations do not account for the multiplet structure.

All theoretical approaches to calculate and analyse XAS spectra have their strengths and limitations and none is capable to account for all effects. So far, the multiplet approach has been used to interpret XPS, Kα and Kβ line as well as K and

L absorption pre-edge structures in RIXS spectroscopy. The DFT method is required for Kβ satellite lines and account well for valence to core XES and resonant valence to core RIXS, as can be seen from the literature describing these experiments (see above and e.g. [46]).

Developments in all programs are currently being pursued. For example, a new version of FEFF9.0 enables non-resonant inelastic X-ray spectroscopy as well as RIXS (core to core and valence to core) [37, 38]. It is clear that, ideally, a program is required which combines full potentials with multiplet theory, based on full structural models.

12.4 Finally

Ultimately, one wants to combine time and energy resolution in one experiment to obtain the greatest information density on your system possible. This has already been achieved in some cases. For example, ultrafast spin-state pump-probe studies have been performed [47]. A method to obtain single shot XES and time-resolved RIXS spectra is the Von Hamos spectrometer [48], which has been recently installed at the SuperXAS beamline of the SLS. It is clear that the application of these advanced high energy resolution XAS and XES techniques, very much depends on the system under investigation and the questions to be answered. Depending on the system, different techniques are more or less suitable or informative.

References

1. Many textbooks exist on catalysis. Look also at http://www.youtube.com/user/proftromp?feature¼watch
2. Koningsberger DC, Prins R (eds) (1988) For example, X-ray absorption, principles, applications and techniques of EXAFS, SEXAFS and XANES. Wiley, New York
3. Newton MA, Dent AJ, Evans J (2002) Bringing time resolution to EXAFS: recent developments and application to chemical systems. Chem Soc Rev 31:83
4. Some examples out of a large amount of literature: Bordiga S, Groppo E, Agostini G, van Bokhoven JA, Lamberti C (2013) Reactivity of surface species in heterogeneous catalysts probed by in situ X-ray absorption techniques. Chem Rev 113:1736; Evans J, Puig-Molina A, Tromp M (2007) In situ EXAFS characterization of nanoparticulate catalysts. MRS Bull 32:1038–1043; Evans J, Tromp M (2008) Interaction of small gas phase molecules with alumina supported rhodium nanoparticles: an in-situ spectroscopic study. J Phys Condens Matter 20:184020; Dent AJ, Evans J, Fiddy SG, Jyoti B, Newton MA, Tromp M (2007) Rhodium dispersion during NO/CO conversions. Angew Chem Int Ed 46 (28):5356; Dent AJ, Evans J, Fiddy SG, Jyoti B, Newton MA, Tromp M (2008) Structure-performance curriculum vitae relationships of Rh and RhPd alloy supported catalysts using combined EDE/DRIFTS/MS. Faraday Discuss 138:287
5. Some selected examples: Tromp M, Sietsma JRA, van Bokhoven JA, van Strijdonck GPF, van Haaren RJ, van der Eerden AMJ, van Leeuwen PWNM, Koningsberger DC (2003)

Deactivation processes of homogeneous Pd catalysts using in situ time resolved spectroscopic techniques. Chem Commun 7(1):128; Tromp M, van Bokhoven JA, van Haaren RJ, van Strijdonck GPF, van der Eerden AMJ, van Leeuwen PWNM, Koningsberger DC (2002) Structure-performance relations in homogeneous Pd catalysis by in situ EXAFS spectroscopy. J Am Chem Soc 124(50):14814; Smolentsev G, Guilera G, Tromp M, Pascarelli S, Soldatov A (2009) Local structures of reaction intermediates probed by time-resolved XANES spectroscopy. J Chem Phys 130:174508

6. Tromp M, van Berkel SS, van den Hoogenband A, Feiters MC, de Bruin B, Fiddy SG, van Bokhoven JA, van Leeuwen PWNM, van Strijdonck GPF, Koningsberger DC (2010) Multi-technique approach to reveal the mechanism of Cu(II) catalyzed arylation reactions. Organometallics 29:3085
7. Chen LX (2005) Probing transient molecular structures in photochemical processes using laser-initiated time-resolved X-ray absorption spectroscopy. Ann Rev Phys Chem 56:221; Chen LX (2004) Taking snapshots of photoexcited molecular structures in disordered media using pulsed X-rays. Angew Chem Int Ed 43:2886; Chen LX (2002) Excited state molecular structure determination in disordered media using laser pump/X-ray probe time-domain X-ray absorption spectroscopy. Faraday Discuss 122:315; Tromp M et al (2013) Energy dispersive XAFS: characterisation of electronically excited states of Cu(I) complexes. Phys Chem B 117 (24):7381
8. Glatzel P, Bergmann U (2005) High resolution 1s core hole X-ray spectroscopy in 3d transition metal complexes – electronic and structural information; and references therein. Coord Chem Rev 249:65
9. Brinkgreve P, Koningsberger DC (1983) Patent application NL8300927; Thulke W, Haensel R, Rabe P (1983) Versatile curved crystal spectrometer for laboratory extended X-ray absorption fine structure measurements. Rev Sci Instrum 54:277; Williams A (1983) Laboratory X-ray spectrometer for EXAFS and XANES measurements. Rev Sci Instrum 54:193; Yamashita S, Taniguchi K, Nomoto S, Yamagucchi T, Wakita H (1992) A new laboratory XAFS spectrometer for X-ray absorption spectra of light elements. X-Ray Spectrom 21:91
10. Tromp M, van Bokhoven JA, van Strijdonck GPF, van Leeuwen PWNM, Koningsberger DC, Ramaker DE (2005) Probing the molecular orbitals and charge redistribution in organometallic (PP)Pd(XX) complexes. A Pd K-edge study. J Am Chem Soc 127:777; Wilke M, Farges F, Petit PE, Brown GE, Martin F (2001) Oxidation state and coordination of Fe in minerals: an Fe K XANES spectroscopic study. Am. Mineral 86:714; DeBeer S, Randall DW, Nersissian AM, Valentine JS, Hedman B, Hodgson KO, Solomon EI (2000) X-ray absorption edge and EXAFS studies of the blue copper site in stellacyanin: effects of axial amide coordination. J Phys Chem B 104:10814; Wong J, Lytle FW, Messmer RP, Maylotte DH (1984) K-edge absorption spectra of selected vanadium compounds. Phys Rev B 30:5596
11. Koningsberger DC, Mojet BL, van Dorssen GE, Ramaker DE (2000) XAFS spectroscopy; fundamental principles and data analysis. Top Catal 10:143
12. Richwin M, Zaeper R, Lützenkirchen-Hecht D, Frahm R (2001) Piezo-QEXAFS: advances in time-resolved X-ray absorption spectroscopy. J Synchrotron Radiat 8:354
13. Frahm R, Richwin M, Lützenkirchen-Hecht D (2005) Recent advances and new applications of time-resolved x-ray absorption spectroscopy. Phys Scr T115:974
14. Matsushita T, Phizackerley RP (1981) A Fast X-ray absorption spectrometer for use with synchrotron radiation. Jpn J Appl Phys 20:2223; Flank AM, Fontaine A, Jucha A, Lemonnier M, Williams C (1982) Extended X-ray absorption fine structure in dispersive mode. J Phys Lett 43: 315
15. Bartlett SA, Wells PP, Nachtegaal M, Dent AJ, Cibin G, Reid G, Evans J, Tromp M (2011) Insights in the mechanism of selective olefin oligomerisation catalysis using stopped-flow freeze-quench techniques: a Mo K edge QEXAFS study. J Catal 284:247; Bartlett SA, Cibin G, Dent AJ, Evans J, Hanton MJ, Reid G, Tooze RP, Tromp M (2013) Sc(III) complexes with neutral N3- and SNS-donor ligands – a spectroscopic study of the activation of ethene polymerisation catalyst. Dalton Trans 42:2213–2223

16. Chen LX (2005) Probing transient molecular structures in photochemical processes using laser-initiated time-resolved X-ray absorption spectroscopy. Annu Rev Phys Chem 56:221; Chen LX (2004) Taking snapshots of photoexcited molecules in disordered media by using pulsed synchrotron X-rays. Angew Chem Int Ed 43:2886; Chen LX (2002) Excited state molecular structure determination in disordered media using laser pump/X-ray probe time-domain X-ray absorption spectroscopy. Faraday Discuss 122:315–329
17. Gawelda W et al (2007) Structural determination of a short-lived excited Iron(II) complex by picosecond X-ray absorption spectroscopy. Phys Rev Lett 98:057401
18. Chen LX et al (2003) The MLCT state structure and dynamics of a Cu(I) diimine complex characterized by pump-probe X-ray and laser spectroscopies and DFT calculations. J Am Chem Soc 125:7022; Bressler C, Chergui M (2004) Ultrafast X-ray absorption spectroscopy. Chem Rev 104:1781
19. Lockard JV et al (2010) Triplet excited state distortions in a pyrazolate bridged platinum dimer measured by X-ray transient absorption spectroscopy. J Phys Chem A 114:12780
20. Lima FA et al (2011) A high-repetition rate scheme for synchrotron-based picosecond laser pump/x-ray probe experiments on chemical and biological systems in solution. Rev Sci Instrum 82:063111
21. Tromp M et al (2013) Energy dispersive XAFS: characterisation of electronically excited states of Cu(I) complexes. J Phys Chem B 117:7381–7387
22. Martin IPS, Rehm G, Thomas C, Bartolini R (2011) Experience with low-alpha lattices at the diamond light source. Phys Rev Spec Top Accel Beams 14:040705
23. Glatzel P, Bergmann U (2005) High resolution 1s core hole X-ray spectroscopy in 3d transition metal complexes—electronic and structural information; and references therein. Coord Chem Rev 249:65; Glatzel P, Sikora M, Eeckhout SG, Safonova OV, Smolentsev G, Pirngruber G, van Bokhoven JA, Grunwaldt J-D, Tromp M (2007) Hard X-ray photon-in-photon-out spectroscopy with lifetime resolution – of XAS, XES, RIXSS and HERFD. AIP CP879:1731; Glatzel P, Weng TC, Kvashnina K, Swarbrick J, Sikora M, Galo E, Smolentsev N, Mori RA (2012) Reflections on hard X-ray photon-in/photon-out spectroscopy for electronic structure studies. J Electron Spec Relat Phenom 188:17–25
24. For example Tromp M, van Bokhoven JA, van Strijdonck GPF, van Leeuwen PWNM, Koningsberger DC, Ramaker DE (2005) Probing the molecular orbitals and charge redistribution in organometallic (PP)Pd(XX) complexes. A Pd K-edge study. J Am Chem Soc 127:777; Wilke M, Farges F, Petit PE, Brown GE, Martin F (2001) Oxidation state and coordination of Fe in minerals: an Fe K XANES spectroscopic study. Am Miner 86:714; DeBeer S, Randall DW, Nersissian AM, Valentine JS, Hedman B, Hodgson KO, Solomon EI (2000) X-ray absorption edge and EXAFS studies of the blue copper site in stellacyanin: effects of axial amide coordination. J Phys Chem B 104:10814; Wong J, Lytle FW, Messmer RP, Maylotte DH (1984) K-edge absorption spectra of selected vanadium compounds. Phys Rev B 30:5596
25. Fuggle JC, Inglesfield JE (1992) In: Fuggle JC, Inglesfield JE (eds) Unoccupied electronic states: fundamentals for XANES, EELS, IPS and BIS, vol 69, Topics in applied physics. Springer, Berlin/Heidelberg, pp 348–351, 464
26. Knop-Gericke A et al (1998) New experimental technique: X-ray absorption spectroscopy detector for in situ studies in the soft X-ray range (250 eV $\leq$ hv $\leq$ 1000 eV) under reaction conditions. Nucl Instrum Method A 406:311; Knop-Gericke A et al (2000) High-pressure low-energy XAS: a new tool for probing reacting surfaces of heterogeneous catalysts. Topics Catal 10:187; Wagner JB et al (2003) In situ electron energy loss spectroscopy studies of gas-dependent metal–support interactions in Cu/ZnO catalysts. J Phys Chem B 107:7753; Hansen PL et al (2002) Atom-resolved imaging of dynamic shape changes in supported copper nanocrystals. Science 295:2053
27. Kortright JB, Thompson AC (2001) In: Thompson AC, Vaughan D (eds) X-ray data booklet, 2nd edn. Lawrence Berkeley National Laboratory, Berkeley

28. Hämäläinen K et al (1992) Spin-dependent x-ray absorption of MnO and MnF_2. Phys Rev B 46:14274
29. Doonan CJ et al (2005) High-resolution x-ray emission spectroscopy of molybdenum compounds. Inorg Chem 44:2579
30. Smolentsev G et al (2009) X-ray emission spectroscopy to study ligand valence orbitals in Mn coordination complexes. J Am Chem Soc 131:13161–13167 ; Bergmann U, Horne CR, Collins TJ, Workman JM, Cramer SP (1999) Chemical dependence of interatomic X-ray transition energies and intensities – a study of Mn K_b'' and $K_{b2,5}$ spectra. Chem Phys Lett 302:119–124; Eeckhout SG et al (2009) Cr local environment by valence-to-core X-ray emission spectroscopy. J Anal Atom Spectrom 24:215; Delgado-Jaime MU et al (2011) Identification of a single light atom within a multinuclear metal cluster using valence-to-core X-ray emission spectroscopy. Inorg Chem 50:10709
31. Safonov VA, Vykhodtseva LN, Polukarov YM, Safonova OV, Smolentsev G, Sikora M, Eeckhout SG, Glatzel P (2006) Valence-to-core X-ray emission spectroscopy identification of carbide compounds in nanocrystalline Cr coatings deposited from Cr(III) electrolytes containing organic substances. J Phys Chem B 110:23192
32. Hämäläinen K et al (1991) Elimination of the inner-shell lifetime broadening in x-ray-absorption spectroscopy. Phys Rev Lett 67:2850; de Groot FMF et al (2002) Spectral sharpening of the Pt L edges by high-resolution x-ray emission. Phys Rev B 66:195112; van Bokhoven JA, Tromp M et al (2006) Activation of oxygen on gold-alumina catalysts: in-situ high energy resolution detection and time-resolved X-ray spectroscopy. Angew Chem Int Ed 45(28):4651
33. Safonova OV, Tromp M et al (2006) Identification of CO adsorption sites in supported Pt catalysts using high energy resolution fluorescence detection X-ray absorption spectroscopy. J Phys Chem B 110:16162; Frenkel AI, Small MW, Smith JG, Nuzzo RG, Kvashnina KO, Tromp M (2013) An in-situ study of bond strains in 1nm Pt catalysts and their sensitivities to cluster-support and cluster-adsorbate interactions. J Phys Chem C 117(44):23286–23294
34. de Groot FMF (2001) High-resolution x-ray emission and x-ray absorption spectroscopy. Chem Rev 101:1779; Butorin SM (2000) Resonant inelastic X-ray scattering as a probe of optical scale excitations in strongly electron-correlated systems: quasi-localized view. J Elec Spec 110:213; Kotani A, Shin S (2001) Resonant inelastic x-ray scattering spectra for electrons in solids. Rev Mod Phys 73:203; Gel'mukhanov F, Agren H (1999) Resonant X-ray Raman scattering. Phys Rep 312:91
35. Matsubara M et al (2002) Polarization dependence of resonant X-ray emission spectra in $3d^n$ transition metal compounds with $n = 0,1,2,3$. J Phys Soc Jpn 71:347
36. Hobbs S, Thomas RJ, Grattage J, Dhesi S, de Groot FMF, Rehr JJ, Glatzel P, Tromp M (2013) An investigation into the electronic structure of Cr systems using complementary X-ray absorption and emission techniques. J Phys Chem B, submitted for publication
37. Thomas R, Kas J, Glatzel P, De Groot FMF, Mori RA, Kavčič M, Zitnik M, Bucar K, Rehr JJ, Tromp M. Resonant Inelastic X-ray Scattering of molybdenum oxides and sulfides. J Phys Chem C, submitted for publication
38. Hobbs S, Thomas RJ, Samson J, Grattage J, Kvashnina K, Dhesi S, de Groot FMF, Rehr JJ, Glatzel P, Tromp M (2013) In situ XAS and XES techniques revealing important insights in Cr (salen) catalysed epoxidation reaction performances. J Organomet Chem, submitted for publication
39. Glatzel P, Singh J, Kvashnina KO, Van Bokhoven JA (2010) In situ characterization of the 5d density of states of Pt nanoparticles upon adsorption of CO. J Am Chem Soc 132:2555; Small MW, Kas J, Kvashnina KO, Rehr JJ, Nuzzo RG, Tromp M, Frenkel AI (2013) Effects of adsorbate coverage and bond length disorder on the d-band center in carbon-supported Pt catalysts. Angew Chem Int Ed, accepted
40. Garino C, Gallo E, Smolentsev N, Glatzel P, Gobetto R, Lamberti C, Sadler PJ, Salassa L (2012) Resonant X-ray emission spectroscopy reveals d–d ligand-field states involved in the self-assembly of a square-planar platinum complex. Phys Chem Chem Phys 14:15278

41. Gallo E et al (2013) Preference towards five-coordination in Ti Silicalite-1 upon molecular adsorption. ChemPhysChem 14:79
42. FEFF8: Ankudinov AL, Ravel B, Conradson SD (1998) Real space multiple scattering calculation of XANES. Phys Rev B 58:7565; Ankudinov AL, Bouldin C, Rehr JJ, Sims J, Hung H (2002) Parallel calculation of electron multiple scattering using Lanczos algorithms. Phys Rev B 65:104107. FEFF9: Kas JJ, Rehr JJ, Soininen JA, Glatzel P (2011) Real-space Green's function approach to resonant inelastic x-ray scattering. Phys Rev B 83(23):235114
43. de Groot FMF (2005) Multiplet effects in X-ray spectroscopy. Coord Chem Rev 249:31; de Groot F, Kotani A (2008) Core level spectroscopy of solids. Taylor & Francis/CRC Press, Boca Raton
44. Carravetta V et al (2005) StoBe Software
45. Kohn W, Sham LJ (1965) Self-consistent equations including exchange and correlation effects. Phys Rev 140:A1133; Kohn W et al (1996) Density functional theory of electronic structure. J Phys Chem 100:12974
46. Smolentsev N, Sikora M, Soldatov AV, Kvashnina KO, Glatzel P (2011) Spin-orbit sensitive hard x-ray probe of the occupied and unoccupied 5d density of states. Phys Rev B 84 (23):235113
47. Vanko G et al (2012) Spin-state studies with XES and RIXS: from static to ultrafast. J Electron Spec Relat Phenom 188:166
48. Szlachetko J, Nachtegaal M, de Boni E, Willimann M, Safonova O, Sa J, Smolentsev G, Szlachetko M, van Bokhoven JA, Dousse JCl, Hoszowska J, Kayser Y, Jagodzinski P, Bergamaschi A, Schmitt B, David C, Luecke A (2012) A von Hamos x-ray spectrometer based on a segmented-type diffraction crystal for single-shot x-ray emissionspectroscopy and time-resolved resonant inelastic x-ray scattering studies. Rev Sci Instrum 83:103105

Chapter 13
Dynamical Aspects of Biomacromolecular Multi-resolution Modelling Using the UltraScan Solution Modeler (US-SOMO) Suite

Mattia Rocco and Emre Brookes

13.1 Introduction

New and more refined 3D structures, from isolated domains to entire proteins, nucleic acids, and their complexes, are being produced at an ever increasing rate, spurred also by the contributions coming from structural genomics projects (e.g., [1, 2]; see also [3]). Notwithstanding the continuous technological advancements in the current high-resolution methods (X-ray crystallography/NMR), ample bottlenecks remains in several areas, and an exhaustive list of every relevant structure and of its complexes with every partner remains a far fetched goal. Furthermore, dynamical aspects, from local or large scale flexibility, to conformational changes following interactions/binding, to supramolecular structures formation, are not easily amenable to high-resolution analysis. A host of intermediate-resolution techniques are, however, available to complement the higher resolution data. For instance, stable, large scale complexes can be studied with cryo-electron microscopy [4] and electron tomography [5], while small-angle X-ray scattering (SAXS) and, on a slower time scale, small-angle neutron scattering (SANS) [6] are in addition capable to monitor the evolution of structural changes in solution. Since the above-mentioned techniques can provide 3D envelopes at ~10–20 Å resolution, a typical task involves placing the atomic structures of the components inside the envelope [7–9] or to optimize their arrangement to fit experimental scattering data [10].

Single-valued parameters provided by low-resolution techniques, such as the radius of gyration R_g, the translational diffusion coefficient D_t, the sedimentation coefficient s, the Stokes radius R_s, the rotational correlation time τ_c, and the intrinsic

M. Rocco (✉)
Biopolymers and Proteomics Unit, IRCCS AOU San Martino-IST, Genoa, Italy
e-mail: mattia.rocco@hsanmartino.it

E. Brookes
Department of Biochemistry, University of Texas Health Science
Center at San Antonio, San Antonio, TX, USA

J.A.K. Howard et al. (eds.), *The Future of Dynamic Structural Science*,
NATO Science for Peace and Security Series A: Chemistry and Biology,
DOI 10.1007/978-94-017-8550-1_13,

viscosity [η], can be also utilized to screen/confirm potential spatial arrangements of modules/domains, or to monitor their overall conformational changes. In its most efficient implementation, this can be done by computing the same properties from atomistic 3D models. However, this is not an easy task. The most widely used methods involve bead modelling approximations, since nearly exact expressions exist for the computation of the hydrodynamic properties of ensembles of beads of equal or different size, the latter only if they do not overlap [11, 12]. These procedures involve solving by matrix inversion a system of N equations with 3 N unknowns, and thus they rapidly become computationally intensive ($\propto N^3$). Three principal different bead modelling methods are available, all implemented in public-domain computer programs. A "grid" method, where the protein is subdivided into equally sized cubes, and beads corresponding to the each residue falling into a particular cube are then generated, was implemented in the program AtoB [13]; the resolution of the final model depends on the spacing of the initial cubic grid. A radically different approach is taken in the program Hydropro [14], where all residues are first replaced by beads having equal radii, optimized to also take into account a uniform layer of hydration, and then this "primary" model is covered with a shell of small, equal beads on which the computation of the hydrodynamic parameters is carried out. The operation is repeated a number of times decreasing the radii of the shell beads, and each parameter is then extrapolated to zero bead size. Since both methods presented several problems, most notably in the treatment of the hydration contribution, a third method was developed by the Rocco and Byron groups, originating the so called SOMO (SOlution MOdeller) approach [15]. In SOMO, each residue can be represented by a user-defined number of beads, whose initial volume is calculated from the anhydrous volume of the atoms assigned to it, plus that of the theoretically bound solvent (water) molecules [16]. In its standard implementation, for proteins two beads are used for each residue, one for the main chain peptide bond segment and one for the side chain, whose initial position is also under user control (by default, it's the centre of mass for the peptide bond, hydrophobic and ring-containing side chains, and at the end of the side chain for polar/charged side chains). An accessible surface area (ASA) screening is performed on the starting structure, and beads representing exposed side chains are placed first. The overlap between these beads are then removed, either in a sequential or synchronous manner, by first fusing together beads whose overlap exceeds a pre-set threshold ("popping"), and then reducing the radii of the remaining overlapping beads but at the same time translating by the same amount their centres outwardly along a line connecting them to the centre of mass of the protein. This effectively preserves the original hydrated surface in the resulting bead model. The exposed main chain segments beads are then placed and their overlaps reduced without outward translation (OT), and finally the beads corresponding to buried main chain segments and side chains are then placed/overlaps removed, again without OT. After a re-check for accessibility, the hydrodynamic computations are then carried out using only the non-buried beads, thus greatly reducing the computational load.

The SOMO and an improved AtoB methods have been subsequently incorporated in the analytical ultracentrifugation data analysis open source program UltraScan [17], putting them under a sophisticated GUI interface and vastly enhancing their performance and capabilities (US-SOMO; [18–20]; http://somo.uthscsa.edu/). Since its initial release, US-SOMO has been already cited in ~35 publications, and the current user base includes ~700 registered researchers and 56 registered laboratories worldwide. Recognizing the greater potential of small-angle scattering (SAS) techniques in multi-resolution modelling, an initial nucleus of a SAS data analysis and modelling module was implemented in late 2010, and has undergone a significant expansion in recent times [21]. Furthermore, a Discrete Molecular Dynamics (DMD; [22, 23]) tool running on several supercomputer clusters has been implemented, allowing the exploration of conformational space for flexible or partially disordered structures. In this brief review, the use of the US-SOMO suite with respect to dynamical aspects in multi-resolution modelling of biomacromolecules is discussed.

13.2 US-SOMO Suite Overview

In Fig. 13.1, the main US-SOMO panel is shown. The left side is divided into three parts, dealing with (a) the loading and direct processing of PDB-formatted structures, (b) bead models generation, loading and visualization functions, and

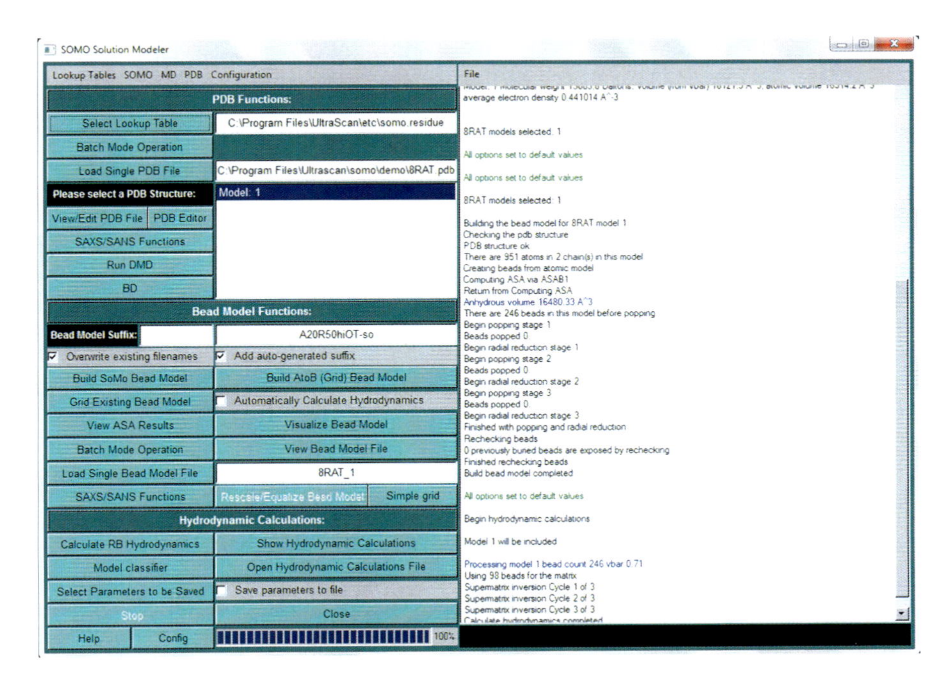

Fig. 13.1 The main US-SOMO panel

(c) hydrodynamic computations and related operations, respectively. The right side is a progress window (shown are the final stages of the bead model generation and hydrodynamic computations for the 8RAT.pdb RNase A structure).

In the PDB functions area, the reference files containing all the information necessary to properly recognize each residue and the atoms within it can be changed from the default ones by pressing "Select Lookup Tables". These files can be edited from the top bar pull-down menu ("Lookup Tables"), and the coding of each residue is a fundamental step both in the hydrodynamic bead models generation as well as for SAS computations, as they contain the atomic radii, hydration, SAXS coefficients, and atoms to beads conversion and beads positioning rules. Since atoms/residues coding and assignment to beads is not a simple operation (advanced editors are nevertheless available), skipping non-coded atoms/residues or approximate methods to represent them are provided from the "PDB" pull-down menu in the top bar. However, it is important to note that for best results all residues/atoms present in the sample on which experimental data are gathered must be also present in the structural model on which the computations are carried out. In other words, approximate location and coding are better than total absence. Loading PDB files can be done either on single or batch mode (the latter will open a new windows with advanced functions not described here). When NMR-style files are opened, either individual or multiple/all models can be selected for further operations. Upon loading, the structure is automatically visualized using RasMol ([24]; http://www.bernstein-plus-sons.com/software/rasmol/). An advanced PDB editor is also provided, including cut/splice capabilities and the possibility to extract individual models from NMR-style files of to create NMR-style files from single models. Both functions are particularly useful when using the DMD utility (see below), for instance to splice multiple conformation say of a connecting segment to one or two static domains. A panel for the SAXS/SANS functions with computations directly on the atomic structure can be also accessed from here (see below). A Brownian Dynamics (BD) module (in preparation) will be also available in the future, allowing a more reliable computation of the hydrodynamic parameters for flexible or partially disordered structures.

The Bead Model Function part allows generating either a SoMo- or AtoB-type bead model(s) from the loaded structure(s), and to visualize it again using RasMol. The ASA screening results can be also accessed from here. Bead models previously generated by US-SOMO, or coming from other sources like DAMMIN/DAMMIF [25] can be uploaded and further processed here.

In the Hydrodynamic Computation part the hydrodynamic properties in the rigid body approximation are carried out ("Calculate RB Hydrodynamics"), and a list of the more widely-used calculated parameters can be accessed by pressing "Show Hydrodynamics Calculations". A complete list is produced by default as a text file (accessible via the "Open Hydrodynamic Calculations File" button). A selected list of calculated parameters can be saved in a csv-style file for further processing, either with user-provided programs or with the built in "Model Classifier", where the results coming from multiple models can be compared and ranked against user-provided experimental parameters.

13.3 US-SOMO SAS Module

In Fig. 13.2 the US-SOMO SAS module is shown. It can perform operation both in reciprocal and real space, notably allowing the comparison through non-negative least squares (NNLS) procedures between experimental and calculated data. In reciprocal space, the latter include the computation of scattering intensities as a function of scattering vector, $I(q)$ *vs.* q ($q = 4\pi \sin(\theta)/\lambda$, with 2θ the scattering angle and λ the incident radiation wavelength), for both SAXS and SANS, starting from atomic-level structures. In case of explicit hydration (which should be externally provided, see e.g., Poitevin et al. [26]), it uses the Debye equation [27] and its variant computed with spherical harmonics [28]. For implicit hydration, Crysol [29], Cryson [30], and a fast Debye method based on the FoXS concept [31] are internally available. Guinier analyses, providing the z-average overall, cross-sectional (for rod-like molecules), and transverse (for disk-like molecules) radii of gyration ($<R_g>_z$, $<R_c>_z$, $<R_t>_z$) and the w- and w/z-averages molecular weight, mass/unit length and mass/unit area, respectively ($<M>_w$, $<M/L>_{w/z}$, and $< M/A>_{w/z}$) can be performed in manual or semi-automatic mode. A primary data reduction utility is also present. In addition, a novel module for the processing of HPLC-SAXS data was recently added. It allows to correct for spurious background intensity arising from capillary fouling, and to apply Gaussian decomposition to non-baseline-resolved SAXS peaks [21]. In real space, pair-wise distance distribution function $P(r)$ *vs.* r curves can be computed directly from atomic-level structures, and compared with

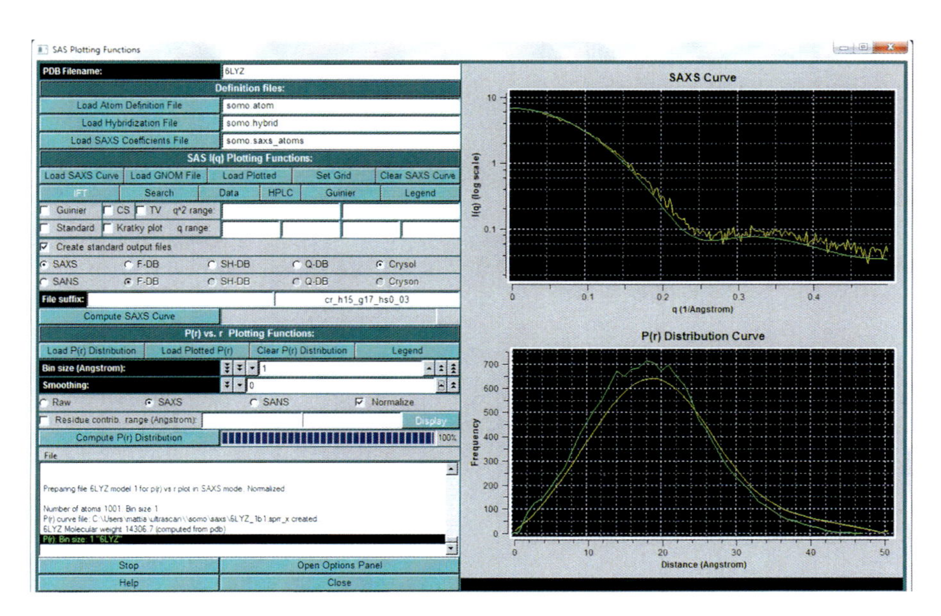

Fig. 13.2 The US-SOMO SAS module. Experimental $I(q)$ *vs.* q and its derived $P(r)$ *vs.* r curves for lysozyme (*yellow*) are compared with calculated curves for the 6LYZ.pdb structure (*green*)

data derived by inverse Fourier transformation of reciprocal-space data. To better understand the way the location of residues in a macromolecule affects the $P(r)$ *vs.* r distribution, a novel tool was developed: the structure is visualized using RasMol with its residues colour-coded according to their contribution to a particular distance range.

13.4 US-SOMO DMD Module

DMD is utilized within US-SOMO primarily to expand conformational space. Starting from an atomic structure, DMD provides a fast method of physically accurate molecular dynamics by discretizing the continuous potential functions, which alleviates the need to have explicit time steps in the differential equation time evolution. Instead, a collision-to-collision jump step occurs, where a collision is defined as a step change in a potential function. The results of DMD provide a time series (trajectory) of alternate conformations, thus enabling the researcher to screen each conformation against hydrodynamic or SAS experimental data. DMD can be run for user-specified durations and temperatures. Low temperatures are recommended for structured regions, but fully unstructured regions can be run at higher temperatures to more rapidly cover possible conformational space. Residue ranges can be defined as static during a DMD run, which means that those residues will not move. This capability allows known domains to remain unchanged while flexible linkers or other unstructured regions remain dynamic during the run. Unfortunately, multiple static regions will also remain mutually static. To move the domains relative to each other, the structure must be cut into pieces before a DMD run. A cut-and-splice tool is provided by the included PDB Editor enabling efficient rejoining of these pieces. Furthermore, presently only standard amino acids/nucleic acids bases are recognized by the available DMD software. Therefore, prosthetic groups like carbohydrates must be cut before a DMD run and spliced back afterwards. All DMD runs must be submitted to cluster facilities, as the underlying DMD code is proprietary and only available for certain computer architectures. Cluster accounts are readily available to interested researchers.

Currently, DMD in conjunction with SAXS and hydrodynamic data is used in several laboratories to study a variety of dynamical structural problems. For instance, we are applying it to generate random conformations of the unstructured N- and C-terminal regions of the human fibrinogen chains to model the molecule starting from incomplete crystal structures; it is used to model the flexible disordered arms of *B. subtilis* single stranded-DNA binding protein (D. Scott, Nottingham University, UK, personal communication); and it will be used to explore the conformational space associated with molecular flexibility, mainly from the long C-terminal helix, of the smac/diablo apoptosis regulator protein (P. Vachette, CNRS-Université Paris-Sud, Orsay, FR, personal communication).

13.5 A US-SOMO Application Case: Integrin $\alpha_{IIb}\beta_3$ Conformational States

As an example of the potential of a multi-resolution approach in determining the conformational states and transitions of a biomacromolecule, we briefly describe an overview of previous work on the integrin $\alpha_{IIb}\beta_3$ [32, 33]. Integrins are a class of heterodimeric transmembrane receptors playing fundamental roles in cell anchoring to the extracellular matrix, and in two-way signal transduction [34]. Each subunit is composed of several modules making up the large extracellular portion (ectodomain), a single pass transmembrane (TM) region, and a usually small intracellular C-terminal domain. Several crystal structures of the ectodomain of various integrins exist, and they all show a severely bent structure (reviewed in Arnaout et al. [35]; Luo et al. [36]). The N-terminal β-propeller module of the α subunit interacts with the βA module of the β subunit, forming a pocket for the binding of RGD-type ligands which is unavailable in the resting state. In the α subunit, three modules (thigh, calf-1 and calf-2) link the β-propeller to the TM helix. In the β subunit, the N-terminal starts with a PSI module, and then the chain enters a special "hybrid" module, exits forming the βA module, from which it enters again in the hybrid module. It then proceeds with four EGF-like modules and a β-terminal domain before forming the TM helix. Elegant crystallographic work with a construct containing only a few N-terminal modules of both chains has suggested that upon activation/priming the hybrid module "swings out" allowing the opening of the pocket for RGD binding [37]. This movement is believed to induce chain separation at the TM level allowing signal transduction. Furthermore, the bent conformation is believed to represent the resting state, with a transition to an extended conformation required for the integrin to become competent for binding/activation.

We have built bent and extended models of the integrin $\alpha_{IIb}\beta_3$ starting from the crystal structure of the ectodomain [38] and adding NMR-based models of the TM and cytoplasmic domains [39, 40], with either an explicit octyl-glucoside (OG) micelle or embedded into nanodiscs models [41] (see Fig. 13.3a, b).

The extended model, both in a "closed" and in an "open" conformation differing by the hybrid module swing-out, was fitted into an electron tomography-derived density map [42] obtained for a primed (activated) $\alpha_{IIb}\beta_3$ (Fig. 13.3c, d). In the open conformation, the TM helices were either kept in contact just changing their relative orientation (Fig. 13.3d), or allowed to fully separate (not shown). We have then compared the US-SOMO-calculated hydrodynamic parameters for all the conformation examined with a series of experimental data (in Table 13.1 only the *w*-average sedimentation coefficients reduced to standard conditions values, $<s^0_{20,H_2O}>_w$, are reported).

Furthermore, we have computed the $P(r)$ *vs.* r profiles for the closed/bent and closed/extended $\alpha_{IIb}\beta_3$ models and compared them with an experimentally-derived profile from SANS data collected on $\alpha_{IIb}\beta_3$ solubilized in Triton and contrast matched to render the micelle invisible [43]. As can be seen in Fig. 13.4, while neither model fits the data well, a 1:1 mixture closely approximates them.

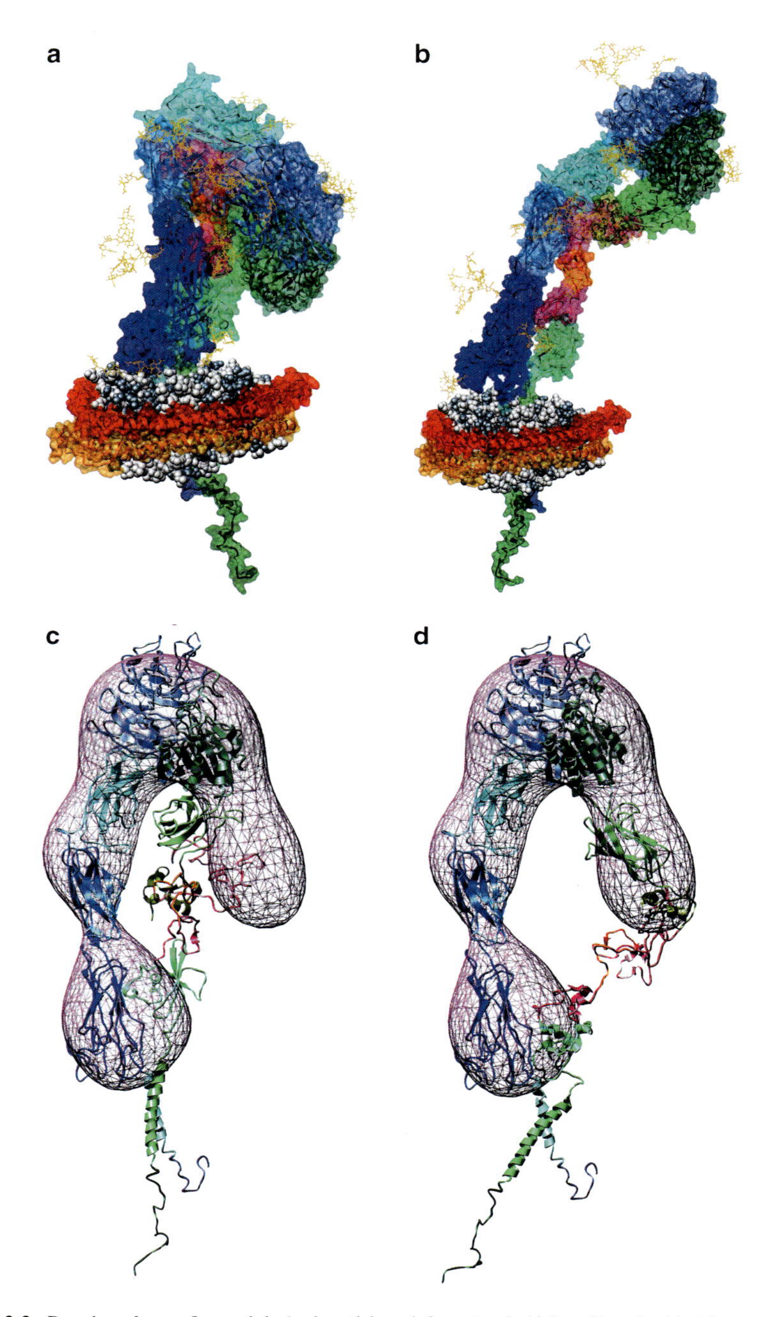

Fig. 13.3 Panels **a**, **b**, $\alpha_{IIb}\beta_3$ models (**a**, bent/closed; **b**, extended/closed) embedded in a nanodisc. Panels **c**, **d**, $\alpha_{IIb}\beta_3$ models fitted into an electron tomography-derived density map (**c**, extended/closed; **d**, extended/open-tails in contact). In all panels the α_{IIb} subunit modules are in *blue shades*, those of the β_3 subunit in *green* or *red shades*. Carbohydrates are *yellow*; they and the OG micelle are omitted for clarity in **c**, **d** (Adapted from Rosano and Rocco [33])

Table 13.1 Comparison between experimental and computed sedimentation coefficients for full-length, OG-solubilized and nanodiscs-embedded $\alpha_{IIb}\beta_3$ and the various models

Experimental/models	$<s^0_{20,H_2O}>_w$, S (% difference from resting values)
$\alpha_{IIb}\beta_3$, resting, OG-solubilized	8.18 ± 0.07 (–)
Bent/closed model	9.08 (+11.0 %)
Extended/closed model	8.20 (+0.2 %)
$\alpha_{IIb}\beta_3$, primed, OG-solubilized	7.65÷7.97 (−6.5÷−2.6 %)
Extended/open-tails crossed model	7.81 (−4.5 %)
Extended/open-tails separated model	7.41 (−9.4 %)
$\alpha_{IIb}\beta_3$, resting, nanodiscs-embedded	9.02 (–)
Bent/closed models	10.1÷10.3 (+12÷+14 %)
Extended/closed models	9.13÷9.31 (+1.2÷+3.2 %)

From Rosano and Rocco [33]

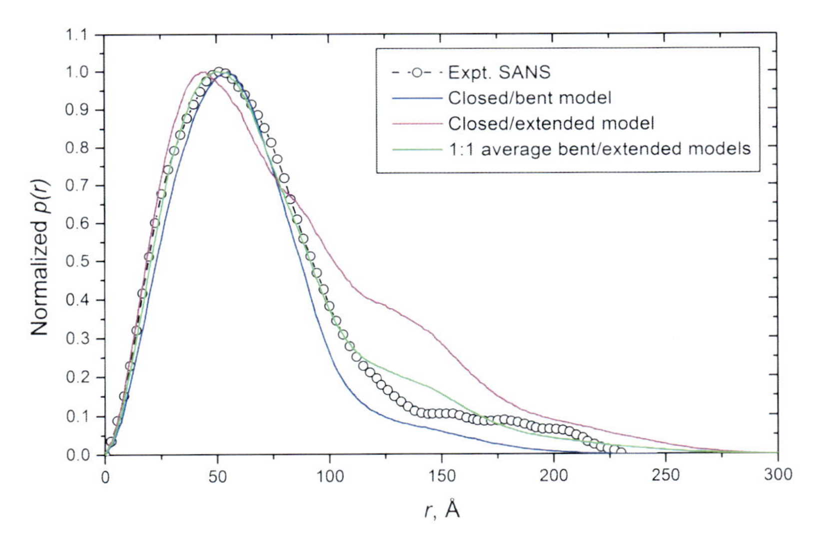

Fig. 13.4 Comparison of $P(r)$ *vs.* r profiles between SANS experimental data for $\alpha_{IIb}\beta_3$ solubilized in Triton and those calculated by US-SOMO for closed/bent and closed/extended $\alpha_{IIb}\beta_3$ models (Adapted from Rosano and Rocco [33])

From our modelling efforts, we suggested [32, 33] that: a-in a small micelle-forming detergent (OG) or nanodiscs-solubilized, the majority of resting $\alpha_{IIb}\beta_3$ appears to be in the extended conformation, while in a larger micelle (Triton) there is an almost equimolecular distribution; b-the transition appears to be snap-like, without the presence of intermediates, ruling out a flexible model in which the integrin continuously shifts from bent to extended until it is either activated from the cell interior, or is primed by a small extracellular ligand; c-this suggests that the bent conformation could be used mainly to transport the newly formed integrins to the cell surface, after which they become extended; d-interactions at the cytoplasmic domain level could be important for the transition from bent to extended; e-the activation via the hybrid domain swing-out does not necessarily immediately lead to the TM helices separation.

References

1. Burley SK, Almo SC, Bonanno JB et al (1999) Structural genomics: beyond the human genome project. Nat Genet 23:151–157
2. Todda AE, Marsdena RL, Thornton JM, Orengo CA (2005) Progress of structural genomics initiatives: an analysis of solved target structures. J Mol Biol 348:1235–1260
3. Smith JD, Clayton DA, Fields S et al (2007) Report of the Protein Structure Initiative Assessment Panel. http://www.nigms.nih.gov/News/Reports/PSIAssessmentPanel2007.htm
4. van Heel M, Gowen B, Matadeen R et al (2000) Single-particle electron cryo-microscopy: towards atomic resolution. Q Rev Biophys 33:307–369
5. McEwen BF, Marko M (2001) The emergence of electron tomography as an important tool for investigating cellular ultrastructure. J Histochem Cytochem 49:553–564
6. Svergun DI, Koch MHJ (2003) Small-angle scattering studies of biological macromolecules in solution. Rep Prog Phys 66:1735–1782
7. Wriggers W, Milligan RA, McCammon JA (1999) Situs: a package for docking crystal structures into low-resolution maps from electron microscopy. J Struct Biol 125:185–195
8. Suhre K, Navaza J, Sanejouand Y-H (2006) NORMA: a tool for flexible fitting of high-resolution protein structures into low-resolution electron-microscopy-derived density maps. Acta Cryst D62:1098–1100
9. Topf M, Lasker K, Webb B et al (2008) Protein structure fitting and refinement guided by cryo-EM density. Structure 16:295–307
10. Petoukhov MV, Svergun DI (2005) Global rigid body modeling of macromolecular complexes against small-angle scattering data. Biophys J 89:1237–1250
11. García de la Torre J, Bloomfield VA (1981) Hydrodynamic properties of complex, rigid, biological macromolecules: theory and applications. Q Rev Biophys 14:81–139
12. Spotorno B, Piccinini L, Tassara G et al (1997) BEAMS (BEAds Modelling System): a set of computer programs for the generation, the visualization and the computation of the hydrodynamic and conformational properties of bead models of proteins. Eur Biophys J 25:373–384 (Erratum 26:417)
13. Byron O (1997) Construction of hydrodynamic bead models from high-resolution X-ray crystallographic or nuclear magnetic resonance data. Biophys J 72:408–415
14. García de la Torre J, Huertas ML, Carrasco B (2000) Calculation of hydrodynamic properties of globular proteins from their atomic level structure. Biophys J 78:719–730
15. Rai N, Nöllmann M, Spotorno B et al (2005) SOMO (SOlution MOdeler): differences between X-ray and NMR-derived bead models suggest a role for side chain flexibility in protein hydrodynamics. Structure 13:723–734
16. Kuntz ID, Kauzmann W (1974) Hydration of proteins and polypeptides. In: Anfinsen CB, Edsall JT, Richards FM (eds) Advances in protein chemistry, vol 28. Academic, New York, pp 239–345
17. Demeler B (2005) UltraScan. A comprehensive data analysis software package for analytical ultracentrifugation experiments. In: Scott DJ, Harding SE, Rowe AJ (eds) Modern analytical ultracentrifugation: techniques and methods. Royal Society of Chemistry, Cambridge, pp 210–229
18. Brookes E, Demeler B, Rosano C, Rocco M (2010) The implementation of SOMO (SOlution MOdeller) in the UltraScan analytical ultracentrifugation data analysis suite: enhanced capabilities allow the reliable hydrodynamic modeling of virtually any kind of biomacromolecule. Eur Biophys J 39:423–435
19. Brookes E, Demeler B, Rocco M (2010) Developments in the US-SOMO bead modeling suite: new features in the direct residue-to-bead method, improved grid routines, and influence of accessible surface area screening. Macromol Biosci 10:746–753
20. Rocco M, Brookes E, Byron O (2013) US-SOMO: methods for construction and hydration of macromolecular hydrodynamic models. In: Gordon R (ed) Encyclopedia of biophysics. Springer, Berlin/Heidelberg, pp 2707–2714

21. Brookes E, Pérez J, Cardinali B et al (2013) Multiple species resolved after fibrinogen analysis by HPLC-SAXS followed by data processing within the UltraScan SOlution MOdeler (US SOMO) enhanced SAS module. Submitted to the J Appl Crystallogr 46:1823–1833
22. Dokholyan NV, Buldyrev SV, Stanley HE, Shaknovich EI (1998) Discrete molecular dynamics studies of the folding of a protein-like model. Fold Des 3:577–587
23. Ding F, Dokholyan NV (2006) Emergence of protein fold families through rational design. PLoS Comput Biol 2:e85
24. Sayle RA, Milner-White EJ (1995) RasMol: biomolecular graphics for all. Trends Biochem Sci 20:374–376
25. Franke D, Svergun DI (2009) DAMMIF, a program for rapid ab-initio shape determination in small-angle scattering. J Appl Cryst 42:342–346
26. Poitevin F, Orland H, Doniach S et al (2011) AquaSAXS: a web server for computation and fitting of SAXS profiles with non-uniformally hydrated atomic models. Nucl Acids Res 39: W184–W189
27. Glatter O, Kratky O (eds) (1982) Small-angle X-ray scattering. Academic, New York
28. Svergun DI, Stuhrmann HB (1991) New developments in direct shape determination from small-angle scattering 1. Theory and model calculations. Acta Cryst A47:736–744
29. Svergun DI, Barberato C, Koch MHJ (1995) CRYSOL – a program to evaluate X-ray solution scattering of biological macromolecules from atomic coordinates. J Appl Cryst 28:768–773
30. Svergun DI, Richard KMHJ et al (1998) Protein hydration in solution: experimental observation by X-ray and neutron scattering. Proc Natl Acad Sci U S A 95:2267–2272
31. Schneidman-Duhovny D, Hammel M, Sali A (2010) FoXS: a web server for rapid computation and fitting of SAXS profiles. Nucl Acids Res 38:W540–W544
32. Rocco M, Rosano C, Weisel JW et al (2008) Integrin conformational regulation: uncoupling extension/tail separation from changes in the head region by a multiresolution approach. Structure 16:954–964
33. Rosano C, Rocco M (2010) Solution properties of full-length integrin $\alpha_{IIb}\beta_3$ refined models suggest environment-dependent induction of alternative bent/extended resting states. FEBS J 277:3190–3202
34. Hynes RO (2002) Integrins: bidirectional, allosteric signalling machines. Cell 110:673–687
35. Arnaout MA, Goodman SL, Xiong J-P (2007) Structure and mechanics of integrin-based cell adhesion. Curr Opin Cell Biol 19:495–507
36. Luo BH, Carman CV, Springer TA (2007) Structural basis of integrin regulation and signaling. Annu Rev Immunol 25:619–647
37. Xiao T, Takagi J, Coller BS et al (2004) Structural basis for allostery in integrins and binding to fibrinogen-mimetic therapeutics. Nature 432:59–67
38. Zhu J, Luo B-H, Xiao T et al (2008) Structure of a complete integrin ectodomain in a physiologic resting state and activation and deactivation by applied forces. Mol Cell 32:849–861
39. Lau T-L, Kim C, Ginsberg MH, Ulmer TS (2009) The structure of the integrin $\alpha_{IIb}\beta_3$ transmembrane complex explains integrin transmembrane signaling. EMBO J 28:1351–1361
40. Yang J, Ma YQ, Page RC et al (2009) Structure of an integrin $\alpha_{IIb}\beta_3$ transmembrane cytoplasmic heterocomplex provides insight into integrin activation. Proc Natl Acad Sci U S A 106:17729–17734
41. Ye F, Hu G, Taylor D et al (2010) Recreation of the terminal events in physiological integrin activation. J Cell Biol 188:157–173
42. Iwasaki K, Mitsuoka K, Fujiyoshi Y et al (2005) Electron tomography reveals diverse conformations of integrin $\alpha_{IIb}\beta_3$ in the active state. J Struct Biol 150:259–267
43. Nogales A, García C, Pérez J et al (2010) Three-dimensional model of human platelet integrin $\alpha_{IIb}\beta_3$ in solution obtained by small angle neutron scattering. J Biol Chem 285:1023–1031

Chapter 14
Linking Diffraction, XAFS and Spectroscopic Studies on Short Lived Species

Paul R. Raithby

14.1 Background

As we have seen already different spectroscopic and diffraction methods are best suited to studying different processes across a range of timescales. Until recently, time resolved spectroscopy was the preferred option, particularly in solution, while "snapshot" single-crystal crystallographic methods have allowed macromolecular processes to be monitored on the 100 ps timescale [1]. The latter method, of course, requires single crystals of the compound to have been grown. Another method that can be applied successfully to both solution and solid-state time-resolved measurements is EXAFS (Extended X-ray Absorption Fine Structure). EXAFS is part of the more general area of X-ray Absorption Spectroscopy (XAFS) that also includes X-ray Absorption Near Edge Structure (XANES). Thus, for a given chemical or biological system we can probe the structural dynamics over the full range of timescales form picoseconds to the metastable state (Fig. 14.1).

The EXAFS technique requires the use of synchrotron radiation and the experimental method involves scanning a solid or liquid sample with a monochromatic X-ray beam across an energy range. During this process core electrons of atoms of specific elements are excited into continuum states, thus the EXAFS is a measure of the energy dependence of the absorption spectrum. In the spectrum obtained there is a step function centred at the electron binding energy of the element that has been excited. This peak is broadened by the resolution of the measurement and by the inherent lifetime of the core-hole. Above this peak the spectrum decreases monotonically with increase in energy. However, imposed on this higher energy signal is *fine structure* – the EXAFS spectrum. The simple way of thinking about this is to consider the excited photoelectron as an outwardly propagating spherical wave that is scattered by surrounding atoms, and the interference between the outgoing and

P.R. Raithby (✉)
Department of Chemistry, University of Bath, Claverton Down, BA2 7AY Bath, UK
e-mail: p.r.raithby@bath.ac.uk

J.A.K. Howard et al. (eds.), *The Future of Dynamic Structural Science*,
NATO Science for Peace and Security Series A: Chemistry and Biology,
DOI 10.1007/978-94-017-8550-1_14,

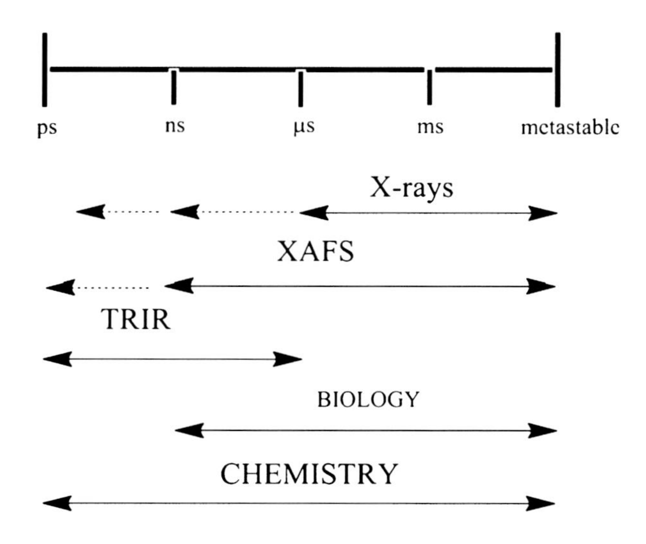

Fig. 14.1 Timescales for spectroscopy and diffraction experiments

backscattered waves causes the *fine structure*. So an EXAFS spectrum provides important information about the environment around the excited atom and is interpreted as a radial distribution function.

In the interpretation of time resolved spectroscopic, XAFS and crystallographic studies of short lifetime species it is essential to have a good theoretical understanding of the processes and of the excited state generated. Therefore, DFT studies, or other computational investigations, should be carried out in conjunction with the time resolved experiments and used in the interpretation of the results.

14.2 Early Multiple Technique Experiments on Short Lifetime Complexes

Among the first molecular species to be studied by time-resolved crystallographic methods was tetrakis(pyrophosphito)diplatinate(II), $[Pt_2(pop)_4]^{4-}$, {pop = $(H_2P_2O_5)^{2-}$} [2], the structure of which is shown in Fig. 14.2. Earlier spectroscopic analysis had indicated that a contraction of the Pt-Pt bond occurred upon photoexcitation, and this was attributed to a $5d\sigma^* \rightarrow 6p\sigma$ transition of an electron from a metal-metal antibonding orbital to one that was bonding with respect to the metal-metal bond. This $^3A_{2u}$ triplet excited state has a lifetime of 9–10 μs in solution at room temperature and exhibits a high phosphorescence quantum yield [3]. The pump-probe, time-resolved single-crystal diffraction experiment was performed on the tetramethylammonium salt, at 17 K, using an Nd/YAG laser and synchrotron radiation, where the lifetime of the excited species was 50 μs. A 2 % conversion to the excited state was observed, and the Pt-Pt bond distance was found to shorten by

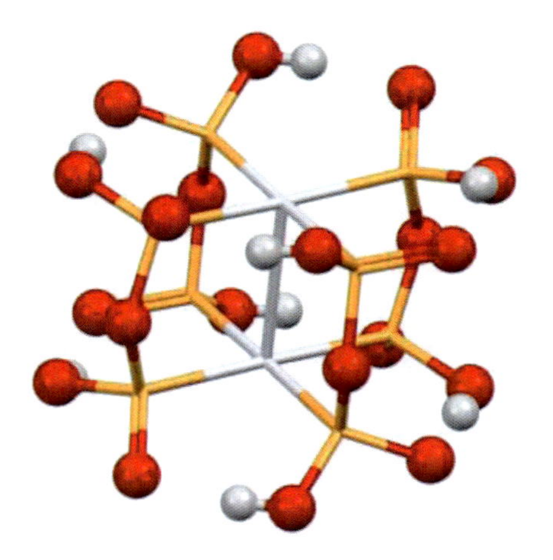

Fig. 14.2 The structure of the $[Pt_2(pop)_4]^{4-}$ anion

0.28(9) Å, compared to the ground state value of 2.9126(2) Å, and there was a 3° molecular rotation of the excited molecules within the crystal [4].

An EXAFS study assigned a shortening of 0.52(13) Å to the distance between the two planes described by the coordinated phosphorus atoms attached to each platinum centre, however, a Pt-Pt excited state distance of 2.75 Å was assumed and combined with other information from the spectrum [5]. The Pt-P distances were measured directly from the EXAFS spectrum and are reported as contracting by 0.047(11) Å.

DFT calculations using quasi-relativistic Pauli and ZORA formalisms both predict a Pt-Pt bond shortening and a slight Pt-P bond lengthening upon excitation to the lowest triplet state. The PW86LYP functional with the ZORA relativistic treatment provides molecular parameters that are in good agreement with the time-resolved crystallographic and earlier spectroscopic results, and indicate that a weak Pt-Pt covalent bonding interaction is present only in the excited state.

The dynamics of the cation $[Cu(dmp)_2]^+$ (dmp = 2,9-dimethyl-1,10-phentharoline) (Fig. 14.3) has been investigated by pump-probe, solution XAFS and femto-second optical transient spectroscopy, supported by DFT calculations [6] while a time-resolved single-crystal X-ray diffraction study has been carried out on the closely related $[Cu(dmp)(dppe)]^+$ cation [7].

For the $[Cu(dmp)_2]^+$ cation the photoexcited metal-to-ligand-charge-transfer (MLCT) state was studied. The XAFS investigation was carried out to a time resolution of 100 ps and it was established that the thermally equilibrated MLCT state had the same oxidation state as the corresponding Cu(II) complex in its ground state, and was pentacoordinate, through an interaction with the donor solvent, with an average Cu–N distance 0.04 Å shorter than in the ground state of the Cu(I) cation. Using femtosecond optical transient spectroscopy, it was also possible to follow the evolution from the photoexcited Franck-Condon MLCT state to the thermally equilibrated MLCT state. In this study three time constants were obtained: 500–700 fs, 10–20 ps and 1.6–1.7 ns. These were considered to be related to the kinetics

Fig. 14.3 The structure of the $[Cu(dmp)_2]^+$ cation

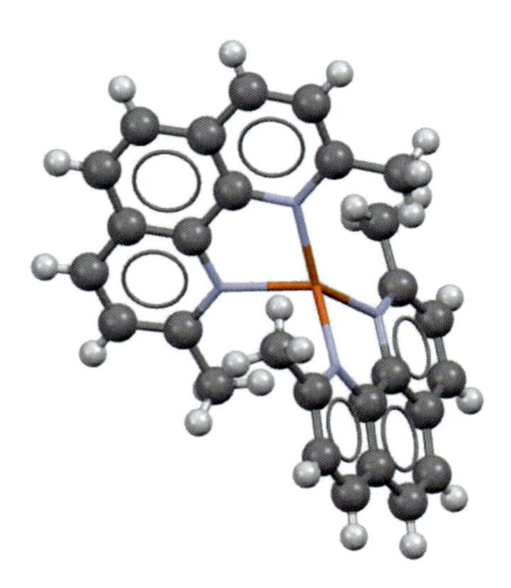

of the formation of the MLCT triplet state, structural relaxation, and the MLCT state decay to the ground state, respectively. These assignments were supported by DFT calculations.

The pump-probe, time-resolved photocrystallographic study on the triplet state of the $[PF_6]^-$ salt of the $[Cu(dmp)(dppe)]^+$ cation was carried out at 16 K using 50 ns laser pulses ($\lambda = 355$ nm). The crystal structure contained two independent cations in the asymmetric unit and at 16 K the lifetime of the excited state species is 85 μs. In the excited state the two independent molecules behave differently, with one molecule showing a significant flattening of 3.2(5)° from the ground state tetrahedral geometry while the second molecule shows no significant change. This difference is attributed to the slightly different constraints placed on the two cations in the crystal environment. However, the observed flattening is consistent with changes predicted by DFT calculations.

As is apparent from the foregoing discussion time-resolved spectroscopic and XAFS studies have generally been carried out in solution. There have, however, been attempts to carry out time-resolved infrared vibrational studies of the photodynamics of crystalline materials [8]. Photochemical investigations of two classes of molecules in the crystalline state have been undertaken using a highly sensitive time-resolved IR instrument, with the samples either suspended in a nujol mull or spin coated. The first system to be studied was $[Re(bpy)(CO)_3Cl]$ because its time-resolved solution behaviour was well known [9]. It was found that in the solid state the relaxation pathway for $[Re(bpy)(CO)_3Cl]$ (Fig. 14.4) was strongly perturbed by comparison to the solution state. A simultaneous population of intraligand and MLCT states along with a considerable reduction in the excited state lifetime was observed, and there was also a nonlinear pump power dependence. The latter feature may be attributed to triplet-triplet annihilation.

The second class of compounds are represented by 1,5-dihydroxyanthraquinone (1,5-DHAQ) (Fig. 14.4) which undergo excited state intramolecular proton transfer

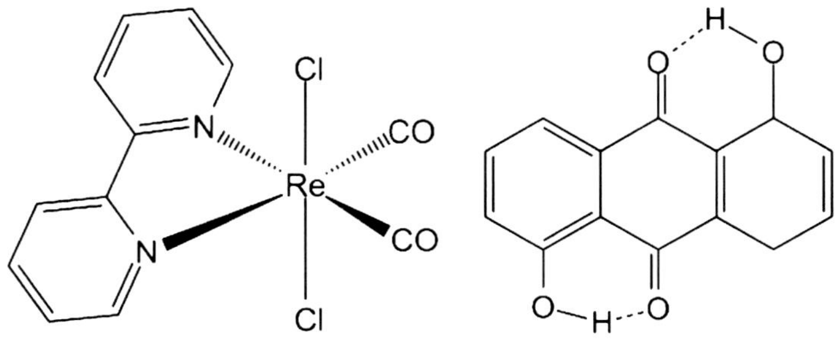

Fig. 14.4 [Re(bpy)$(CO)_3$Cl] and 1,5-DHAQ

(ESIPT). The solid state time-resolved study showed that the ESIPT process is quenched. In the solid 1,5-DHAQ undergoes sub-picosecond internal conversion back to the ground state. The results also demonstrate that thermal changes in the solid are relatively long lived and have the potential to disturb ps or ns studies.

These investigations show that the move from time-resolved solution studies to solid-state studies is not trivial, and a deeper understanding of the process involved and further developments of the spectroscopic technique are required before the method becomes more generally applicable.

14.3 Pump-Probe X-ray and Laser Spectroscopic Studies on Metal-Metal Multiply Bonded Systems

14.3.1 Introduction

An area of chemistry to which time-resolved photocrystallographic and spectroscopic studies can be applied to good effect is the investigation of metal-metal bond length changes in dimetal multiply bonded systems. The presence of the heavy metals, with the associated spin-orbit coupling, lengthens the lifetime of the triplet states through the relaxation of the selection rules that prevent intersystem crossing. The heavy metals dominate the scattering in the X-ray experiments making it relatively easy to identify small changes in metal-metal distances upon excitation. Similarly, XAFS can also be used to identify changes in bond length around heavy metal centres, by scanning the appropriate absorption edge for the given metallic element. Metal-metal bond stretches can also be identified spectroscopically and time-resolved Raman studies should be able to probe the metal-metal interactions directly while time-resolved IR can be used to study changes in metal-ligand or ligand-ligand interactions.

Coppens carried out one of the first studies on the dirhodium complex $[Rh_2(1,8\text{-diisocyano-p-menthane})_4]^{2+}$ cation [10]. At 23 K the triplet lifetime of the $[PF_6]^-$ salt of the cation is 11.7 μs in the crystalline state, and in a pump-probe

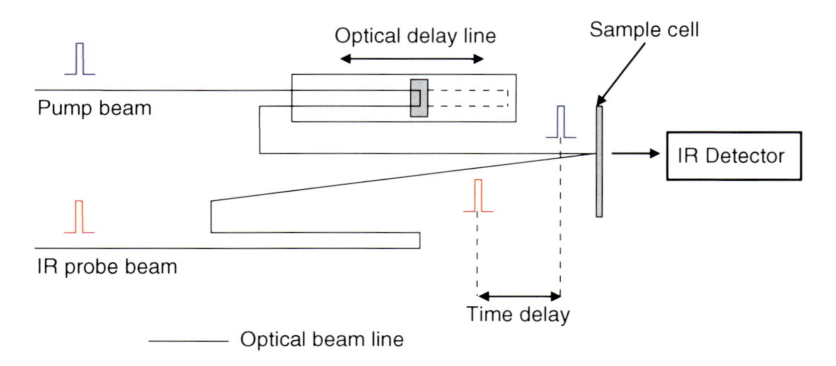

Fig. 14.5 Diagram showing the basic principle of a "pump-probe" TRIR spectrometer

experiment with synchronised laser and X-ray pulses the excited state structure was obtained in which the Rh–Rh distance reduced to 3.64 Å from 4.496(1) Å.

Following on from this we embarked on a series of time-resolved spectroscopic and crystallographic studies on molybdenum and tungsten dimers containing multiple metal-metal bonds. Upon excitation, removal of electrons originating in a $M_2\delta$-based orbital would be expected to bring about a lengthening of the M-M bond length. We are also interested in the dynamics of the change in bond lengths within ligands that bridge the electronically communicating metal fragments in molybdenum and tungsten "dimer of dimers". This will aid the greater understanding of the mechanism for electronic communication in these complexes.

Before investigating complicated bridged complexes it is important to gain an understanding of the dynamics in simple model complexes and we have therefore commenced preliminary investigations into the TRIR spectroscopy of several molybdenum carboxylate and amidinate dimers. From these studies we hope to compliment the data obtained by Chisholm et al. [11] and also that the observed excited states are long-lived enough to be applicable to time-resolved crystallographic studies. The change in stretching frequencies observed will also give information regarding the changes in bonding interactions in the excited state.

14.3.2 Technical and Experimental Information

The TRIR spectroscopy is performed using a pump-probe technique in which the sample is pumped periodically by a laser beam tuned to a certain wavelength to access the excited states. The sample is then probed using a broad band IR laser at given time delays after excitation. The exciting laser beam travels along an optical beam line containing an electronic delay system which can be moved to effect different optical path lengths and thus changing the difference between the incidence of the pump and probe beams. A simple diagram is shown in Fig. 14.5 to

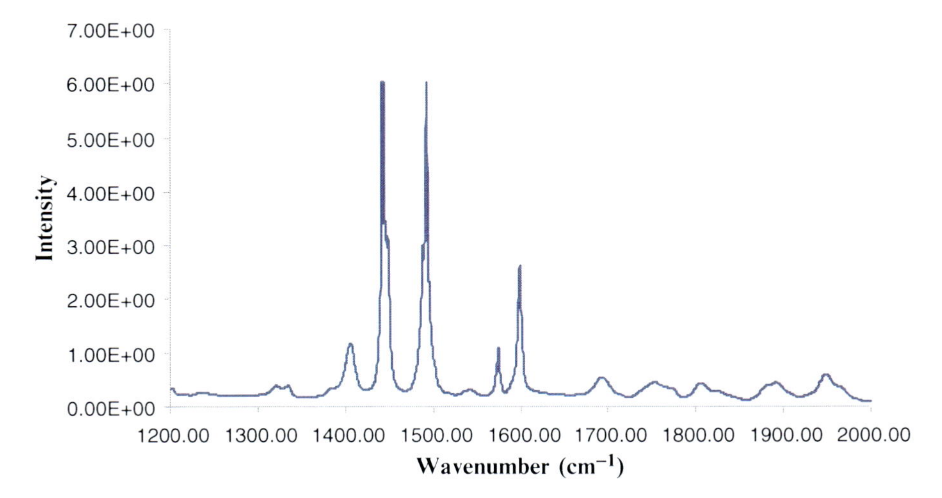

Fig. 14.6 FTIR spectrum of *cis*-stilbene

illustrate this. The IR spectrum of the sample is also measured at the ground state to allow pump on – pump off normalisation which formulates a difference spectrum showing only the IR dynamics relating to the excited state. The spectra therefore display negative bleaches due to the loss of ground state vibrations and positive transient signals corresponding to the vibrations unique to the excited state.

Both picosecond and nanosecond time delays were used to study a range of electronic dynamics. Either a titanium-sapphire (ps) or Nd/YAG (ns) laser was used as the pump. An excitation wavelength of 400 nm was implemented for the ps experiments representing the second harmonic of the titanium-sapphire laser source. For the ns experiments an excitation wavelength of 355 nm was used which is the third harmonic of the Nd-YAG laser source. Pump beam energies were tuneable between 0.1 and 10 μJ and were adjusted to maximise sample absorption without decomposition. All samples are rastered during data collection to prevent localised heating in the sample window.

The spectra obtained in these studies are recorded with two separate IR detectors which measure independent spectral windows which overlap slightly with each other. For this reason the raw data that is obtained is not referenced against wavenumber but against pixels (128 pixels for each spectrometer). The overlap in the spectral windows also means that some vibrations may be present in both windows and thus will appear twice in the raw data. To deal with this a reference sample must be used which exhibits known vibrational frequencies. Using the reference spectra the raw data from each spectrometer can be converted to wavenumber and the two spectra can be stitched together in the correct place. In our studies we have used *cis*-stillbene as reference sample. To show how the TRIR spectra appear before stitching a FTIR spectrum for *cis*-stilbene and an unstiched spectrum obtained on the TRIR setup for the same species is shown in Figs 14.6 and 14.7.

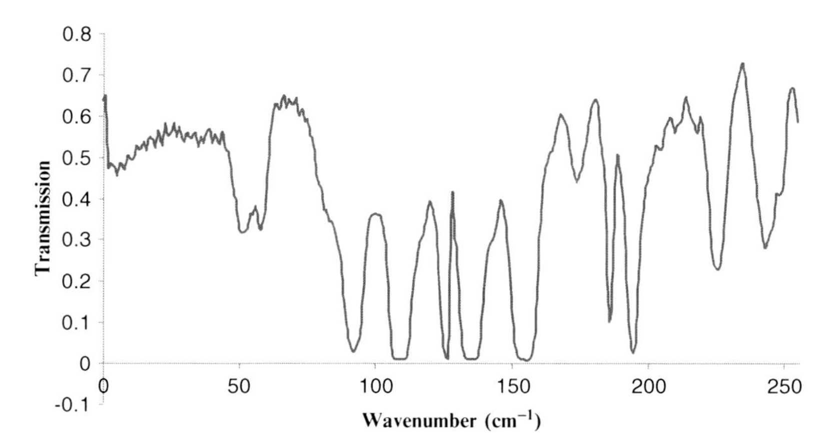

Fig. 14.7 Raw IR spectrum obtained from the TRIR setup

14.3.3 Studies on Metal Dimers

Our initial studies investigated $Mo_2(OAc)_4$ which, as one of the most fundamental quadruply bonded systems, should provide relatively simple signals in the IR spectrum. An immediate problem encountered was that $Mo_2(OAc)_4$ offers very low solubility in all common solvents. For this reason solid samples were prepared for the IR analysis. Several methods of sample preparation were trialled to try and maximise the homogeneity of the solid dispersion. Although dispersive agents such as nujol and perfluoro oils gave initially regular dispersions of the finely ground sample, after a short period of time, clustering of the particles was observed. This inhomogeneity led to anomalous changes in intensity as the probe beam passes through areas of different sample density. In addition to the uniformity of the sample, the IR absorption of the media also proved a problem allowing only a very restricted spectral window to be analysed. To avoid these problems samples were prepared without a suspending medium. This was done by applying drops of a hexane slurry of finely ground $Mo_2(OAc)_4$ onto the CaF_2 plates and allowing the solvent to evaporate. Whilst anisotropic effects were still observed in the measured spectra, the effects were significantly reduced and a greater spectral window was possible in the oil mulls.

Our first experiments looked at the ultrafast picosecond dynamics of $Mo_2(OAc)_4$ after excitation with a 400 nm wavelength pump. The spectra obtained from this study are shown in Fig. 14.8.

An initial analysis of the plot shows two well resolved transient signals at 1,350 and 1,479 cm^{-1} corresponding to excited state vibrations, and one bleach at 1,354 cm^{-1} corresponding to the decay of the ground state. It is difficult to read anything into the region between 1,390 cm^{-1} and the transient at 1,479 cm^{-1} as it appears to be composed of overlapping transient and bleach signals which have

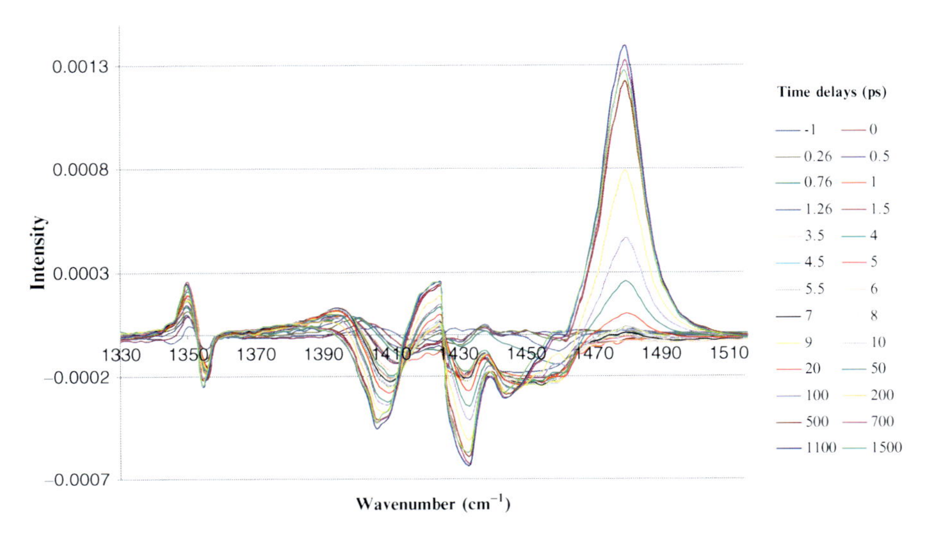

Fig. 14.8 TRIR spectra of a solid dispersion of $Mo_2(OAc)_4$ irradiated at 400 nm at a pulse energy of 0.6 μJ. Time delays are given in picoseconds

partially cancelled each other out. It is likely that this region contains a bleach signal which is coupled to the transient at 1,479 cm^{-1}. The transient and bleach pair at 1,350 and 1,354 cm^{-1} grow in over approximately 200 ps and no decay of this state is observed on the timeframe used for this experiment. The signal is also observed at t = −1 (before the pump beam arrives). This suggests that the excited state does not fully decay before the following pump beam arrives. The relative energies of the transient and bleach signals indicate that the corresponding molecular vibration is of lower energy in the excited state relative to the ground state.

The transient at 1,479 cm^{-1} grows in between 6 and 1,100 ps indicating slower population of the excited state than was observed from the signals at 1,350 and 1,354 cm^{-1}. Once again this signal shows negligible decay within the timeframe being used. The fact that no signal is observed at negative time shows that the excited state has fully decayed prior to the following pump pulse arriving. The corresponding excited state is therefore shorter lived than the excited state represented by the low wave number transient and bleach pair. This suggests that the two transients correspond to different states and the relationship between these is yet to be assigned.

To try and further explore the dynamics of the excited states that were observed on the picosecond timescale, the experiment was repeated with nanosecond time delays in order to potentially observe decay of the transients. The spectrum obtained (Fig. 14.9) unfortunately showed a lot more noise around the baseline however the bleach and transients described for the picosecond experiment were once again observed.

All signals were fully formed after the lowest time delay which was expected from the observations in the picosecond study. Using the longer timeframe we were

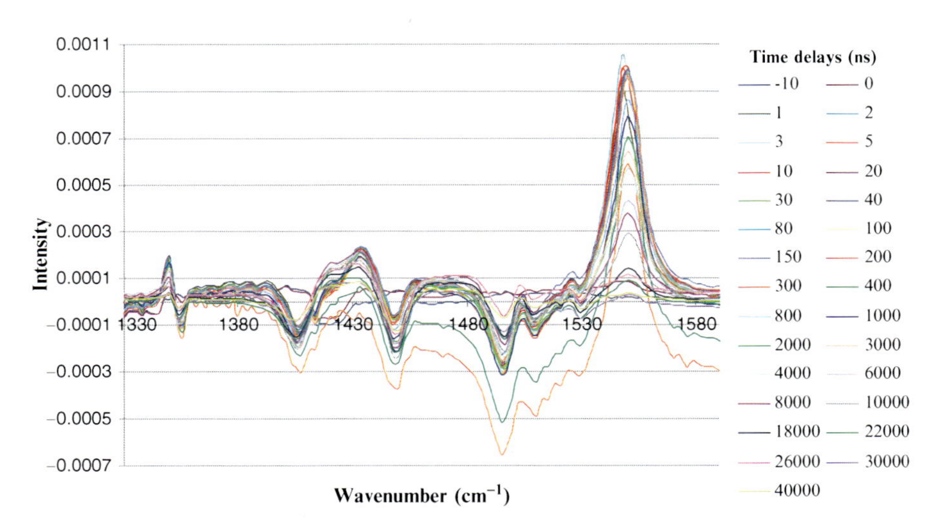

Fig. 14.9 TRIR spectra of solid $Mo_2(OAc)_4$ collected with nanosecond time delays with an excitation pulse of 355 nm and 0.4 μJ

able to observe the full decay of the intense transient at 1,550 cm^{-1} over about 50 μs and partial decay of the second transient at 1,350 cm^{-1}. The latter had not fully returned to the ground state within the time available in this study. Whilst the dynamics of this signals at 1,350 and 1,354 cm^{-1} have not yet been modelled, an exponential fit was applied to the decay of the transient at 1,550 cm^{-1}. The corresponding excited state was shown to have an approximate lifetime of 10 μs and decayed with a rate constant of 1×10^5 s^{-1}. On the grounds of the relatively long lifetime this excited state has been assigned to a triplet state and is likely to be the ^{3}MLCT state reported by Chisholm et al. [11, 12]. A comparison with the solution state arylcarboxylate molybdenum dimer data reported by Chisholm, the lifetime of $Mo_2(OAc)_4$ excited state is significantly shorter. Considering that our measurements were collected in the solid state the lifetimes for the different species are not directly comparable.

References

1. Perman B, Srajer V, Ren Z, Teng TY, Pradervand C, Ursby T, Bourgeois D, Schotte F, Wulff M, Kort R, Hellingwerf K, Moffat K (1998) Energy transduction on the nanosecond time scale: early structural events in a xanthopsin photocycle. Science 279(5358):1946–1950; Schotte F, Lim MH, Jackson TA, Smirnov AV, Soman J, Olson JS, Phillips GN, Wulff M, Anfinrud PA (2003) Watching a protein as it functions with 150-ps time-resolved X-ray crystallography. Science 300(5627):1944–1947
2. Novozhilova IV, Volkov AV, Coppens P (2003) Theoretical analysis of the triplet excited state of the $[Pt_2(H_2P_2O_5)_4]^{4-}$ ion and comparison with time-resolved x-ray and spectroscopic results. J Am Chem Soc 125(4):1079–1087

3. Steigman AE, Rice SF, Gray HB, Miskowski VM (1987) Electronic spectroscopy of D8–D8 diplatinum complexes – 1A2U(DSIGMA STAR- P-SIGMA), 3EU(DXZ, DYZ- P-SIGMA), and 3,1B2U(DSIGMA STAR- DX2–Y2) excited-states of $Pt_2(P_2O_5H_2)_4^{4}$-. Inorg Chem 26 (7):1112–1116
4. Coppens P, Novozhilova I, Kovalevsky A (2002) Photoinduced linkage isomers of transition-metal nitrosyl compounds and related complexes. Chem Rev 102(4):861–883
5. Thiel DJ, Livins P, Stern EA, Lewis A (1993) Microsecond-resolved XAFS of the triplet excited-state of $PT_2(P_2O_5H_2)_4^{4-}$. Nature 362(6415):40–43
6. Chen LX, Shaw GB, Novozhilova I, Liu T, Jennings G, Attenkofer K, Meyer GJ, Coppens P (2003) MLCT state structure and dynamics of a copper(I) diimine complex characterized by pump-probe X-ray and laser spectroscopies and DFT calculations. J Am Chem Soc 125 (23):7022–7034
7. Coppens P, Vorontsov II, Graber T, Kovalevsky AY, Chen YS, Wu G, Gembicky M, Novozhilova IV (2004) Geometry changes of a Cu(I) phenanthroline complex on photoexcitation in a confining medium by time-resolved x-ray diffraction. J Am Chem Soc 126 (19):5980–5981
8. Towrie M, Parker AW, Ronayne KL, Bowes KF, Cole JM, Raithby PR, Warren JE (2009) A time-resolved infrared vibrational spectroscopic study of the photo-dynamics of crystalline materials. Appl Spectrosc 63(1):57–65
9. George MW, Johnson FPA, Westwell JR, Hodges PM, Turner JJ (1993) Excited-state properties and reactivity of $ReCl(CO)_3$(2,2′-BIPY) (2,2′-BIPY = 2,2′-BIPYRIDYL) studied by time-resolved infrared-spectrzoscopy. J Chem Soc-Dalton Trans 19:2977–2979
10. Coppens P, Gerlits O, Vorontsov II, Kovalevsky AY, Chen YS, Graber T, Gembicky M, Novozhilova IV (2004) A very large Rh-Rh bond shortening on excitation of the $[Rh_2(1,8\text{-diisocyano-p-menthane})_4]^{2+}$ ion by time-resolved synchrotron X-ray diffraction. Chem Commun 19:2144–2145
11. Byrnes MJ, Chisholm MH, Gallucci JA, Liu Y, Ramnauth R, Turro C (2005) Observation of (MLCT)-M-1 and (MLCT)-M-3 excited states in quadruply bonded Mo_2 and W_2 complexes. J Am Chem Soc 127(49):17343–17352
12. Burdzinski GT, Ramnauth R, Chisholm MH, Gustafson TL (2006) Direct observation of a (MLCT)-M-1 state by ultrafast transient absorption spectroscopy in $Mo_2(O_2C\text{-9-anthracene})_4$. J Am Chem Soc 128(21):6776–6777

Chapter 15
Structure Analyses of Unstable Reaction Intermediates Using the Technique of Acid-Base Complex or Polymorphic Crystal Formation

Yuji Ohashi

15.1 Introduction

Although a variety of reaction intermediates in gaseous and solution states have been proposed from the spectroscopic results and synthetic work, it has been very rare that the structures of reaction intermediates have been analyzed by X-ray crystallography because they are very unstable and difficult to isolate. We found that the crystal of a cobaloxime complex gradually changed its unit cell dimensions without degradation on exposure to visible light or X-rays. From the crystal structure analyses at seven stages from the initial state to the converted state of the reaction, it was found that the chiral alkyl group bonded to the cobalt atom is racemized keeping the single crystal form [1]. Such a reaction was called crystalline-state reaction [2]. Similar examples have been observed in the crystalline-state racemization and isomerization of cobaloxime complexes [3].

The intermediate structures observed in the reaction described above, in the crystalline-state reactions, indicated to us that the unstable reaction intermediate structures which have been assumed only from spectroscopic measurement or theoretical calculation can be made clear by crystal structure analysis. We succeeded in analysing the light-induced unstable structures of lophyl radicals [4–6], carbenes [7, 8] and unstable species of salicylideneanilines [9] with retention of their single crystal forms.

However, it became clear that molecules in a crystal, in general, cannot react only by photoirradiation. It is necessary that the reactive group has enough void space for reaction or the reaction cavity has sufficient volume. How to design the size of reaction cavity is the most important factor if reaction intermediates are to be observed. One of the most promising methods is acid-base complex formation. We introduced a carboxyl group, –COOH, into the reactant molecule and tried to

Y. Ohashi (✉)
Ibaraki Quantum Beam Research Center (IQBRC),
162-1 Shirakata, Tokai, Ibaraki, 319-1106, Japan
e-mail: yujiohashi@jcom.home.ne.jp

J.A.K. Howard et al. (eds.), *The Future of Dynamic Structural Science*,
NATO Science for Peace and Security Series A: Chemistry and Biology,
DOI 10.1007/978-94-017-8550-1_15, © Springer Science+Business Media Dordrecht 2014

make acid-base complexes with the several amines which have no light absorption at wavelengths that cause the photoreaction to the reactant molecules. Some complex crystals thus obtained may have enough volume in the reaction cavity for the reactive group. Such crystals may be good candidates for the observation of the reaction intermediates. Another method is the formation of polymorphic crystals. Typical example will be shown in this article.

15.2 Structures of Nitrenes and Various Products from the Nitrenes

A reaction intermediate consisting of a phenyl group and an electrically neutral nitreno group, phenylnitrene, was proposed by Bertho about 90 years ago in the thermal or photo dimerization process of phenylazide as shown in Fig. 15.1 [10].

Among various phenylazide crystals, the crystal of 2-azidobiphenyl produced 2-biphenylnitrene after 5 h photoirradiation. The molecular structures before and after the photoirradiation is shown in Fig. 15.2. This clearly indicates that the 2-azidobiphenyl molecule is partly transformed to dinitrogen and 2-biphenylnitrene molecules. The disordered model is composed of the product and starting material and was refined with severe constraints. However, it was very difficult to discuss

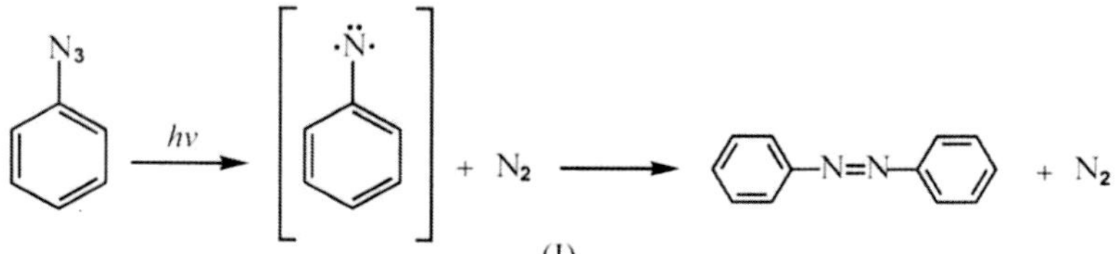

N_3 $\xrightarrow{h\nu}$ [$\cdot\ddot{N}\cdot$] + N_2 ⟶ N=N + N_2

(I)

Fig. 15.1 Proposed reaction scheme of intermediate phenylnitrene by Berto

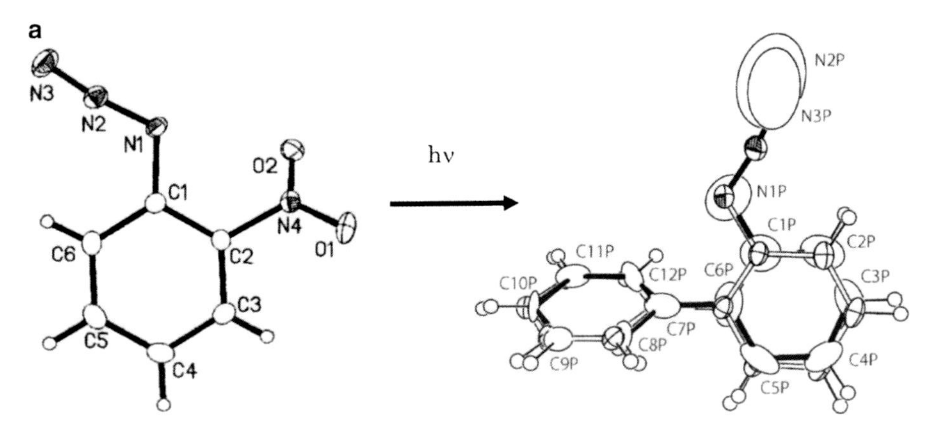

Fig. 15.2 Structural change of 2-azidobipnenyl on photoirradiation

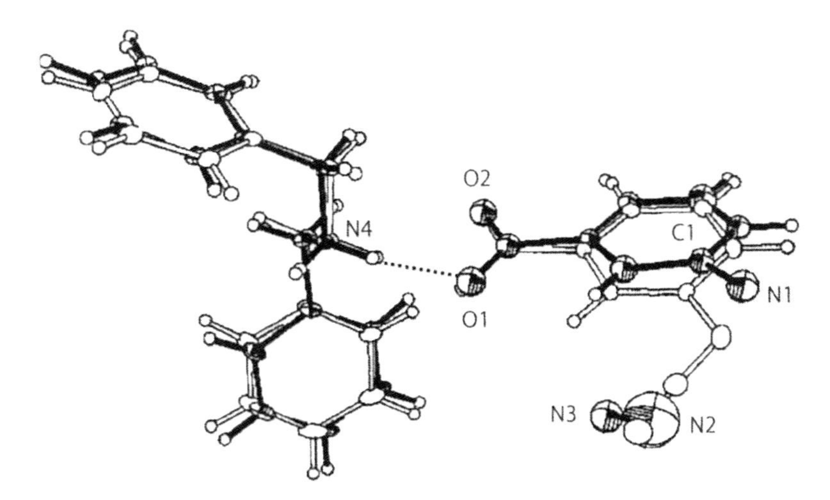

Fig. 15.3 Structure of **I** after photoirradiation

its precise structure since the heavy constraints were applied to separate the photoproducts from the overlapped azide molecule. The dinitrogen molecule is captured in the cavity around the N2 and N3 atoms [11].

Then, a variety of acid-base complex crystals were prepared, using 2-azidobenzoic acid, 3-azidobenzoic acid and 4-azidobenzoic acid as acids and dibenzylamine, N-benzyl-2-phenylethylamine and dicyclohexylamine as bases [12]. Among the combinations of acid-base complexes, suitable crystals for X-ray work were obtained for six complexes. A typical example is the complex between 3-azidobenzoic acid and dibenzylamine, which has two crystal forms **I** and **II**.

Figure 15.3 shows the molecular structure of **I** after photoirradiation. The molecules with white bonds and black bonds are the structures before and after photoirradiation. The nitrogen molecule, N2–N3, is produced from the azido group. The occupancy factor of the dinitrogen molecule converged to 0.25(2).

The intermolecular hydrogen bonds of N4...O1 and N4′...O2 fix the carboxyl group of the benzoic acid tightly to the dibenzylamines, the product 3-carboxyphenylnitrene slightly shifts from the original position of the 3-azidobenzoic acid due to the steric repulsion with the dinitrogen molecule. The C1–N1(nitrene) bond was successfully analyzed to be 1.34(4) Å.

Figure 15.4 shows the molecular structure of **II** after the photoirradiation. New residual peaks appeared in the vicinity of the azido group, which were assigned to dinitrogen molecules. Moreover, new peaks appeared between the 3-azidobenzoic acid molecules related by an inversion centre, N1P = N1P′. This indicates that two photoproduced 3-carboxyphenylnitrenes come close to each other and a dimer of 3,3′-dicarboxy-*trans*-azobenzene, is formed. This is the direct evidence that the azobenzene derivatives are produced from arylazides through arylnitrenes, which was proposed by Bertho. It seems adequate that the arylnitrene is kept intact in the crystalline lattice at low temperatures if the product nitreno groups of the neighbouring molecules are more than 5.5 Å apart, but that they are dimerized if they lie within 5.5 Å of each other.

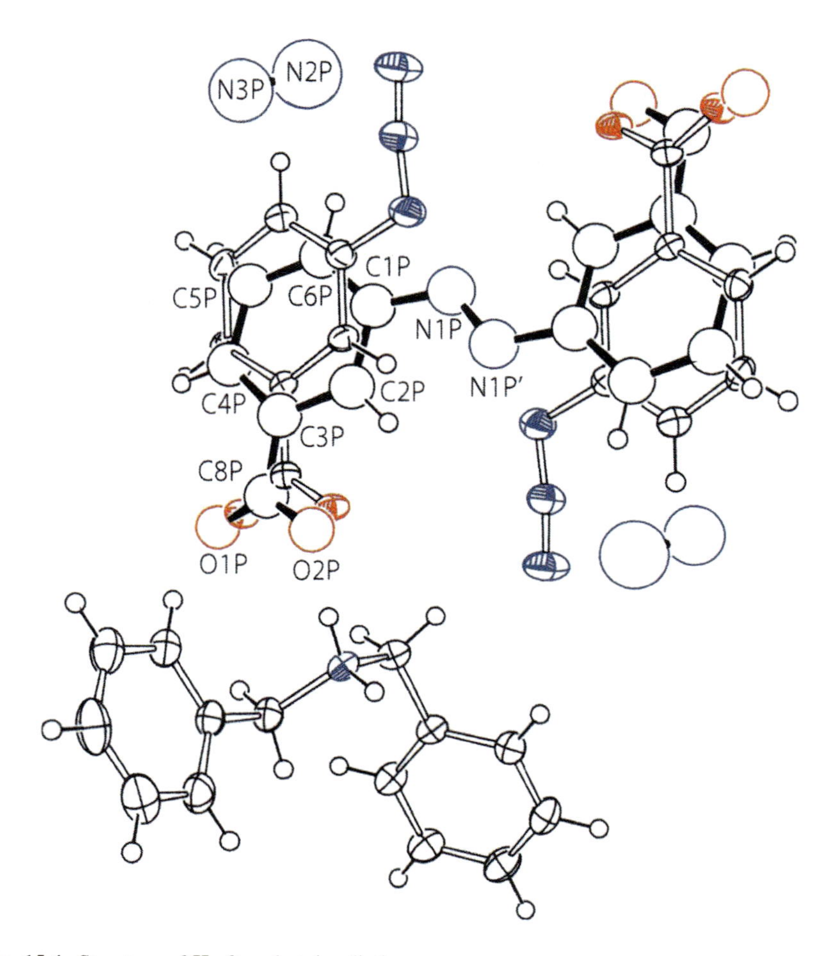

Fig. 15.4 Structure of **II** after photoirradiation

15.3 Photochromism of Salicylideneanilines and the Different Decoloration Rates

Light-induced reversible color change of substances is known as photochromism. *N*-Salicylideneanilines, which are the condensation products of salicylaldehydes and anilines, exhibit photochromism both in solution and in the solid state. Photochromic salicylideneanilines are usually pale yellow and exist in the *enol* form in crystals. It was proposed from spectroscopic data that the hydrogen atom of the hydroxyl group is translocated to the imine nitrogen atom on irradiation with ultraviolet light. A subsequent geometrical rearrangement in the excited state is assumed to form a red photoproduct. The deep red color can be erased by irradiation with visible light or by thermal fading in the dark. Although many studies have been carried out over three decades, there remains a controversy about the structure of the colored species.

The first example was selected to be *N*-3,5-di-*tert*- butylsalicylidene-3-nitroaniline, **1**, as shown in Fig. 15.5, because the photo-product of **1** was reported

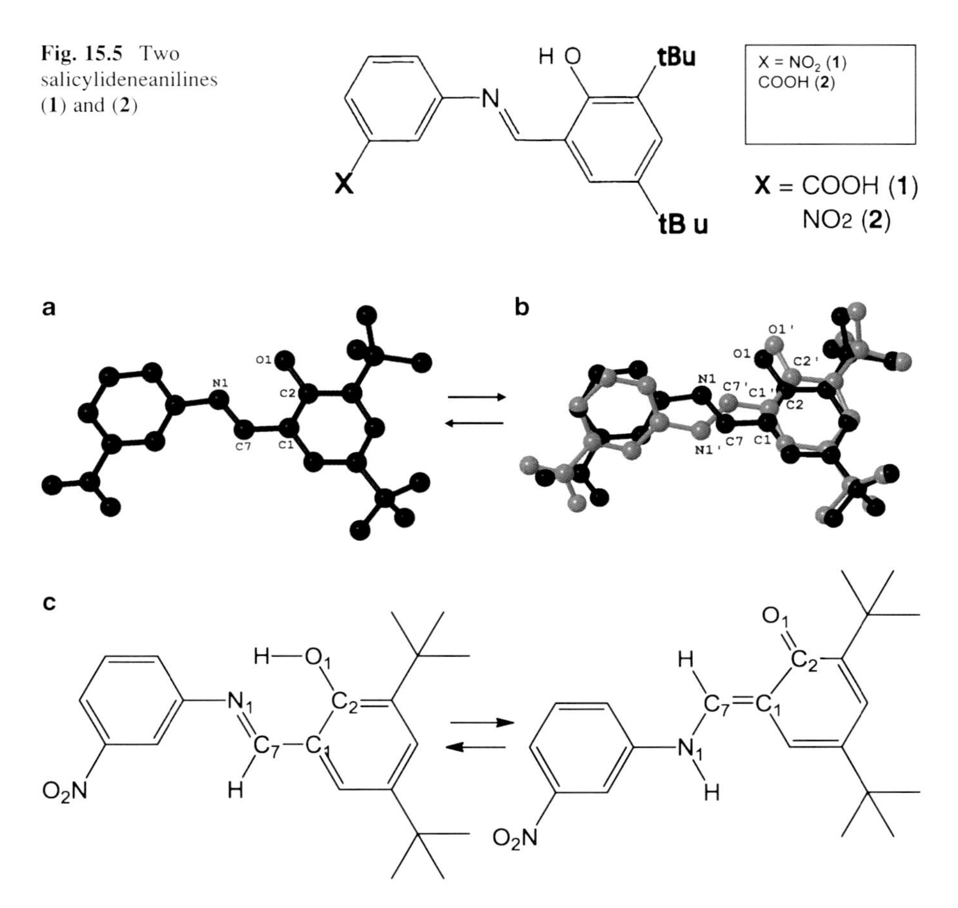

Fig. 15.5 Two salicylideneanilines (**1**) and (**2**)

Fig. 15.6 Structural change in the photochromism of **1**

to have the longest lifetime among a number of salicylideneanilines [9]. The crystal structure of **1** before photoirradiation was analyzed by X-rays at 90 K. The molecular structure is shown in Fig. 15.6a. It is clear that the molecule **1** exists in an *enol* form before photoirradiation. Using two-photon excitation techniques, the crystal was irradiated with laser light of 730 nm at room temperature. New peaks were assigned and successfully refined as another molecule which coexists with the original enol form in the crystal as shown in Fig. 15.6b. The occupancy factor of the newly produced molecule is 0.104(2). It is clear that the newly produced red species is a *trans*-keto form of **1**. When the same crystal was irradiated with a xenon lamp with a filter ($\lambda > 530$ nm) at room temperature, the dark-red color returned to the original pale yellow. The crystal structure analyzed under the same conditions is essentially the same as that of the enol form before photo-irradiation. These results clearly indicate that the photo-reversible color changes in the crystals are caused by an inter-conversion between the *cis*-enol and trans-keto forms as shown in Fig. 15.6c.

These results clearly show why photochromism occurs in the crystals of salicylideneanilines. However, it remained unclear why the metastable coloured species of

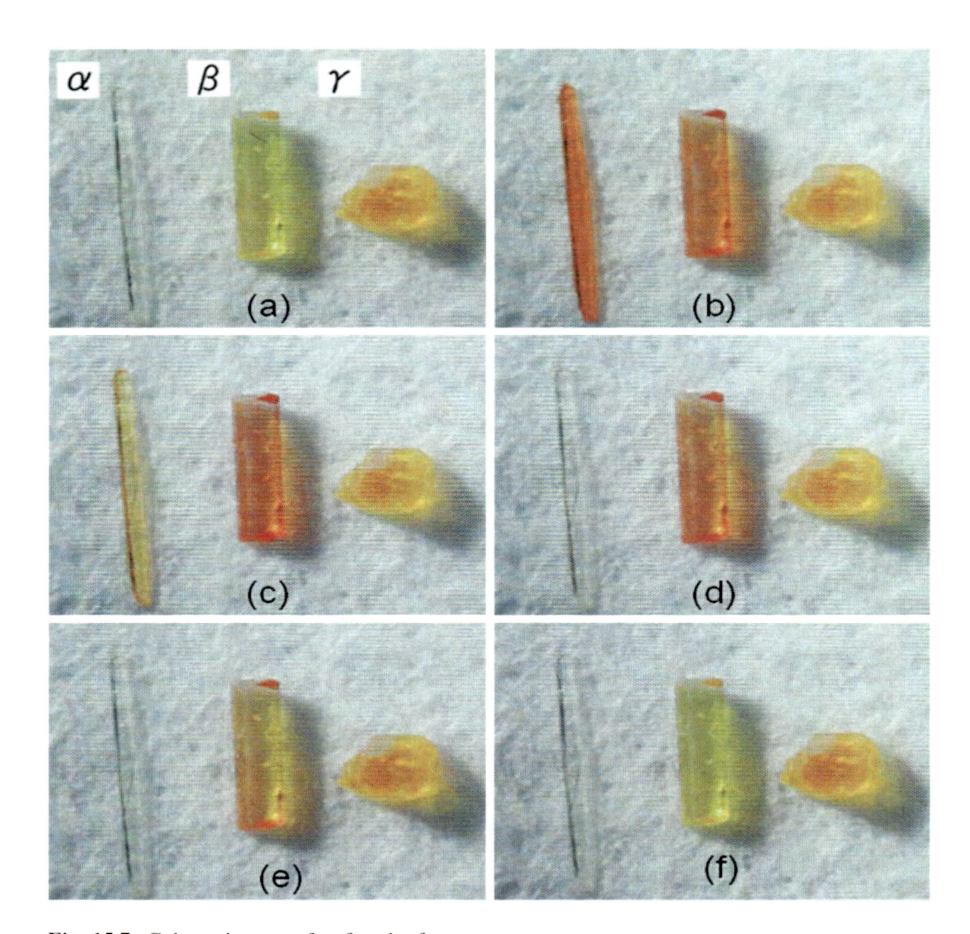

Fig. 15.7 Colour changes of α, β and γ forms

different crystals have different photochromic lifetime. Recently three polymorphic crystals, α (colorless needle), β (pale yellow parallelepiped) and γ (orange block), were found in the crystallization of *N*-3,5-di-*tert*-butylsalicylidene-3-carboxy-aniline, **2**. On exposure to a high-pressure Hg lamp for more than 10 h at room temperature, the crystals of **2α** and **2β** turned red whereas **2γ** was unchanged as shown in Fig. 15.7a, b [13]. When the three colored crystals kept in the dark at room temperature for 30, 60, 80 min and 72 h, the color of each crystal is shown in Fig. 15.7c–f, respectively. The color of **2α** faded after 60 min, whereas the crystal of **2β** showed red colour after 24 h and the colour finally faded after 72 h. However, the colour of **2γ** remained unchanged.

In order to examine the relation between the lifetime and the crystal structure, the crystal structures of the three forms before the photo-irradiation were analyzed. The molecular structures are very similar to each other, except that the torsion angle of aniline ring in **2α** is different from the other two. It was proposed that the dihedral angle between two phenyl rings of the salicylideneaniline molecule plays an important role in showing photochromic properties. Crystals with the dihedral angles of less

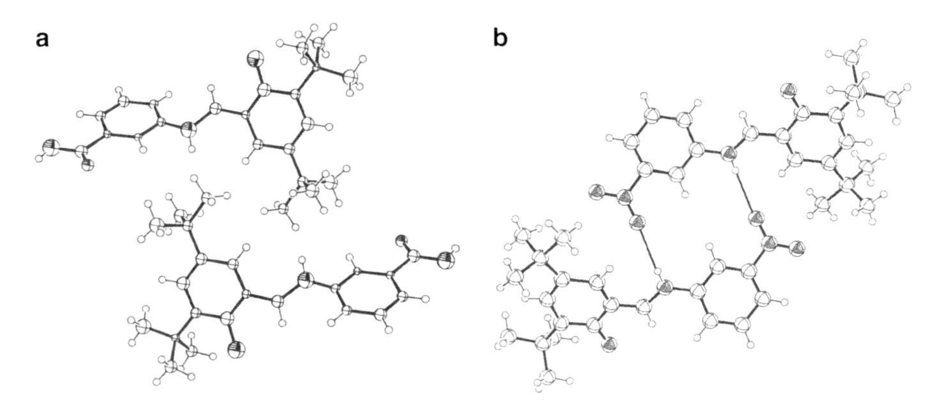

Fig. 15.8 Intermolecular interaction of the *trans-keto* forms of (**a**) **2α** and (**b**) **1**

than 20° were non-photochromic whereas those with the dihedral angles of more than 30° were photochromic. Crystals with the dihedral angle between 20° and 30° were either photochromic or non-photochromic. The dihedral angles for **2α**, **2β** and **2γ** are 60.95, 37.34 and 28.90°, respectively. This indicates that **2α** and **2β** would show photochromic property but that **2γ** may be non-photochromic. This clearly explains the photochromic properties for the three polymorphic crystals [13].

In order to examine the lifetime of the coloured species of **2α**, **2β** and **1** more quantitatively, KBr pellets (0.5 wt %) of the three crystals were prepared and were irradiated with the UV light at room temperature. The observed lifetime τ values for **2α**, **2β** and **1** are 17, 780 and 1,200 min, respectively.

Using a very thin crystal of **2α** (0.03 mm), the crystal structure after photoirradiation was analyzed. New peaks appeared which were assigned to the atoms of *trans-keto* form similar to that observed in the crystal of **1**. The occupancy factor of the *keto*-form is 0.117(2). The two photo-products of **2α** around an inversion center are shown in Fig. 15.8a. The N–H group of the imino group is surrounded by the methyl of *tert*-butyl groups of the neighbouring molecule. No hydrogen bond is formed between the two molecules. In the structure of the photo-coloured crystal of **1**, on the other hand, the two intermolecular hydrogen bonds were made between the N–H group of the *trans-keto* imine group and the oxygen atom of the nitro group around an inversion center as shown in Fig. 15.8b. It is clear that the longest lifetime of the photo-coloured species (*trans-keto* form) of **1** comes from the stabilization energy due to such hydrogen-bond formation.

For the crystal of **2β**, it was not possible to identify additional peaks in the difference electron density map after photoirradiation. It seems adequate to assume that the *trans-keto* form has essentially the same structure as those of **2α** and **1**, which are produced using the pedal motion. The molecule with the *trans-keto* form of **2β** was present instead of the enol form in the crystal of **2β** before photo-irradiation. As shown in Fig. 15.9, the N–H group of the *trans-keto* form of **2β** makes a hydrogen bond with the oxygen atom of the carboxyl group of the neighbouring molecule. This may be the reason why **2β** has a longer lifetime than **2α**. On the other hand, the dimer of **1** with the *trans-keto* form may easily be formed

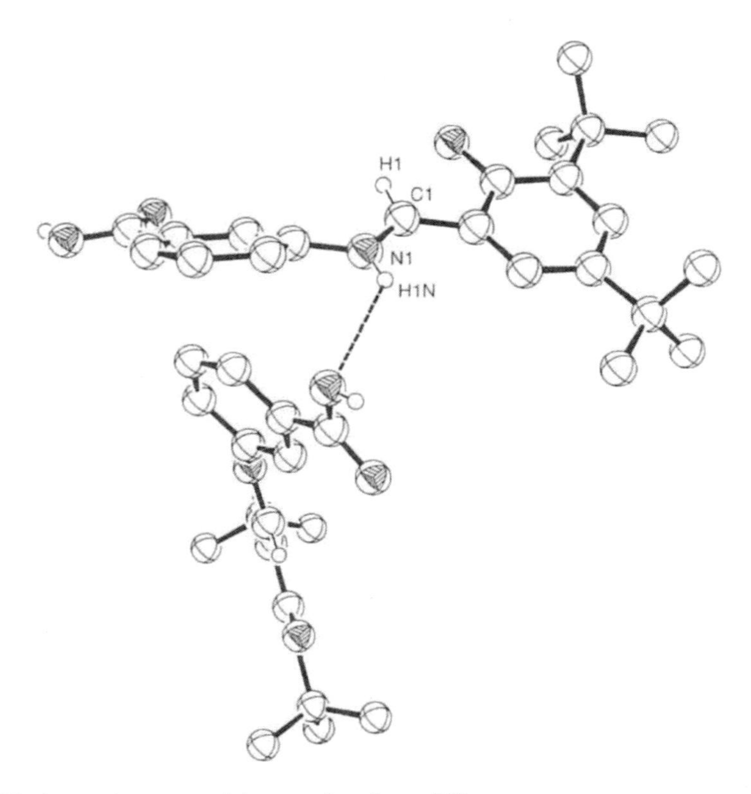

Fig. 15.9 Assumed structure of the *trans-keto* form of **2β**

and has two hydrogen bonds. This is why the lifetime of **2β** (780 min) is significantly shorter than that of **1** (1,200 min).

More interesting results were obtained when the complex crystals between the salicylideneanilines with carboxyl group and the dibenzylamine or dicyclohexylamine were examined. *N*-Salicylidene-3-carboxyaniline, **3**, makes acid-base complexes with dibenzylamine and dicyclohexylamine, whereas *N*-salicylidene-4-carboxyaniline, **4**, forms an acid-base complex with dibenzylamine. Although the crystals of **3** and **4** showed no photochromism, the complex crystals turned red when they were irradiated with UV light. The similar results to the above polymorphic crystals have been observed for the complex crystals [14].

15.4 Structural Change of Platinum Complexes at Excited State

It was reported that the diffraction peaks of $[N(C_4H_9)_4]_4[Pt_2(pop)_4]^{4-}$ (pop = diphosphite) in the X-ray powder pattern shifted to the higher angles on exposure to a xenon lamp at room temperature and returned to the original angles when the

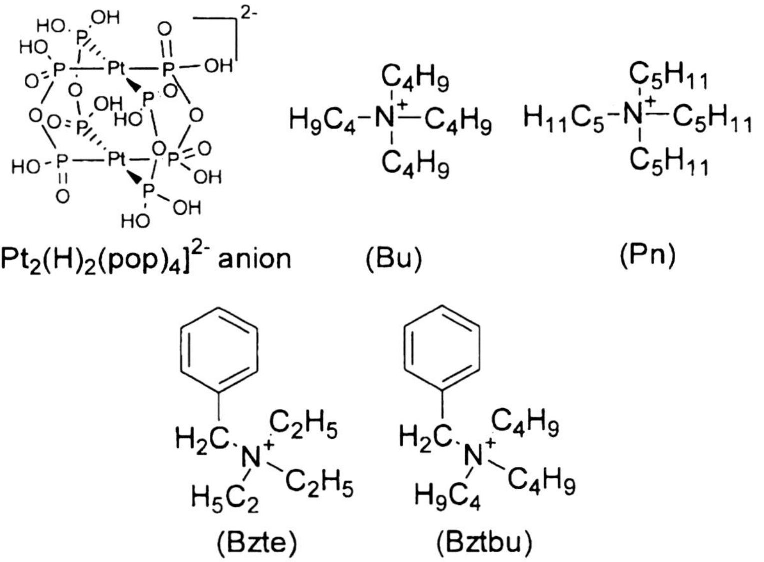

$[Pt_2(H)_2(pop)_4]^{2-}$ anion (Bu) (Pn)

(Bzte) (Bztbu)

Fig. 15.10 The Pt anion and four alkyl ammonium cations

light was turned off [15]. This indicated that the cell dimensions contracted probably due to the photo-excitation of the platinum complex in the crystal. Since the platinum complex is an anion, several different alkyl ammonium ions were selected as cations to make an acid-base complex. The platinum anion and the four alkyl ammonium cations, tetrabutyl- (Bu), tetrapentyl- (Pn), benzyltriethyl- (Bzte) and benzyltributylammoniums (Bztbu) are shown in Fig. 15.10. The crystal structure analyses of the four complex crystals revealed that the platinum anion has two extra protons. This indicated that the Pt anion should be represented as $[Pt_2(pop)_2(popH)_2]^{2-}$, not $[Pt_2(pop)_4]^{4-}$ and there are two ammonium cations per one platinum complex anion [16].

Each crystal was mounted on the diffractometer and irradiated with a xenon lamp. A combination of filters was inserted between the lamp and the crystal to select wavelengths between 400 and 550 nm. The temperature was set at 173 K for the Bu and Pn complex crystals and 103 K for the Bzte and Bztbu complex crystals. Such temperatures were selected for which the largest volume change of the unit cell was obtained before and during photo-irradiation. The changes of the unit-cell volumes for the four crystals under light-off and light-on conditions are summarized in Table 15.1.

To examine whether the unit-cell change is affected by the crystal size, the powder diffraction patterns of the Pn crystal at the light-off and light-on stages were measured. The diffraction features shift to angles greater than those with without illumination. The shifted pattern is similar to that observed for the $[N(C_4H_9)_4]_4[Pt_2(pop)_4]^{4-}$ crystal. The light-on stage indicates that the photo-excited and ground-state molecules reach an equilibrium state and that a crystalline lattice

Table 15.1 Change of unit-cell volume (Å^3) at light-off (V_{off}) and light-on (V_{on}) stages for Bu, Pn, Bzte, Bztbu complex crystals

Crystals	V_{off} (Å^3)	V_{on}(Å^3)	$\Delta(V_{on} - V_{off})$ (Å^3)	$\Delta|V_{on} - V_{off}|/V_{off}$ (%)
Bu	1,320.18(3)	1,312.16(5)	−8.02(49)	0.61
Pn	3,046.45(7)	3,026.12(17)	−20.33(12)	0.67
Bzte	2,108.78(12)	2,094.92(11)	−13.86(11)	0.66
Bztbu	2,696.42(16)	2,691.27(12)	−5.15(14)	0.19

Table 15.2 The differences in the bond distances of Pt-Pt, Δ(Pt-Pt), (Å) and Pt-P, Δ(Pt-P), (Å) between the light-off and light-on stages

	Bu	Pn	Bzte	Bztbu
Δ(Pt-Pt)	−0.0038(3)	−0.0081(3)	−0.0127(5)	−0.0031(3)
Δ(Pt-P)	−0.0069(10)	−0.0053(11)	−0.0085(14)	−0.0019(12)

corresponding to equilibrium state produced. The powder pattern was calculated based on the structural change of the single crystal. The pattern is essentially the same as the observed one. This suggests that the equilibrium structure is independent of the number of incident photons and that the number of excited molecules reaches a maximum at the equilibrium state. The structural change at the excited state may be obtained from that at the equilibrium state if the concentration of the excited molecules exceeds the threshold value.

Although there are no significant structural differences in the cation molecules between light-off and light-on stages, the bond distances in the Pt complex anions are significantly altered from the corresponding ones at the light-off stage. It may be adequate to assume that the Pt complex has D_{4h} (4/m) symmetry except for the exocyclic P=O or P–OH groups. There are significant differences in the Pt-Pt and Pt-P bonds between the light-off and light-on stages. The changes of the other bonds are within the experimental error. The differences for the four crystals are shown in Table 15.2.

The contractions of the Pt-Pt and Pt-P bond distances have ranges 0.0031 (3)–0.0127(5) Å and 0.0019(12)–0.0085 Å, respectively. Such differences are caused by the varied concentrations of excited molecules in the four crystals. The difference in the packing around the Pt complex might be responsible for the varied concentrations of the excited molecules. Among the four crystals, the bond contractions of Pn and Bzte structures are significant. The ratios of |Δ(Pt-Pt)|/|Δ (Pt-P)| for the Pn and Bzte crystals are 1.53 and 1.49, respectively, the excited molecular structures for the two crystals are essentially the same as each other.

The schematic drawing of the excited molecule is shown in Fig. 15.11. It may be possible to refine the structure of the excited molecule assuming its model structure obtained from the theoretical calculation [17]. However, the refined structure probably depends strongly on the model structure because the structural difference between ground and excited states is too small. Moreover, the excited molecular

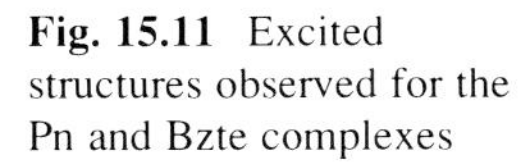
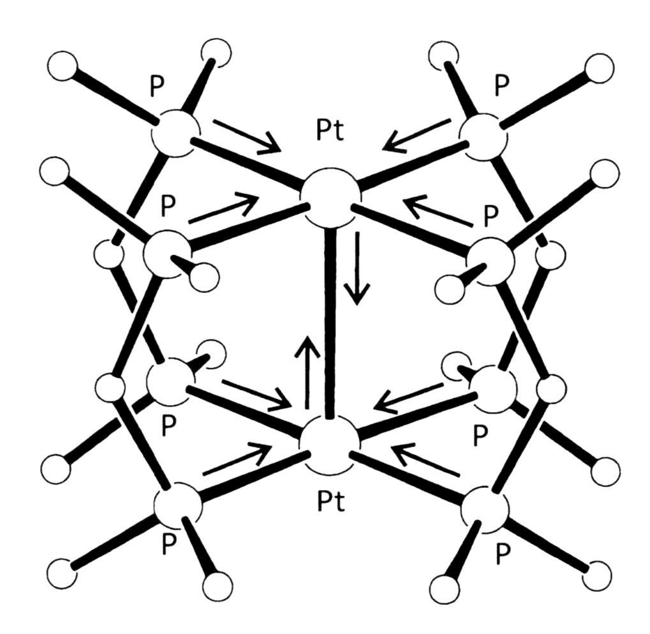

Fig. 15.11 Excited structures observed for the Pn and Bzte complexes

structure should depend significantly on the crystal environment, because the excited molecule is not in the minimum energy, different from the ground state molecule. This means that we must consider the environment effect around the excited molecule to get the quantitative lengths of arrows in Fig. 15.11 quantitatively. Recently the method of the stroboscopic structure analysis was reported [18]. The method is very useful for observing the change of the diffraction pattern within a nano or a pico second after the photo-irradiation. However, if we want to obtain the change of the electron density or crystal structure, it is very important to confirm the periodic structure in such a short time. Another question is why the method should be applied to the structural change in a reversible process, such as excited-state structure. If there are only two structures in the crystal during the light-on stage, that is, ground and excited structures, the diffraction should be brought about by the periodic structures composed of ground and excited molecules. If the concentration of the excited molecules exceeds the threshold value in the equilibrium state, the structure of the excited molecule can be analyzed by using the conventional diffraction data, no matter how quickly the excitation occurs.

Although it appears to be very difficult to observe the unstable structures by X-rays, it is not so difficult if we can make an appropriate crystal environment to keep the unstable species in a crystal, such as acid-base complex formation or polymorphic crystals. The rapid data collection is not so important since more than 1 s may be necessary before the crystal will reach the equilibrium state at room temperature. However, the precise structure analysis is most important to separate the original and produced structures. Recently such results are reviewed [19].

References

1. Ohashi Y, Sasada Y (1977) X-ray analysis of Co-C bond cleavage in the crystalline state. Nature 267:142–144
2. Ohashi Y, Yanagi K, Kurihara T, Sasada Y, Ohgo Y (1981) Crystalline state reaction of cobaloxime complexes by X-ray exposure. 1. A direct observation of Co-C bond cleavage in [(*R*)-1-cyanoethyl][(*S*)-(−)-α-methylbenzylamine]bis(dimethylglyoximato)-cobalt(III). J Am Chem Soc 103:5805–5812
3. Ohashi Y (2013) Dynamic motion and various reaction paths of cobaloxime complexes in crystalline-state photoreaction. Cryst Rev 19(Suppl 1):2–146
4. Kawano M, Sano T, Abe J, Ohashi Y (1999) The first in situ direct observation of the light-induced radical pair from a Hexaarylbiimidazolyl derivative by X-ray crystallography. J Am Chem Soc 121:8106
5. Kawano M, Ozawa Y, Matsubara K, Imabayashi H, Mitsumi M, Toriumi K, Ohashi Y (2002) Synchrotron radiation structure analyses of the light-induced radical pair of a Hexaarylbiimidazolyl derivative. Origin of the spin-multiplicity change. Chem Lett 1130–1131
6. Kawano M, Sano T, Abe J, Ohashi Y (2000) In situ observation of molecular swapping in a crystal by X-ray analysis. Chem Lett 1372–1373
7. Kawano M, Hirai Y, Tomioka H, Ohashi Y (2001) Structure analysis of a triplet carbene trapped in a crystal. J Am Chem Soc 123:6904–6908
8. Kawano M, Hirai K, Tomioka H, Ohashi Y (2007) Structure determination of triplet diphenylcarbenes by in situ X-ray crystallographic analysis. J Am Chem Soc 129:2383–1391
9. Harada J, Uekusa H, Ohashi Y (1999) X-ray analysis of structural changes in photochromic salicylideneaniline crystals. Solid-state reaction induced by Tweo-photon excitation. J Am Chem Soc 121:5809–5810
10. Bertho A (1924) Der Zerfall des Phenylazids in Benzol und in p-Xylol. Chem Ber 57:1138–1142
11. Takayama T, Mitsumori T, Kawano M, Sekine A, Uekusa H, Ohashi Y, Sugawara T (2010) Direct observation of arylnitrene formation in the photoreaction of arylazide crystals. Acta Cryst B66:639–646
12. Mitsumori T, Sekine A, Uekusa H, Ohashi Y (2010) Direct observation of various reaction pathways of arylnitrenes in different crystal environments caused by acid–base complex formation. Acta Cryst B66:647–661
13. Johmoto K, Sekine A, Uekusa H, Ohashi Y (2009) Elongated lifetime of unstable colored species by intermolecular hydrogen bond formation in photochromic crystals. Bull Chem Soc Jpn 82:50–57
14. Johmoto K, Sekine A, Uekusa H (2012) Photochromism control of salicylideneaniline derivatives by acid–base co-crystallization. Cryst Growth Des 12:4779–4786
15. Ikegawa T, Okumura T, Otsuka T, Kaizu Y (1997) Distortion of the unit cell of platinum(II) complexes under light irradiation. Chem Lett 829–830
16. Yasuda N, Uekusa H, Ohashi Y (2004) X-ray analysis of excited-state structures of the diplatinum complex anions in five crystals with different cations. Bull Chem Soc Jpn 77:933–944
17. Ozawa Y, Terashima M, Mitsumi M, Toriumi K, Yasuda Y, Uekusa H, Ohashi Y (2003) Photoexcited crystallography of diplatinum complex by multiple-exposure IP method. Chem Lett 32:62–63
18. Kim CD, Pillet S, Wu G, Fullagar K, Copens P (2002) Excited-state structure by time-resolved X-ray diffraction. Acta Cryst A58:133–137
19. Ohashi Y (2013) Direct observation of unstable reaction intermediates by acid–base complex formation. Chem Rec 13:303–325

Chapter 16
Molecular Mechanisms of Light-Harvesting and Photo-Protection by Carotenoids Explaining Configurational Selections by Antenna and Reaction Centre Complexes

Yasushi Koyama and Yoshinori Kakitani

16.1 Summary

1. *Singlet states of carotenoids:* Singlet states of carotenoids consist of the optically-allowed $1B_u^+$ and the optically-forbidden $1B_u^-$ and $2A_g^-$ states which are involved in Car → BChl singlet-energy transfer. These energies of Car decrease with the number of conjugated double bonds, n, as linear functions of $1/(2n + 1)$.
2. *Configurational selections:* The all-*trans* configuration is selected by antenna complexes, whereas the 15-*cis* configuration, by reaction centres. The Car → BChl singlet-energy transfer of $1B_u^+$-to-Q_x is always allowed, but $1B_u^-$-to-Q_x and $2A_g^-$-to-Q_y can be allowed or forbidden, depending on the length of Car conjugated chain. The Car-to-BChl singlet-energy transfer efficiencies through the three channels and the sum ($1B_u^+/1B_u^-/2A_g^-$/sum) become (48/19/22/88 %) for $n = 9$, (46/18/20/84 %) for $n = 10$ and (48/2/1–4/51–54 %) for $n = 11$.
3. *Mechanism of photo-protection by 15-cis Car ($n = 10$) bound to the RC:* After excitation of BChl triplet (^{3}P), a sequence of Car triplet species, ^{3}Car(I) → ^{3}Car(R) → ^{3}Car(II), is generated. Raman spectroscopy showed that the twisted conformation of ^{3}Car(II) around *cis* C15 = C15′, *trans* C13=C14 and *trans* C11=C12 to be (+45°, −30°, +30°), whereas EPR spectroscopy, to be (+45°, −40°, +40°). The time constants of the ^{3}Car(I) → ^{3}Car(R) and ^{3}Car(R) → ^{3}Car(II) conformational changes were determined to be 3.4 and 2.0 μs. Very fast (1.0 μs) intersystem crossing from ^{3}Car(R) plays the most important role in the triplet-energy dissipation.

Y. Koyama (✉) • Y. Kakitani
Faculty of Science and Technology, Kwansei Gakuin University, 2-1 Gakuen, Sanda, Japan
e-mail: ykoyama@kwansei.ac.jp

J.A.K. Howard et al. (eds.), *The Future of Dynamic Structural Science*,
NATO Science for Peace and Security Series A: Chemistry and Biology,
DOI 10.1007/978-94-017-8550-1_16,

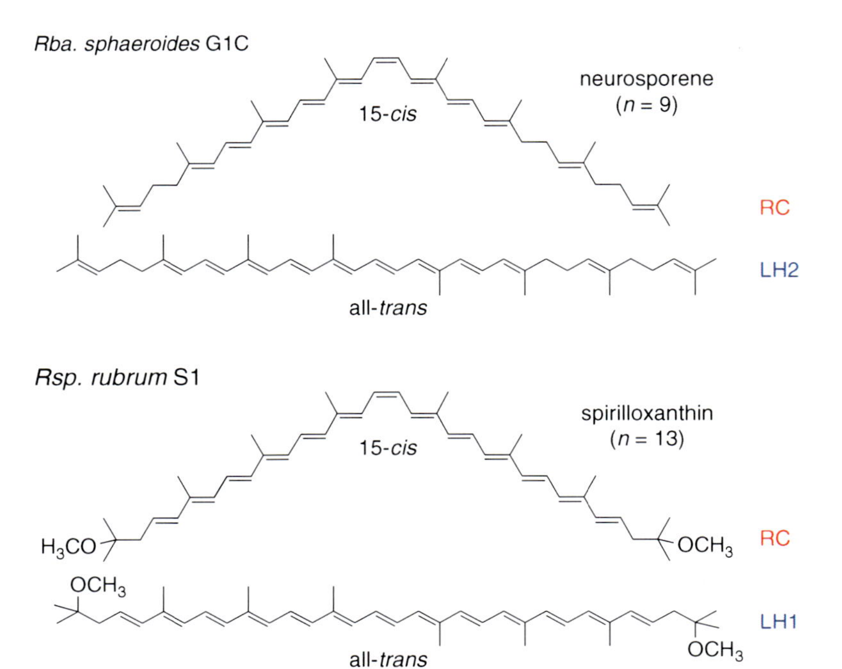

Fig. 16.1 The 15-*cis* configuration of Cars bound to the RCs and the all-*trans* configuration of Cars bound to the LH2 and LH1; in neurosporene ($n = 9$) of *Rba. sphaeroides* G1C and in spirilloxanthin ($n = 13$) of *Rsp. rubrum* S1, respectively

16.2 Configurational Selections

In both neurosporene ($n = 9$) in *Rba. sphaeroides* G1C and spirilloxanthin ($n = 13$) in *Rsp. rubrum* S1, the all-*trans* configuration is selected by the antennas (LH2 and LH1), whereas the 15-*cis* configuration is selected by the reaction centres (RCs) (Fig. 16.1).

16.3 The Mechanism of Light-Harvesting by All-*trans* Cars in LH2 and LH1 Complexes

The mechanisms of Car-to-BChl singlet-energy transfer in LH2 and LH1 complexes are illustrated in Fig. 16.2. The mechanism of light-harvesting, i.e., the absorption of light energy by Cars followed by singlet-energy transfer to BChls has been established. The importance of the all-*trans* configuration, in facilitating the plural channels of Car-to-BChl singlet-energy transfer during the stepwise internal conversion through the singlet states of different symmetries located close-by in energy is also described.

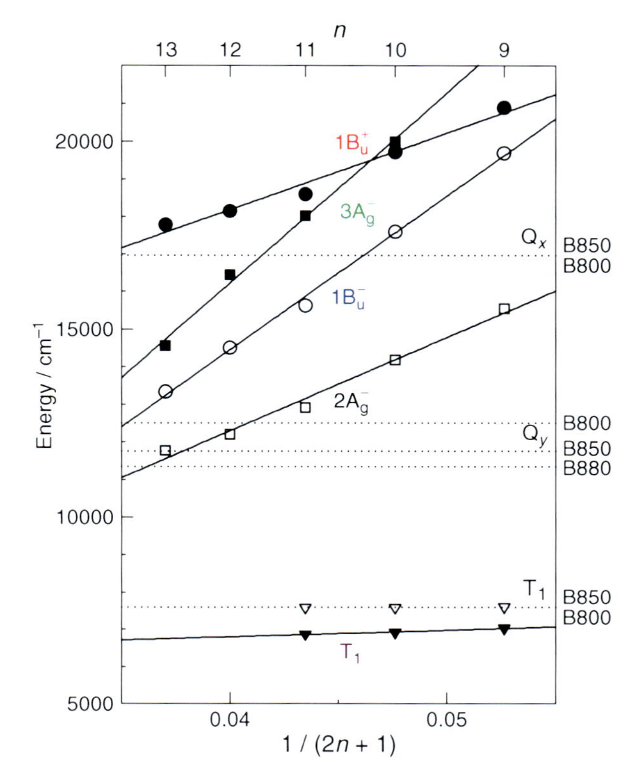

Fig. 16.2 The energies of the $1B_u^+$, $3A_g^-$, $1B_u^-$ and $2A_g^-$ singlet-excited states determined by measurement of resonance-Raman excitation profiles for crystalline all-*trans*-mini-9-β-carotene ($n = 9$), spheroidene ($n = 10$), lycopene ($n = 11$), anhydrorhodovibrin ($n = 12$) and spirilloxanthin ($n = 13$) and those of the T_1 (1^3B_u) state determined by high-sensitivity emission spectroscopy for neurosporene ($n = 9$) and spheroidene ($n = 10$) bound to the LH2 complexes from *Rba. sphaeroides* G1C and *Rba. sphaeroides* 2.4.1 and rhodopin + lycopene ($n = 11$) bound to the LH1 complex from *Rsp. molischianum*. For comparison, the energies of the Q_x and Q_y transitions in LH2 (B800 & B850) and LH1 (B880) complexes and of the T_1 states of BChls are shown in *dotted lines*

An energy diagram summarizes the energies of low-lying singlet states, i.e., $1B_u^+$, $3A_g^-$, $1B_u^-$ and $2A_g^-$, that were determined by measurement of resonance-Raman excitation profiles for crystalline Cars ($n = 9$–13) and the energies of the T_1 states of BChl *a* and Cars ($n = 9$–11) bound to the LH2 complexes that were determined by high-sensitivity phosphorescence spectroscopy. The energies of all the electronic states of Cars decrease with n, as functions of $1/(2n + 1)$. The Q_x, Q_y and T_1 energies of BChl *a* absorbing at 800 and 850 nm (called 'B800' and 'B850') in LH2 and at 880 nm ('B880') in LH1 are shown in horizontal lines.

This energy diagram predicts that the $1B_u^+ \rightarrow Q_x$, Car-to-BChl singlet-energy transfer channel is always open, but the $1B_u^-$-to-Q_x singlet-energy transfer should be closed in Cars with longer conjugated chains ($n > 10$).

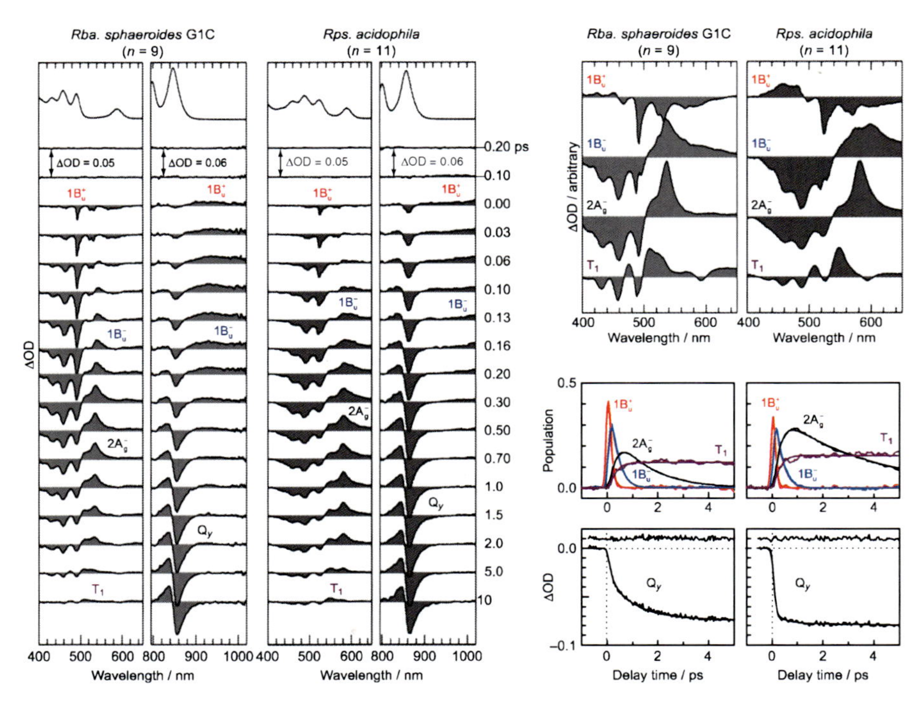

Fig. 16.3 *Left:* Sub-picosecond time-resolved stimulated-emission and transient-absorption spectra of neurosporene (n = 9) and BChl *a* in *Rba. sphaeroides* G1C (LH2) and of rhodopin glucoside (n = 11) and BChl *a* in *Rps. acidophila* (LH2) that were pumped to the $1B_u^+(0)$ level and probed with a white continuum in the visible (left-hand-side) and the near-infrared (right-hand-side) regions. *Right:* The results of the SVD and global-fitting analysis, including SADS (*upper panels*) and time-dependent changes in population in the visible and near-infrared region (*middle and bottom columns*). Only the T_1 (^3B_u) levels generated from the $1B_u^-$ state were observed. Only the two cases of *Rba. sphaeroides* G1C (n = 9) and *Rps. acidophila* (n = 11) are presented here. SVD and SADS indicate singular value decomposition and species-associated difference spectra, respectively

Taking into account the energy diagram of Cars and BChl (Fig. 16.2) and the relaxation scheme of Cars (Fig. 16.4), we are going to determine the Car → BChl singlet energy-transfer efficiencies through three channels ($1B_u^+ \rightarrow Q_x$, $1B_u^- \rightarrow Q_x$ and $2A_g^- \rightarrow Q_y$) and the efficiency of singlet-to-triplet fission within Car: Here, we choose Car (n = 9) and Car (n = 11) as the representative shorter-chain and longer-chain Cars, to explain the procedure (Fig. 16.3): (i) Sub-picosecond time-resolved, stimulated-emission and transient-absorption spectra in the visible and near-infrared regions were recorded to probe the excited-state dynamics of Cars and BChl *a*, respectively (left-hand-side). Here, the contribution of the T_2 (1^3A_g) state of Cars is also seen. (ii) As shown in Fig. 16.4, by fixing the time constants of internal conversion within Car and BChl *a* transferred from those determined in solution, the time constants of Car-to-BChl singlet-energy transfer through the

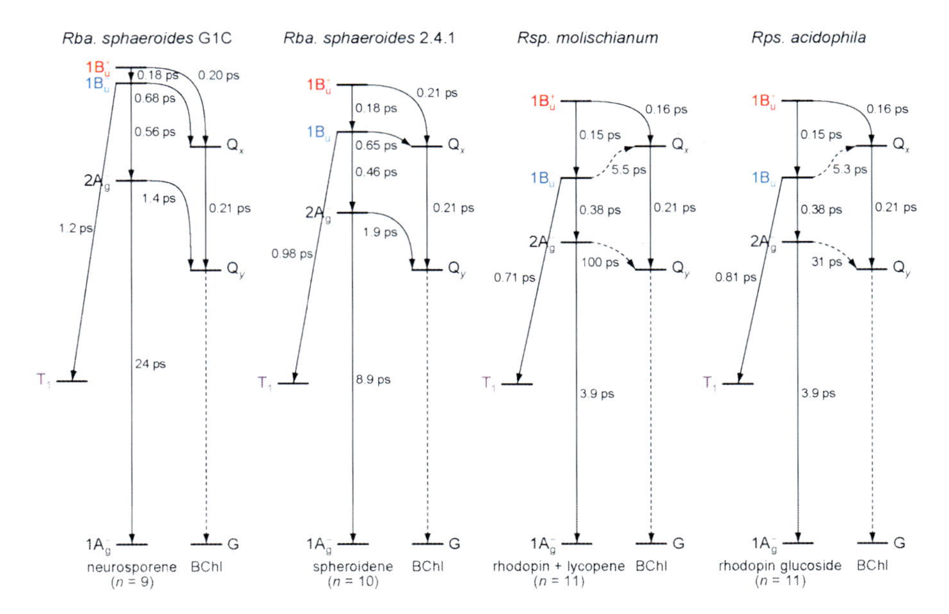

Fig. 16.4 Relaxation dynamics after excitation to the $1B_u^+(0)$ level of a set of Cars, including neurosporene ($n = 9$), spheroidene ($n = 10$), rhodopin + lycopene ($n = 11$) and rhodopin glucoside ($n = 11$), and BChl *a* are shown for *Rba. sphaeroides* G1C, *Rba. sphaeroides* 2.4.1, *Rsp. molischianum* and *Rps. acidophila*, respectively. The same type of branching in LH2 and LH1 as described in Fig. 16.3. The $1B_u^+$, $1B_u^-$, $2A_g^-$ and T_1 (1^3B_u) states of Cars are shown in *red*, *blue*, *black* and *magenta*, respectively

$1B_u^-$ and $2A_g^-$ channels and the time constant of $1B_u^-$ to T_1 (1^3B_u) transformation were determined. (iii) Again, singular-value-decomposition (SVD) followed by global fitting of spectral data matrices was used, the results of which are shown in Fig. 16.3 (right-hand-side). The spectral patterns of three singlet-excited and T_1-excited states of Cars are reproduced as species-associated difference spectra (SADS), whereas the branching relaxation processes in Car, as time-dependent changes in population.

Figure 16.4 compares the time constants of singlet-energy transfer through the $1B_u^+ \rightarrow Q_x$, $1B_u^- \rightarrow Q_x$ and $2A_g^- \rightarrow Q_y$ channels and those of the $1B_u^- \rightarrow T_1$ transformation, among neurosporene ($n = 9$) in *Rba. sphaeroides* G1C, spheroidene ($n = 10$) in *Rba. sphaeroides* 2.4.1, rhodopin & lycopene ($n = 11$) in *Rsp. molischianum* and rhodopin glucoside ($n = 11$) in *Rps. acidophila* 10050. They can be characterized as follows: (i) The $1B_u^+$ channel is open in all the cases and the energy transfer speeds up only slightly on going from $n = 10$ to $n = 11$. (ii) The $1B_u^-$ channel is open in $n = 9$ and 10, but it gets practically closed in $n = 11$, because it becomes an up-hill energy transfer. (iii) The $2A_g^-$ channel slows down on going from $n = 9$ to $n = 10$. This channel practically becomes closed in $n = 11$, even though this channel can apparently perform down-hill energy transfer.

Table 16.1 Efficiencies (Φ in %) of Car-to-BChl singlet-energy transfer through the $1B_u^+$, $1B_u^-$ and $2A_g^-$ channels and those of the $1B_u^-$-to-T_1 singlet-to-triplet conversion in Cars as determined by the SVD and global-fitting analyses of sub-picosecond time-resolved, spectral data matrices. The overall efficiencies, determined by comparison of the fluorescence-excitation and electronic-absorption spectra, Φ(obs), are also shown

Carotenoids (n)	$\Phi(1B_u^+)$	$\Phi(1B_u^-)$	$\Phi(2A_g^-)$	Φ(sum)	$\Phi(T_1)$	Φ(obs)
(a) Native LH2 complexes						
neurosporene (9)[a]	48	19	22	88	10	92
spheroidene (10)[b]	46	18	20	84	12	89
rhodopin + lycopene (11)[c]	48	2	1	51	17	53
rhodopin glucoside (11)[d]	48	2	4	54	16	56
(b) Reconstituted LH1 complexes[e]						
neurosporene (9)	47	11	20	78	19	
spheroidene (10)	47	9	19	75	20	
lycopene (11)	41	2	3	46	27	
anhydrorhodovibrin (12)	39	0	1	40	29	
spirilloxanthin (13)	35	0	1	36	31	

[a] *Rba. sphaeroides* G1C
[b] *Rba. sphaeroides* 2.4.1
[c] *Rsp. molischianum*
[d] *Rps. acidophila*
[e] Incorporated into the LH1 complex from *Rsp. rubrum* G9

As a result, the Car to BChl singlet-energy transfer efficiencies through $1B_u^+$/$1B_u^-$/$2A_g^-$/sum are 48/19/22/88 % for neurosporene, 46/18/20/84 % for spheroidene, 48/2/1/51 % for rhodopin + lycopene and 48/2/4/54 % for rhodopin glucoside (see Table 16.1).

Since the set of time constants predicts a sudden decrease in the overall efficiency of the Car-to-BChl singlet-energy transfer on going from $n = 10$ to $n = 11$, we tried to determine these values by an independent experiment. This experiment is based on the comparison of fluorescence-excitation spectrum to the electronic-absorption spectrum of each Car bound to each antenna complex, normalizing at the Q_x fluorescence/absorption peaks. The overall energy-transfer efficiencies were determined to be 92, 89 and 53–56 %, for the bound Cars ($n = 9$, 10 and 11, respectively). The set of values seem to nicely correspond to the closing of both the $1B_u^- \rightarrow Q_x$ and $2A_g^- \rightarrow Q_y$ channels on going from $n = 10$ to $n = 11$. The results of both experiments are listed in Table 16.1 in the last column.

16.4 The Mechanism of Photo-Protective Function by 15-*cis*-Spheroidene Bound to the RC from *Rba. sphaeroides* 2.4.1

Sets of stretching force constants were determined for the C=C and C–C bonds in the conjugated chain. Changes in the bond orders in the triplet-excited region are now evaluated by the use of stretching force constants as a scale: Double bonds become more single bond-like, whereas single bonds become more double bond-like. Note that the centre of the triplet-excited region is located at the centre of *the conjugated chain*, i.e., C12–C13 in spheroidene.

Then, we proceeded to the case of RC-bound 15-*cis*-spheroidene: By the use of the set of force constants determined above for all-*trans*-spheroidene, we tried to determine the rotational angles around the central double bonds by means of test calculations, changing the rotational angles by a 5° interval.

(a) The S_0 Raman line at 1,239 cm^{-1} shifts to the T_1 Raman line at 1,335 cm^{-1}, while the S_0 Raman line at 954 cm^{-1} shifts to the T_1 Raman line at 935 cm^{-1}, when the C15=C15′ bond is rotated by 45°. This is due to the decoupling of the C15–H and C15′–H in-plane bending from the C15=C15′ stretching in the former, and the decoupling of the C15–H and C15′–H our-of-plane waggings from the C15=C15′ torsion in the latter.
(b) The closely overlapped S_0 Raman lines at 1,534 and 1,532 cm^{-1} widely split into T_1 Raman lines at 1,526 and 1,505 cm^{-1}. Such a large down shift of the C13=C14 stretching Raman line from 1,532 to 1,505 cm^{-1} could be realized, when the C13=C14 bond is twisted by 30°.
(c) In the S_0 state, the C14–C15 and the C10–C11 stretches appear at 1,168 and 1,159 cm^{-1}, whereas in the T_1 state, they appear at 1,156 and 1,182 cm^{-1}, respectively, with the reversed order in frequency. Twisting around the C11=C12 bond by 30° caused the decoupling among the C10–C11 stretching and the C11–H and C12–H bendings and, as a result, the high-frequency shift of the C10–C11 stretching to 1,182 cm^{-1} took place.

Thus, the rotational angles around the *cis* C15=C15′, *trans* C13=C14 and *trans* C11=C12 bonds have been determined to be (+45°, −30°, +30°); the set of signs (+, −, +) was chosen to fit the model into the peptide binding pocket of the Car. Thus, the structure of the RC-bound T_1 Car as well as the changes in Raman frequencies can also be rationalized.

After fixing the rotational angles to the above set of values, the relevant stretching force constants and the cross terms were re-adjusted by the least-square fitting.

Sub-microsecond time-resolved EPR spectra of the RC from *Rba. sphaeroides* 2.4.1 recorded at 100 K were obtained. Immediately after excitation (0.0 μs) at 532 nm, the EPR signal of ^{3}P (BChl pair) appeared. It was followed by three different T_1 Car species, which we named '^{3}Car(I)', '^{3}Car(R)' and'^{3}Car(II)'. We applied SVD followed by global fitting to the data matrices. It was necessary to add the intermediate, ^{3}Car(R), to obtain a set of three SAS for the triplet Cars with similar intensities, showing a reasonably good fit between the observed and the simulated time-dependent changes in population (see each pair of smooth and noisy lines in Fig. 16.5).

The conformation of each intermediate can be determined by the zero-field splitting parameters, $|D|$ and $|E|$, read from the pattern of EPR signal obtained as SAS. Briefly, $|D|$ and $|E|$ values represent the deviation of the molecular structure from the spherical and spheroidal symmetry, respectively. The relative $|D|$ and $|E|$

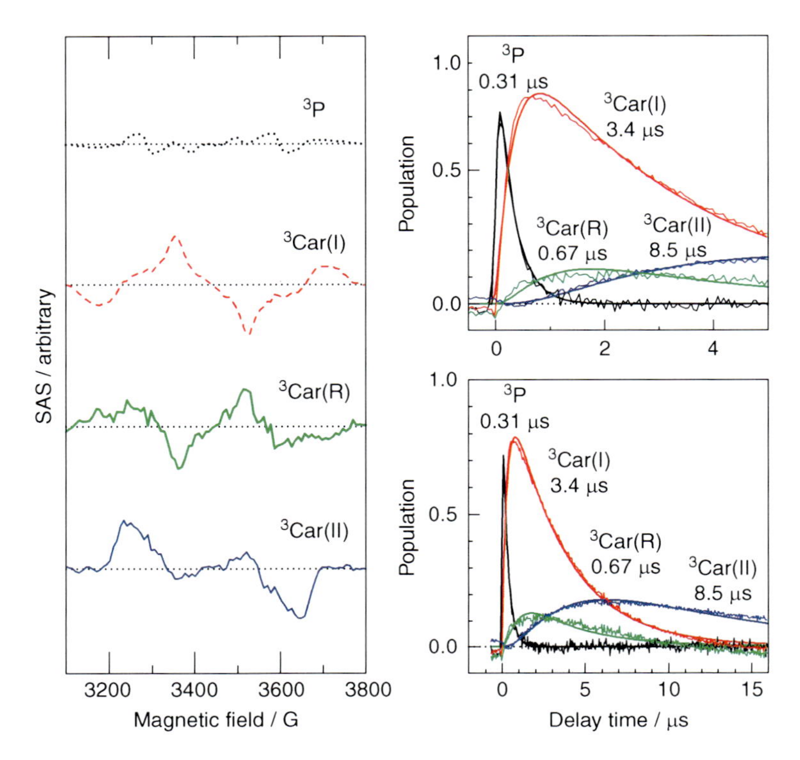

Fig. 16.5 The result of four-component SVD and global-fitting analysis of time-resolved EPR data matrix at 100 K. *Left:* The SAS spectral patterns of the special-pair BChl in the T_1 state (^{3}P) and three sequential ^{3}Car species (^{3}Car(I), ^{3}Car(R) and ^{3}Car(II)) (see Fig. 16.8 for the relaxation scheme used). *Right:* Time-dependent changes in population of the four triplet species shown in two different time scales. The decay time constants are indicated for each triplet species

Table 16.2 Zero-field splitting parameters observed in ^{3}Car(I), ^{3}Car(R) and ^{3}Car(II) and those simulated in models with the rotational angles around (C15=C15′, C13=C14, C11=C12) bonds for the RC-bound 15-*cis*-spheroidene

	^{3}Car(I)	^{3}Car(R)	^{3}Car(II)
|*D*|	1[a]	0.90	0.73
|*E*|	1	0.89	0.50
	(0°, 0°, 0°)	(+20°, −20°, +20°)	(+45°, −40°, +40°)
|*D*|	1	0.98	0.82
|*E*|	1	0.89	0.50

[a]The |*D*| and |*E*| values are normalized to those of ^{3}Car(I)

values thus obtained and then normalized to those values of ^{3}Car(I), are obtained (Table 16.2). On the other hand, the |*D*| and |*E*| values simulated based on the structures and then normalized to the values for the (0°, 0°, 0°) planar conformation that was assumed for ^{3}Car(I). In the simulation of the |*D*| and |*E*| values for ^{3}Car

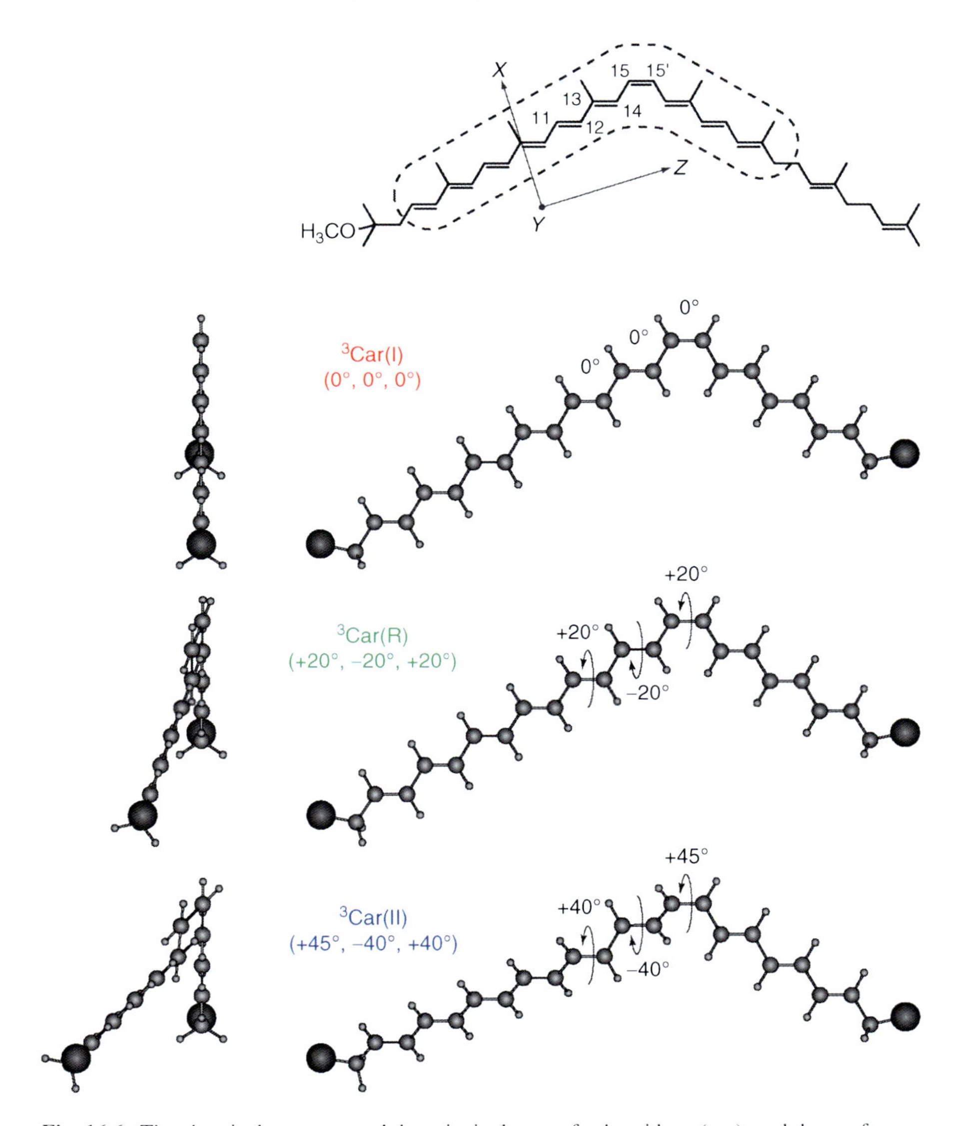

Fig. 16.6 The chemical structure and the principal axes of spheroidene (*top*), and the conformation of ^{3}Car(I) assumed to be planar 15-*cis* and those conformations of ^{3}Car(R) and ^{3}Car (II) determined by comparison between the observed and the calculated zero-field splitting parameters, |*D*| and |*E*| (see Table 16.2). Both side and front views are shown for the structures of the three ^{3}Car species

(R) and ^{3}Car(II), the rotational angles around the *cis* C15=C15′, *trans* C13=C14 and *trans* C11=C12 bonds were changed stepwise by a 5° interval.

The simulated models around the C15=C15′, C13=C14 and C11=C12 angles, (+20°, −20°, +20°) and (+45°, −40°, +40°), perfectly reproduced the normalized, relative |*E*| values of ^{3}Car(R) and ^{3}Car(II). However, these models tend to give rise to slightly higher |*D*| values than the observed (see Table 16.2 and Fig. 16.6).

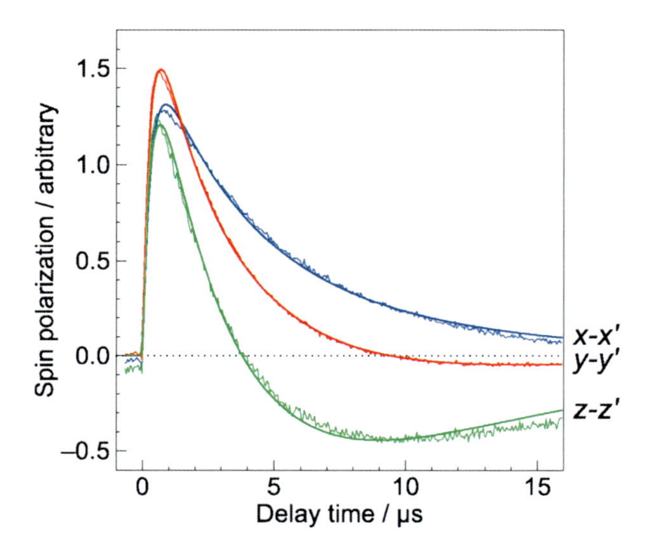

Fig. 16.7 Time-dependent changes in spin polarization for the x–x', y–y' and z–z' components; see the top panel of Fig. 16.6 for the definition of the principal axes, X, Y and Z. The ordinate scale represents the difference in the triplet population, N_0 minus N_{+1} (N_{-1}). The inversion of the spin polarization reflects the presence of the external magnetic field applied to the spin system

Importantly, the rotational angles (+45°, −40°, +40°) for the final stationary-state T_1 species, ^{3}Car(II), determined here by EPR spectroscopy are *in excellent agreement* with the rotational angles (+45°, −30°, +30°) for the final stationary-state T_1 species determined above by resonance-Raman spectroscopy and normal-coordinate analysis, even such a complete difference in the two spectroscopic techniques was taken into account.

Figure 16.7 shows the rise and decay time profiles of spin polarization along the x–x', y–y' and z–z' triplet principal axes. Under ordinary conditions, there is *no chance* for spin polarization (along the x–x', y–y' and z–z' axes) to become *negative*.

However, we did observe the following: (i) The x–x' component decays toward zero, keeping the same sign of spin polarization (as usual), (ii) the y–y' component slightly reverses the sign of spin polarization while relaxing toward zero, and (iii) the z–z' component exhibits rapid decay, a strong inversion of sign, and, then, slow recovery to zero in spin polarization.

The anomalous changes in spin polarization, i.e., (iii) and (ii), suggests a strong influence of *external magnetic field* on the pure spin system. The following observations suggest the influence of the orbital angular momentum that is generated by the rotational motions around the central double bonds: (i) As shown on the top of Fig. 16.6, the directions of the three central C=C bonds, around which the rotational motions takes place, in the transformation, ^{3}Car(I) → ^{3}Car(R) → ^{3}Car(II), are approximately around the z–z' principal axis. The later conformational changes (see figures on the left-hand-side) may give rise to changes in the angular momentum in the y–y' direction. (ii) The timing of the inversion of the spin polarization in the z–z' component (3.8 µs) is slightly later than the opening of the leaking channel of the triplet population (3.4 µs) (compare Figs. 16.7 and 16.8). These observations lead us to an idea that changes in the orbital angular momentum due to the

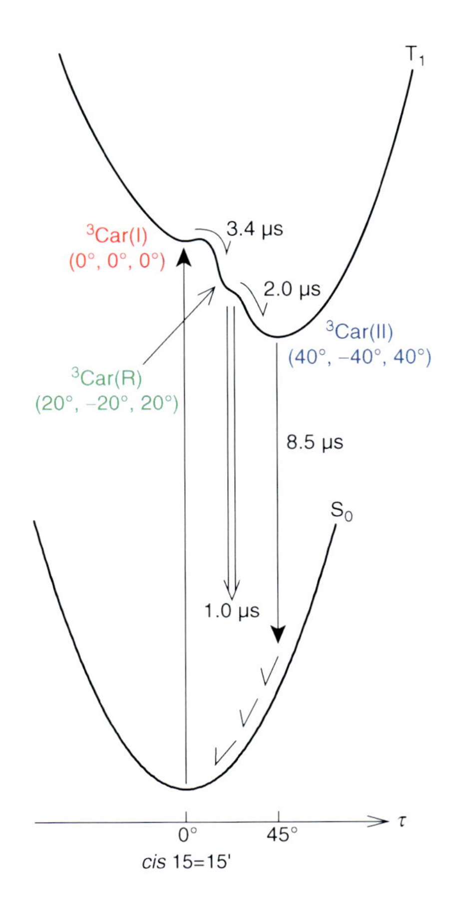

Fig. 16.8 A kinetic scheme of structural changes and triplet-energy dissipation by three ^{3}Car components, in the order, ^{3}Car(I) → ^{3}Car (R) → ^{3}Car(II). After excitation to the T_1 state, the dissipation of the triplet energy takes place through two channels; one, from ^{3}Car(R) (major component) and the other from ^{3}Car (II) (minor component). Cyclic structural changes take place along the T_1 and S_0 potential curves

conformational changes of triplet Cars is responsible for the inversion of spin polarization.

As depicted in Fig. 16.8, the SVD followed by global-fitting analysis determined the time constants of the two steps of conformational changes to be 3.4 and 2.0 μs, and those of the two channels of $T_1 \rightarrow S_0$ intersystem crossing (triplet-energy dissipation) to be 1.0 and 8.5 μs. After intersystem crossing, a reverse structural change to the original conformation, (0°, 0°, 0°), is supposed to take place along the S_0 potential. This is a hypothetical mechanism of triplet-energy dissipation, including the cyclic conformational changes in the T_1 and S_0 states. The most efficient leaking channel (1.0 μs) from the intermediate, ^{3}Car(R), plays the most important role in the triplet-energy dissipation.

This is obviously the reason for the selection of the 15-*cis* Car by the RC. EPR spectroscopy of RC suggests that the rotational motion around the central double bonds must play the most important role in the $T_1 \rightarrow S_0$ intersystem crossing, or in other words, the triplet-energy dissipation.

Reference

1. Koyama Y, Kakitani Y, Watanabe Y (2008) Photophysical properties and light-harvesting and photoprotective functions of carotenoids in bacterial photosynthesis: structural selections. In: Renger G (ed) Primary processes of photosynthesis, Part 1: Principles and apparatus, vol 8, Comprehensive series in photochemistry & photobiology. RSC Publishing, Cambridge, pp 151–201

Chapter 17
Time-Resolved Macromolecular Crystallography in Practice at BioCARS, Advanced Photon Source: From Data Collection to Structures of Intermediates

Vukica Šrajer

17.1 Introduction

Time-resolved macromolecular crystallography is one of several "kinetic crystallography" methods [2, 3]. In kinetic crystallography experiments, genuine biological function is triggered in the crystal with a goal of capturing molecules in action and determining structures of intermediate states. In time-resolved experiments in particular, short and intense X-ray pulses are used to probe intermediates in real time and at room temperature, in reactions that are initiated synchronously and rapidly in the crystal. In other kinetic crystallography approaches, physical or chemical trapping has been used (such as trapping by freezing or chemical modifications) to extend the lifetimes of intermediate states. These states can then be studied by the more conventional and less technically challenging static crystallography.

During the last two decades, essential advancements have been made in the development of high-intensity synchrotron X-ray sources, synchrotron instrumentation, as well as methodology and software for Laue data processing and analysis. This facilitated successful time-resolved crystallographic experiments with 100 ps time resolution [17, 20, 24]. Time-resolved macromolecular crystallography at synchrotron X-ray sources is in its mature phase today. Structural changes as small as 0.2–0.3 Å are detectable with 100 ps time-resolution and even if the reaction is initiated in a relatively small fraction of molecules in the crystal [10–12, 25, 27–29, 34]. In addition, determination of time-independent structures of intermediate states from measured time-dependent structure factor amplitudes has been successfully demonstrated [14, 22].

V. Šrajer (✉)
Center for Advanced Radiation Sources, The University of Chicago,
Chicago, IL 60637, USA
e-mail: v-srajer@uchicago.edu

J.A.K. Howard et al. (eds.), *The Future of Dynamic Structural Science*,
NATO Science for Peace and Security Series A: Chemistry and Biology,
DOI 10.1007/978-94-017-8550-1_17,

We provide here an overview of time-resolved crystallography as implemented today at the BioCARS beamline 14-ID at the Advanced Photon Source [7], with an overview of future directions and challenges.

17.2 Time-Resolved Crystallographic Experiments: General Considerations

17.2.1 Reaction Triggering

To examine short-lived intermediates in real time, it is critical to initiate the reaction in the molecules in the crystal in a time period that is significantly shorter than the lifetimes of such intermediates [19, 20, 24]. The fastest method of reaction initiation by far is the use of ultra-short laser pulses. This method is readily applicable to molecules that are inherently photosensitive and undergo a reversible reaction, like ligand photo-dissociation in heme proteins [29] or chromophore isomerization in photoactive yellow protein [10]. Alternatively, other molecules can be rendered light sensitive by chemically attaching photosensitive groups to substrates, cofactors or important protein residues [2, 3, 6]. Reactions can also be initiated by diffusion of small molecules into the crystal (such as substrates, cofactors or redox agents) but this is a slow process that requires seconds or minutes (or even longer). It is therefore suitable as a trigger only for slow reactions [2, 3].

Protein crystals contain a very large number of molecules, on the order of 10^{13}–10^{14}, and therefore have high optical density (OD) in the wavelength regions where the chromophore absorbs significantly. In order to photo-initiate the reaction by laser pulses throughout the crystal, smaller (thinner) crystals have to be used and the laser wavelength tuned to a spectrum region where OD $<<1$. Typical laser power densities used for reaction triggering are 2–5 mJ/mm^2.

17.2.2 Pump-Probe Method

In time-resolved X-ray diffraction experiments laser pulses (or other triggering methods) are used as "pump" pulses that trigger the reaction, while X-ray pulses are used to "probe" the reaction, at various time delays following the pump pulse. Due to slow readout time of large area X-ray detectors (10–100 ms at best), single-pump/single-probe is typically employed, unless a reaction is slow and single-pump/multiple-probe is possible, where one collects several diffraction images following a single reaction initiation.

17.2.3 Time-Resolution

Time-resolution of the experiment is determined by the duration of the pump or probe pulses, whichever is longer. At synchrotrons, best time resolution is ultimately limited by the ~100 ps duration of the X-ray pulses, although somewhat better time-resolution is also possible in principle (~10 ps) at the expense of the reduced signal-to-noise ratio [8, 9].

17.2.4 X-ray Source

A critical requirement for time-resolved experiments is the high X-ray flux per pulse, typically available only at third generation synchrotron sources, such as Advanced Photon Source (Argonne National Laboratory, USA), European Synchrotron Radiation Facility (Grenoble, France) and SPring 8 (Japan). Flux greater than 10^{10} photons/pulse is needed when a single 100 ps X-ray pulse is used to record a diffraction image from a protein crystal. If such high flux is not available or crystals are smaller in size (<200 μm) and/or weakly-diffracting, the pump-probe sequence needs to be repeated a number of times prior to the detector readout. Such repeated pump-probe cycles are clearly possible only for fully reversible reactions. For irreversible reactions, single X-ray pulse data acquisition is necessary, where each image requires a single pump-probe sequence and a new crystal volume. For slower reactions with less demanding time resolution requirement, an X-ray pulse train of longer total duration and higher overall intensity can be used as a probe pulse, resulting in diffraction images with improved signal-to-noise ratio.

17.2.5 Single X-ray Pulse Isolation

An X-ray shutter train is necessary to isolate single 100 ps X-ray pulses or longer pulse trains from the continuous stream of synchrotron pulses (see Fig. 17.1 and [7]).

17.2.6 Pump/Probe Synchronization

Accurate and precise synchronization of the arrival of the X-ray probe pulse at the sample with respect to the laser pump pulse has to be achieved for time-resolved experiments [7].

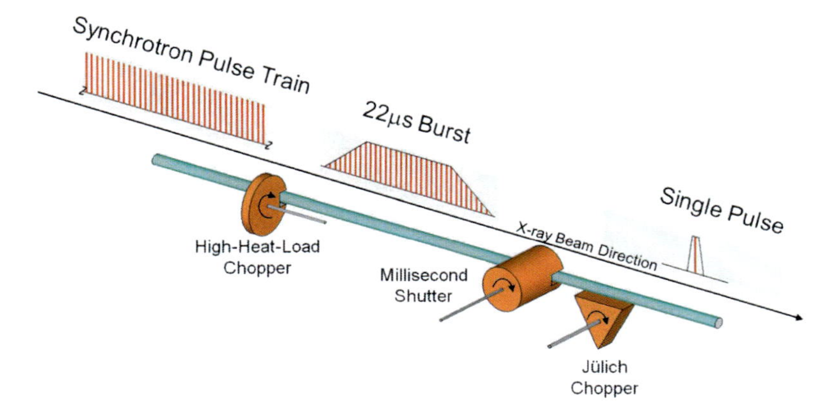

Fig. 17.1 X-ray shutter train for time-resolved experiments at BioCARS 14-ID beamline (Graber et al. [7]. Reproduced with permission of the International Union of Crystallography)

17.2.7 Laue Diffraction

Mechanical crystal rotation used in the standard monochromatic oscillation method is much too slow to probe fast, sub-second reactions. Laue diffraction with stationary crystals therefore has to be used. Laue diffraction employs significantly wider energy bandwidth than the standard monochromatic diffraction. It has been shown that undulators are best X-ray sources for time-resolved Laue diffraction experiments [4, 30]. The typical bandpass of undulators used today for time-resolved experiments is 3–5 % at 12–15 keV. Specialized software for Laue data processing is required, for example CCP4 *Daresbury Laue* Software Suite or Precognition/Epinorm [17].

17.2.8 Crystal Mounting

Crystals for time-resolved crystallography are typically mounted in thin-walled glass or quartz capillaries or using MiTeGen RT style mounts. This is because data has to be collected at or near room-temperature rather than at cryo temperatures to facilitate complete protein mobility as the reaction in the crystal proceeds.

17.3 Time-Resolved Crystallographic Experiments at BioCARS

17.3.1 Upgraded BioCARS 14-ID Beamline

BioCARS 14-ID beamline at the Advanced Photon Source (Argonne National Laboratory, USA) was originally designed for time-resolved crystallography.

It has been completely upgraded in 2007–2008. The new beamline layout, specifications and operation are described in detail in Graber et al. [7]. In short, the X-ray source consists of two in-line undulators with periods of 23 and 27 mm, providing high pink-beam flux at 12 keV, as well as first-harmonic coverage from 6.8 to 19 keV. A high-heat-load chopper (Fig. 17.1) is implemented to reduce the average power load on downstream components to less than 1 % of the original power in the X-ray beam. This chopper typically produces a 22 μs burst of X-rays at a repetition rate of 82.3 Hz (although it can also be operated at 1 kHz). A second, high-speed chopper (Jülich chopper; Fig. 17.1) isolates single 100 ps X-ray pulses at 1 kHz in both hybrid and 24-bunch modes of the APS storage ring. Since these two choppers are continuously rotating, an additional, millisecond shutter is used for on-demand exposures of the samples to X-ray pulses.

In hybrid APS mode, each X-ray pulse delivers up to ~4×10^{10} photons to the sample at 14-ID. X-ray beam is focused by a new KB mirror system to a minimum spot size of 90 μm (h) × 20 μm (v). A new high-power and widely tunable picosecond laser system was also installed, covering the wavelength range of 450–2000 nm. These laser pulses are synchronized to the storage ring RF clock, with long-term stability better than 10 ps rms. A portable nanosecond laser is also available.

17.3.2 Typical Time-Resolved Crystallographic Data Collection

A complete time-resolved data set spans four-dimensions: three traditional reciprocal space dimensions and time. Similar to a standard monochromatic data set, a Laue data set at each time delay contains images collected at a number of different crystal orientations, except that crystal is stationary during each X-ray exposure. Integrated intensities are nevertheless recorded due to the polychromatic nature of the X-ray source. With 14-ID undulator sources, the angular step in crystal orientation needed is typically 2 – 3°. The number of time points required to characterize the reaction kinetics depends on the number of intermediates and their lifetimes. A good starting point is to collect five points per decade in time, equally spaced in logarithmic time.

Since the signal of interest is the difference in intensity of diffraction before and after the photo-activation, it is best to collect both data sets on the same crystal, with interleaving the laser-on and laser-off images for each crystal orientation. To minimize errors across the time domain, time delay is typically a fast variable [20, 24]: time delay is scanned for a given crystal orientation, starting with the laser-off image and collecting a series of laser-on images that span the desired time domain. The time scan is then repeated for a new crystal orientation. Several crystals might be required for a complete space-time data in case of radiation-sensitive samples.

Fig. 17.2 BioCARS 14-ID experimental hutch. Laser light is delivered to the sample via optical fiber (ns laser) or via mirrors (ps laser). The complex ps laser system is located in a separate laser lab. *Left* panel: experimental table is shown. Above the table is a box with ps laser conditioning and pulse stretching optics. Current Mar165 X-ray detector will be replaced in 2013 by a Rayonix MX340 HS detector. Significantly larger area will permit studies of crystals with larger unit cells. Faster readout time of the new detector (0.1–0.01 s) will also permit using single-pump/multiple-probe style of experiments with sub-second time resolution (readout time of the current Mar165 detector is 3–5 s). This is particularly important in studies of irreversible reactions where a new sample is needed after each reaction initiation. A single-pump/multiple-probe method will reduce the number of needed samples

17.3.3 Reaction Initiation

Optimal reaction initiation is critical for the success of time-resolved experiments. Laser light is delivered to the sample either by optical fibers (ns laser) or directly (ps laser) and focused at the sample. Picosecond laser pulses are typically stretched from 1 to ~30 ps for a better match to the X-ray pulse duration (100 ps) and for minimizing laser-induced crystal damage. Variable laser focusing is available to best match the crystal size. Both elliptical (as shown in Fig. 17.3) and circular focus shape is available. Given the fact that crystals are very optically thick, laser wavelength is typically tuned to the part of the spectrum with lower absorption but even then light penetration into the crystal may not be very deep. A perpendicular laser/X-ray geometry is therefore employed (Figs. 17.2 and 17.3), where X-ray beam probes only the laser-illuminated surface layer of the crystal [7]. To properly position the crystal in the X-ray beam, crystal is scanned vertically through the X-ray beam (edge scan) and X-ray diffraction is used for determining correct crystal position. This is repeated for each significantly different crystal orientation. Diagnostics tools for on-line monitoring of the laser beam position and intensity, as well as the laser-to-X-ray time delay, are essential to detect and correct any positional, intensity and timing drifts during the time-resolved data collection.

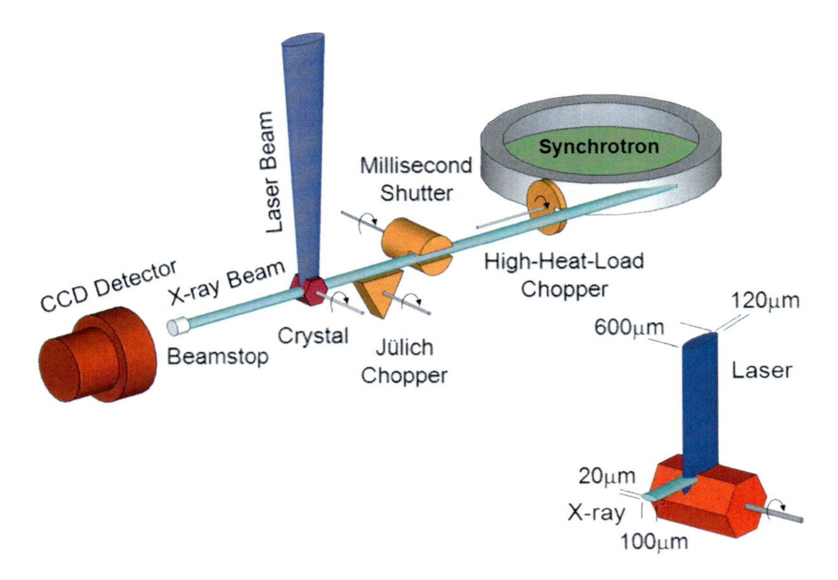

Fig. 17.3 Schematic layout for a typical pump-probe experiment at BioCARS 14-ID beamline (Graber et al. [7]. Reproduced with permission of the International Union of Crystallography)

17.3.4 Data Collection Software

Time-resolved data collection software used at BioCARS, LaueCollect, was developed by Friedrich Schotte (NIH/NIDDK). The emphasis was on data collection automation as well as on flexibility in data collection strategies. The edge scan and crystal translation (to minimize radiation damage) are incorporated into the data collection protocol. Software allows users to log important information such as measured laser and X-ray pulse intensities and time delay for each diffraction image. Laser position and laser-X-ray timing can also be checked and corrected automatically at regular time intervals during data collection.

17.4 From Laue Data to Structures of Intermediates

Analysis of time-resolved crystallographic data involves three steps: (1) Laue data processing to derive time-dependent structure factor amplitudes from recorded diffraction images; (2) calculation of time-dependent difference electron density maps and analysis of such maps; (3) determination of structures of intermediate states.

17.4.1 Laue Data Processing

Several problems specific to the Laue diffraction method are addressed and resolved by software: spatial overlap of diffraction spots in typically crowded Laue diffraction patterns, wavelength normalization and resolving the harmonic overlaps. The so-called wavelength normalization is necessary due to polychromatic X-rays used and associated wavelength dependence of the incident X-ray intensity, scattering power of the crystal and detector sensitivity. A reflection or its symmetry mate may be stimulated by different wavelengths depending on the crystal orientation. Measured intensities therefore need to be brought to a common scale before data can be merged. The resulting λ-curve combines all wavelength-dependent effects in diffraction intensities as recorded at the detector. Harmonic or energy overlaps result from reflections that lie on a radial line (starting at the origin) in the reciprocal space. Such reflections are stimulated by different energies but scatter in exactly the same direction and overlap exactly at the detector. Program *Precognition-Epinorm* used at BioCARS for Laue data processing (Renz Research Inc) deals successfully both with wavelength normalization and harmonic overlaps. Another commonly used Laue package is *Daresbury Laboratory Laue Processing Suite* [1, 5], http://www.ccp4.ac.uk/cvs/viewvc.cgi/laue/doc/laue_install.txt?revision=1.1&view=markup.

17.4.2 Difference Electron Density Maps

Time-resolved crystallography method is based on a difference measurement and time-dependent difference electron density maps $\Delta\rho(t)$ are calculated and analyzed. Experimentally-determined structure factor amplitudes (SF) of the initial, dark state $|\mathbf{F}^{D}(hkl)|$ and corresponding time-dependent structure factor amplitudes $|\mathbf{F}(hkl,t)|$ are used to calculate time-dependent difference amplitudes, $\Delta F(hkl,t) = |\mathbf{F}(hkl,t)| - |\mathbf{F}^{D}(hkl)|$, for each time point t. In addition to measured difference amplitudes $\Delta F(hkl,t)$, phases derived from the known dark structure, φ^{D}_{hkl}, are used for calculating difference maps $\Delta\rho(t)$. In order to improve the signal-to-noise ratio of difference maps, several weighting schemes for difference SF amplitudes have been used [17, 31, 33]. When SF amplitudes are given on the absolute scale, difference electron density in selected regions can be integrated to provide the information about the total number of electrons displaced from or into a particular volume in space as a function of time. The program *Promsk* [25], for example, can be used to integrate difference density in the region of interest within a specified mask, generated by supplying atomic coordinates and a radius of integration around the coordinates. Time evolution of such integrated densities can provide important information on formation and decay of structural intermediates [12, 25].

Measured time-dependent difference electron density maps are very likely to represent a mixture of intermediate states at any point in time and as such can be extremely difficult to interpret (see Fig. 17.4). A method such as Singular Value

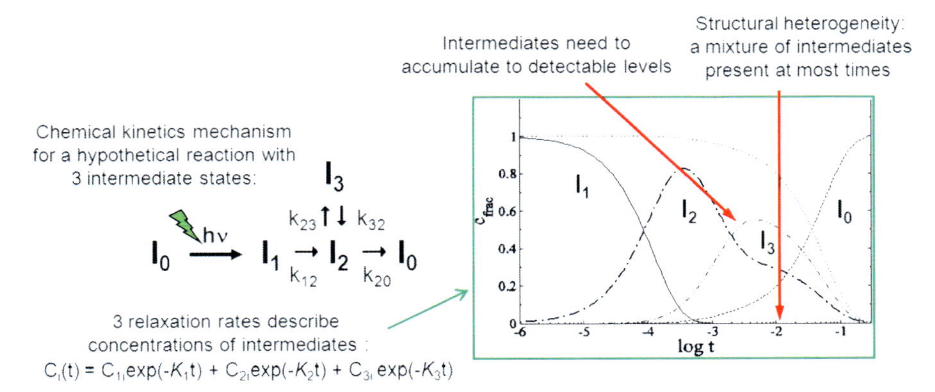

Fig. 17.4 A hypothetical photo-initiated reaction is shown, with three intermediates, I_i (i = 1, 2, 3), and a branched chemical kinetics mechanism. I_0 is the initial state. Three global relaxation rates K_m (m = 1, 2, 3) govern the time dependence of concentrations $C_i(t)$ of all three intermediates. The concentration amplitudes, C_{mi}, as well as these macroscopic, observable rates K_m are functions of four rate coefficients, k_{ij}, which in turn depend on the activation energy barriers between I_i and I_j states. Notice that at most times, a mixture of several states is present. Notice also that, in order for an intermediate state to be detected, rate coefficients need to be such that this intermediate accumulates to a detectable level

Decomposition (SVD) is therefore needed to decompose a series of measured mixed-state, time-dependent maps into time-independent maps corresponding to individual structural intermediates.

17.4.3 Singular Value Decomposition (SVD)

SVD in combination with posterior analysis is a powerful method to analyze time-resolved crystallographic data. The number of intermediates, time-independent structures of such intermediates, rates of their formation and decay, and the overall chemical kinetics mechanism of the investigated reaction can be determined [10, 14, 16, 22, 23]. A comprehensive description of these methods is provided in Schmidt et al. [20, 24]. In short, SVD is a method of global analysis that is applied to a data matrix composed of time-dependent difference electron density maps, $\Delta\rho(t_i)$, i = 1...N, where N is the number of time points. Each map consists of M grid points and constitutes a column in the M × N data matrix. The SVD procedure decomposes such a matrix into N time-independent $\Delta\rho$ maps or left singular vectors (lSV) and corresponding N temporal variations or right singular vectors (rSV). In essence, SVD separates space and time variables. A singular value associated with each vector-pair represents contribution of the lSVs to the experimental difference maps. In matrix form this decomposition can be described as:

$$A = USV^T$$

where A is a M × N data matrix, U is a M × N matrix containing lSVs (maps) as columns, S is an N × N diagonal matrix containing singular values, and V is a N × N matrix, with rows of the V^T matrix containing rSVs (temporal variations).

The actual difference signal is typically contained in only few significant vectors with large singular values. The remaining singular vectors contain only noise. The SVD analysis therefore acts an effective noise filter. The input series of maps can be approximated by S/N improved maps $\Delta\rho'(t)$, reconstituted from significant singular vectors only.

It is important to note that the lSVs (time-independent difference maps) do not represent the actual difference maps of the intermediate states. They are instead their linear combinations. Similarly, the rSVs (temporal variations) are linear combinations of the actual time courses in the observed time-dependent maps. rSVs therefore provide information about relaxation processes and can be used to determine macroscopic, relaxation rates K_m in the reaction (Fig. 17.4). From a global fit of all significant rSVs by a sum of exponential functions, the number of relaxation processes and associated rates K_m are determined. The number of relaxation processes is the lower bound on the number of intermediates.

17.4.4 Structures of Intermediates

With the number of relaxations and associated rates determined from the rSVs, the time-independent difference maps, $\Delta\rho_{Ij}$, corresponding to I_j intermediates can be determined [14, 16, 22, 24]. At this stage, a candidate reaction mechanism has to be assumed (serial, parallel, branched etc.) for the given number of intermediates (Fig. 17.4). For the assumed mechanism, time dependent concentrations are calculated for each intermediate by solving a system of coupled differential equations. The concentrations are functions of the actual rate coefficients k_{ij} that govern the reaction (Fig. 17.4). The rSVs are then fit globally by the sum of concentrations. However, scale factors are necessary since the rSVs are linear combinations of the true time courses and this linear transformation is not known. Both the numerical values of the rate coefficients and the scale factors are varied to fit all significant rSVs jointly. A scale factor E_{nj} represents contribution of the j-th intermediate concentration to the n-th rSV [24].

The time-independent difference maps of intermediates can then be synthesized as a linear combination of significant lSV maps [20, 24]. The scale factors E_{nj} from the rSV fit provide projections of lSVs onto the maps of intermediates: they are contributions of the n-th lSV to the map of the j-th intermediate. The final result is therefore achieved: the mixture of intermediates in the experimental time-dependent difference maps has been separated into the difference maps corresponding to pure intermediates $\Delta\rho_{Ij}$.

Finally, the structures of the reaction intermediates are refined. For this purpose the extrapolated, conventional maps are used instead of difference maps as they are more easily interpretable. For each intermediate, the difference map $\Delta\rho_{Ij}$ obtained from the SVD analysis described above is Fourier transformed and a multiple f of the resulting difference structure factor $\mathbf{\Delta F_{Ij}}$ is added to the dark structure factor $\mathbf{F_D}$ by a vector summation. The resulting structure factor is used to calculate the extrapolated map. The factor f is determined so that the extrapolated map has no contribution from the dark state. The structures of intermediates are then modeled and refined using these conventional maps. Alternatively, difference refinement against $\mathbf{F_D} + \mathbf{\Delta F_{Ij}}$ can be done [32].

To be acceptable, the maps of the intermediate states have to be interpretable by a valid and unique single atomic structure. If not, the particular mechanism must be discarded and another mechanism needs to be evaluated.

17.4.5 Analysis of Kinetic Mechanism

At this stage of the analysis, several candidate kinetic mechanisms can lead to similar, interpretable maps for the intermediates. Plausible mechanisms are then compared by the posterior analysis. Given the models of the initial state and the intermediates, time-dependent difference electron density maps, $\Delta\rho$(t,k)calc, are calculated for each plausible mechanism, where concentrations of intermediates for this mechanism are determined by the set of rate coefficients k_{ij} (Fig 17.4). A fitting procedure is used to adjust mechanism-dependent rate coefficients k_{ij} to best match the measured time-dependent difference maps, $\Delta\rho(t)$. Since electron densities, when on absolute scale, are directly proportional to populations of intermediates, the only scaling factor that needs to be adjusted is the extent of reaction initiation and others scaling factors are not needed. The residual maps $\Delta\Delta\rho(t) = \Delta\rho(t)-\Delta\rho(t,k)_{calc}$ are then inspected. If there are no significant residual densities at any time-points, the mechanism is considered compatible with the data since it has generated concentrations that reproduced the observed difference electron densities. However, if significant residual densities $\Delta\Delta\rho(t)$ exist, the mechanism is considered inconsistent. Although this procedure helps to narrow down possibilities, several mechanisms may still be compatible with the data. To further eliminate inconsistent mechanisms, one can conduct 5D crystallography experiments, where in addition to time, temperature is also a variable parameter [26].

Software for SVD and posterior analysis SVD4TX/GetMech [36] is available and can be obtained from the lab of Marius Schmidt, University of Milwaukee (link to the personal web-page: http://www4.uwm.edu/letsci/physics/staff/marius_schmidt.cfm).

17.5 Future of Time-Resolved Macromolecular Crystallography: Major Challenges

Great advances both in hardware and software in the last two decades enabled successful realization of a challenging goal: watching biological molecules in action with 100 ps time resolution and at the atomic-resolution detail. Never the less, time-resolved crystallography faces a number of challenges in becoming a tool that is used more broadly in studies of dynamic properties of biological macromolecules. A few of major challenges are briefly described below.

Rapid reaction triggering is essential for the success of time-resolved experiments. However, it is specific to each protein and associated reaction. In other words, there is no universal reaction initiation method that can be applied to all molecules. Rather, one needs to determine a suitable method for each particular case. Light can be readily used for initiating fast reactions given widely available pulsed lasers. However, this method can only be applied to naturally photosensitive proteins or when proto-sensitivity can be conferred to otherwise photoinert proteins. In this second category, caged compounds hold a particular promise [2, 3, 6]. They are light-sensitive but biochemically inactivating groups that are chemically attached to substrates, cofactors or catalytically important residues. Activation is then triggered by photo-releasing the active reagent from the caged compound. However, improving characteristics of caged compounds, such as quantum yield and speed of photo-release, is necessary for this method to become more widely applied. Reactions can also be initiated by diffusion of small molecules such as substrates, cofactors, ligands or redox agents into the crystal. This is a simple method in principle but diffusion is a slow process (it could take seconds to days, depending on the crystal size and size of the diffusing molecule). Such method is therefore applicable only to relatively slow reactions. Luckily, many enzyme reactions are slow ($<100\ s^{-1}$ turnover rate). In addition, when using small crystals ($<10\ \mu m$), time resolution of $<$ ms might be achievable in some cases [21]. Also, to facilitate diffusion-based reaction initiation suitable and improved flow cell designs are needed.

Another challenge for time-resolved crystallography is relatively slow readout speed of large-area X-ray detectors. Ideally, one would like to follow a reaction at a number of time delays after a single reaction trigger. However, with current detectors for macromolecular crystallography, only 10–100 Hz frame-rate is possible. This limits the time resolution for such "pump once, probe many" style of experiments. Yet, such style is essential for irreversible reactions where each reaction trigger requires a new crystal or fresh crystal volume. One therefore needs to maximize the data collected following each trigger in order to minimize the number of crystals that is required for a complete data set. In case of reversible reactions, repetition of "pump once, probe once" cycles on the same crystal (or crystal volume) is possible, provided sufficient time is allowed between the cycles for the recovery of the initial state. It is therefore possible to collect complete a complete data set (sampling both reciprocal space and time) using one or only a few crystals. In case of irreversible reactions a large number of crystals is necessary for comprehensive time-resolved data. Any means of reducing the required number of crystals, such as a fast-readout detector, is welcome. Also, when a large number of crystals is required, suitable methods for rapid crystal introduction into the X-ray beam ("serial crystallography") are very important and need to be developed.

Dealing with measured heterogeneous structural states is essential for the analysis of time-resolved data. As mentioned above, at most time points a mixture of state is most likely measured. Both an SVD-based analysis, described above, and a cluster analysis [13, 15] have been used in attempts to determine structures of intermediates from measured mixtures of states. More recently, a new method has been proposed [18], where structural and occupancy refinement of multiple structures across multiple data sets is conducted, based on real space approach and aided by the SVD analysis. In addition to analysis of time-resolved data, such method can be applied to any standard, static crystallographic data whenever several structures are common to multiple data sets. The new method addresses the limitation of current refinement procedures where multiple but common structures cannot be refined against multiple data sets. The method has been applied to a cryo-trapping experiment where a mixture of three common structures was observed at ten temperatures in the range from 100 to 180 K [35].

References

1. Arzt S, Campbell JW, Harding MM, Hao Q, Helliwell JR (1999) LSCALE – the new normalization, scaling and absorption correction program in the Daresbury Laue software suite. J Appl Crystallogr 32:554–562
2. Bourgeois D, Royant A (2005) Advances in kinetic protein crystallography. Curr Opin Struct Biol 15:538–547
3. Bourgeois D, Weik M (2009) Kinetic protein crystallography: a tool to watch proteins in action. Crystallogr Rev 15:87–118
4. Bourgeois D, Wagner U, Wulff M (2000) Towards automated Laue data processing: application to the choice of optimal X-ray spectrum. Acta Crystallogr D56:973–985
5. Campbell JW (1995) LAUEGEN, an X-windows-based program for the processing of Laue diffraction data. J Appl Crystallogr 28:228–236
6. Corrie JET, Katayama Y, Reid GP, Anson M, Trentham DR, Sweet RM, Moffat K (1992) The development and application of photosensitive caged compounds to aid time-resolved structure determination of macromolecules [and discussion]. Philos Trans R Soc Lond Ser A 340:233–244
7. Graber T, Anderson S, Brewer H, Chen Y-S, Cho HS, Dashdorj N, Henning RW, Kosheleva I, Macha G, Meron M et al (2011) BioCARS: a synchrotron resource for time-resolved X-ray science. J Synchrotron Radiat 18:658–670
8. Haldrup K, Harlang T, Christensen M, Dohn A, van Driel TB, Kjær KS, Harrit N, Vibenholt J, Guerin L, Wulff M et al (2011) Bond shortening (1.4 Å) in the singlet and triplet excited states of [Ir2(dimen)4]2+ in solution determined by time-resolved X-ray scattering. Inorg Chem 50:9329–9336
9. Haldrup K, Vankó G, Gawelda W, Galler A, Doumy G, March AM, Kanter EP, Bordage A, Dohn A, van Driel TB et al (2012) Guest–host Interactions investigated by time-resolved X-ray spectroscopies and scattering at MHz rates: solvation dynamics and photoinduced spin transition in aqueous Fe(bipy)32+. J Phys Chem A 116:9878–9887

10. Ihee H, Rajagopal S, Srajer V, Pahl R, Anderson S, Schmidt M, Schotte F, Anfinrud PA, Wulff M, Moffat K (2005) Visualizing reaction pathways in photoactive yellow protein from nanoseconds to seconds. Proc Natl Acad Sci U S A 102:7145–7150
11. Jung YO, Lee JH, Kim J, Schmidt M, Moffat K, Šrajer V, Ihee H (2013) Volume-conserving trans–cis isomerization pathways in photoactive yellow protein visualized by picosecond X-ray crystallography. Nat Chem 5:212–220
12. Knapp JE, Pahl R, Srajer V, Royer WE (2006) Allosteric action in real time: time-resolved crystallographic studies of a cooperative dimeric hemoglobin. Proc Natl Acad Sci U S A 103:7649–7654
13. Kostov KS, Moffat K (2011) Cluster analysis of time-dependent crystallographic data: direct identification of time-independent structural intermediates. Biophys J 100:440–449
14. Rajagopal S, Schmidt M, Anderson S, Ihee H, Moffat K (2004) Analysis of experimental time-resolved crystallographic data by singular value decomposition. Acta Crystallogr D60:860–871
15. Rajagopal S, Kostov KS, Moffat K (2004) Analytical trapping: extraction of time-independent structures from time-dependent crystallographic data. J Struct Biol 147:211–222
16. Rajagopal S, Anderson S, Srajer V, Schmidt M, Pahl R, Moffat K (2005) A structural pathway for signaling in the E46Q mutant of photoactive yellow protein. Structure 13:55–63
17. Ren Z, Bourgeois D, Helliwell JR, Moffat K, Šrajer V, Stoddard BL (1999) Laue crystallography: coming of age. J Synchrotron Radiat 6:891–917
18. Ren Z, Chan PWY, Moffat K, Pai EF, Royer WE, Šrajer V, Yang X (2013) Resolution of structural heterogeneity in dynamic crystallography. Acta Crystallogr D69:946–959
19. Schlichting I, Goody RS (1997) Triggering methods in crystallographic enzyme kinetics. Meth Enzymol 277:467–490
20. Schmidt M (2008) Structure based kinetics by time-resolved X-ray crystallography. In: Braun M, Gilch P, Zinth W (eds) Ultrashort laser pulses in biology and medicine. Springer, Berlin/Heidelberg, pp 201–241
21. Schmidt M (2013) Mix and inject: reaction initiation by diffusion for time-resolved macromolecular crystallography. Adv Condens Mat Phys. Article ID 167276
22. Schmidt M, Rajagopal S, Ren Z, Moffat K (2003) Application of singular value decomposition to the analysis of time-resolved macromolecular X-ray data. Biophys J 84:2112–2129
23. Schmidt M, Pahl R, Srajer V, Anderson S, Ren Z, Ihee H, Rajagopal S, Moffat K (2004) Protein kinetics: structures of intermediates and reaction mechanism from time-resolved X-ray data. Proc Natl Acad Sci U S A 101:4799–4804
24. Schmidt M, Ihee H, Pahl R, Srajer V (2005) Protein-ligand interaction probed by time-resolved crystallography. Methods Mol Biol 305:115–154
25. Schmidt M, Nienhaus K, Pahl R, Krasselt A, Anderson S, Parak F, Nienhaus GU, Srajer V (2005) Ligand migration pathway and protein dynamics in myoglobin: a time-resolved crystallographic study on L29W MbCO. Proc Natl Acad Sci U S A 102:11704–11709
26. Schmidt M, Graber T, Henning R, Srajer V (2010) Five-dimensional crystallography. Acta Crystallogr A66:198–206
27. Schotte F, Lim M, Jackson TA, Smirnov AV, Soman J, Olson JS, Phillips GN, Wulff M, Anfinrud PA (2003) Watching a protein as it functions with 150-ps time-resolved X-ray crystallography. Science 300:1944–1947
28. Schotte F, Cho HS, Kaila VRI, Kamikubo H, Dashdorj N, Henry ER, Graber TJ, Henning R, Wulff M, Hummer G et al (2012) Watching a signaling protein function in real time via 100-ps time-resolved Laue crystallography. Proc Natl Acad Sci U S A 109:19256–19261
29. Srajer V, Teng T, Ursby T, Pradervand C, Ren Z, Adachi S, Schildkamp W, Bourgeois D, Wulff M, Moffat K (1996) Photolysis of the carbon monoxide complex of myoglobin: nanosecond time-resolved crystallography. Science 274:1726–1729
30. Srajer V, Crosson S, Schmidt M, Key J, Schotte F, Anderson S, Perman B, Ren Z, Teng TY, Bourgeois D et al (2000) Extraction of accurate structure-factor amplitudes from Laue data: wavelength normalization with wiggler and undulator X-ray sources. J Synchrotron Radiat 7:236–244

31. Srajer V, Ren Z, Teng TY, Schmidt M, Ursby T, Bourgeois D, Pradervand C, Schildkamp W, Wulff M, Moffat K (2001) Protein conformational relaxation and ligand migration in myoglobin: a nanosecond to millisecond molecular movie from time-resolved Laue X-ray diffraction. Biochemistry 40:13802–13815
32. Terwilliger TC, Berendzen J (1995) Difference refinement: obtaining differences between two related structures. Acta Crystallogr D51:609–618
33. Ursby T, Bourgeois D (1997) Improved estimation of structure-factor difference amplitudes from poorly accurate data. Acta Crystallogr A53:564–575
34. Wöhri AB, Katona G, Johansson LC, Fritz E, Malmerberg E, Andersson M, Vincent J, Eklund M, Cammarata M, Wulff M et al (2010) Light-induced structural changes in a photosynthetic reaction center caught by Laue diffraction. Science 328:630–633
35. Yang X, Ren Z, Kuk J, Moffat K (2011) Temperature-scan cryocrystallography reveals reaction intermediates in bacteriophytochrome. Nature 479:428–432
36. Zhao Y, Schmidt M (2009) New software for the singular value decomposition of time-resolved crystallographic data. J Appl Crystallogr 42:734–740

Chapter 18
Combining Single Crystal UV/Vis Spectroscopy and Diffraction to Structurally Characterise Intermediates and Monitor Radiation Damage

Anna Polyakova and Arwen R. Pearson

18.1 Introduction

Mechanisms in biology are dynamic processes and changes in macromolecular structures are intimately linked to their function. Without insight into the dynamic workings of these molecules, we cannot fully understand them. In an ideal world, a complete time-resolved description of each stage of the reaction pathway, at high resolution, would be obtained. Is this possible in practice?

X-ray crystallography can provide high resolution structural data, but this is averaged over both the time to collect the data and over all the molecules in the X-ray illuminated volume, resulting in electron density where the dynamic detail is smeared out. In order to visualise this dynamic information, the majority of the molecules in the crystal must be doing the same thing, *and* the structural changes that occur must be accommodated in the crystal lattice. If these two conditions are met, and the crystals diffract well, a time-resolved crystallographic experiment can give insights into the molecular process at near atomic resolution.

Fast time-resolved crystallographic experiments are extensively discussed elsewhere in this volume and so we will focus on metastable states for which a number of relatively straightforward strategies for time-resolved studies are available. These 'trapping' methodologies arrest the reaction at an intermediate state, using either chemical or physical traps that prevent full turnover. Chemical traps can include the use of substrate analogues or mutant proteins, while physical traps are predominantly cryo-cooling techniques. In this review we will briefly discuss intermediate trapping strategies, and provide some practical tips on the use of complementary spectroscopic methods. These are important both to guide the trapping experiment and ensure that the X-ray data collected is from the desired

A. Polyakova • A.R. Pearson (✉)
Astbury Centre for Structural Molecular Biology, The University of Leeds, Leeds, UK
e-mail: a.r.pearson@leeds.ac.uk

J.A.K. Howard et al. (eds.), *The Future of Dynamic Structural Science*,
NATO Science for Peace and Security Series A: Chemistry and Biology,
DOI 10.1007/978-94-017-8550-1_18,

structure. We will also discuss some alternate approaches that can be used to structurally characterise time-resolved changes in systems which are not amenable to X-ray crystallography.

18.2 Intermediate Trapping Strategies

Protein crystals normally contain around 50 % solvent, which occupies well-ordered channels in the crystal lattice. Within these channels, diffusion of substrate molecules can occur, and thus protein molecules in crystals often retain their catalytic activity [1]. However, there are often major differences between the kinetics in the crystalline state and in solution due to, amongst other things, viscosity effects and the restraints of the crystal packing [2]. Therefore, the reaction *in crystallo* must be accurately characterised before attempting a trapping experiment to ensure that any off-pathway reactions that occur in the crystalline state are identified [3]. This characterisation will also aid in the identification of conditions under which metastable intermediates can be trapped.

The characterisation of the *in crystallo* reaction is most easily done using spectroscopy. A number of single crystal spectrometers are now available, both in home labs and at synchrotron sources (for a review see Pearson and Owen [4]). These allow a range of spectroscopic measurements, including UV-Visible, fluorescence and fluorescence lifetime, IR and (resonance) Raman. Crystals can either be immobilised in a capillary, through which substrate solutions can be passed, or mounted in a loop (under a humidified vapour stream) and substrate solutions directly pipetted onto the sample [3, 5]. Microscope based systems are also available which allow the reaction to be followed in the crystallization plate after addition of substrate to the droplet [6, 7]. See Box 18.1 for practical tips to help obtain the best possible single crystal spectra.

There are a number of strategies available for altering the reaction conditions in order to halt or slow down reactions, allowing for accumulation of intermediates that are stable or long-lived enough for diffraction data collection.

One of the most common approaches is mechanistic trapping where the reaction conditions can be perturbed in a number of different ways to prevent full turnover. Probably the simplest approach is to withhold a substrate, alternatively reactions involving a proton transfer step can be slowed or stopped by altering the pH [6, 8] Mutant proteins or substrate analogues, in which catalytically important functional groups are removed or modified, are also extremely useful in trapping intermediate states [9]. It is also possible to drive a reaction into steady-state, at which point the intermediate that occurs before the rate determining step should accumulate in the crystal [10]. In most cases, unless the reaction is completely halted by the absence of a required substrate or in steady-state, the intermediate will eventually decay. Therefore, once the desired intermediated has formed, the crystals are normally cryocooled to

Box 18.1: Tips

How to get good quality spectra

- Minimise solvent around the crystal
  - Use a loop that matches the size of the crystal
  - Mesh mounts work fine, but make sure to take a reference on the mesh
  - UV/vis transparent mesh mounts are available
- Optimise cryoprotection – a critical step!
  - Avoid glycerol if looking at features around 500–700 nm and for Raman
  - Ice reduces transparency and scatters light
- Try different parts and orientations of the crystal for the "best" spectrum
  - Avoid icy areas
  - Try finding the thinnest part in high optical density samples
  - Remember the spectrum is anisotropic, there may not be an exact match to the solution spectrum, but you will be able to follow relative changes
- Use a loop that is still big enough to fit the focal spot of the light source
  - Ideal to have a loop big enough to be able to take a reference spectra from the mother liquor
  - This is especially useful for high optical density samples
- Dealing with high optical density samples
  - If you can't see through it, neither can the spectrometer – smaller crystals are better!
  - Bandpass filters can allow you to increase exposure time without saturating the detector
  - Raman and Fluorescence in backscatter mode are not affected by high optical density, but be aware you will only obtain spectral data from the surface layers of the crystal.

100 K for X-ray diffraction data collection. This further stabilises the intermediate and allows long-term storage under liquid nitrogen until beamtime becomes available.

An alternate approach to trap intermediates after initiating the reaction in the crystal at room temperature, is to rapidly flash-cool the crystal to 100 K when it reaches the intermediate of interest. This is best done after the kinetics have been systematically characterised in the crystal, and the time-scales of intermediate state formation and lifetime are known. However, there are numerous examples of intermediate states successfully trapped by flash-cooling on-the-fly [11]. Here, several delays (usually from seconds to minutes or hours) after reaction initiation are used after which the reaction is stopped by rapid cooling to 100 K. The structures are then determined and the electron density, together with any complementary spectroscopic information is used to identify any trapped intermediates.

If the intermediates of interest cannot be mechanistically trapped, or are too short-lived at room temperature for cryo-trapping, intermediate accumulation can also be improved by lowering the temperature. This lowers the reaction rates and may enable stabilisation of an intermediate that is otherwise inaccessible [12]. A detailed description of buffer compositions for < 0 °C experiments can be found in Douzou [13].

18.3 Radiation Damage

Once a crystal containing a trapped intermediate has been obtained the next step is to collect a diffraction dataset. Extreme care must be taken to ensure that radiation damage does not change the state of the crystal during the data collection, even at 100 K. Radiation damage occurs due to absorption of photons from the X-ray beam, by the crystal, giving rise to high-energy electrons that propagate through the crystal and cause chemical changes [14]. Low dose radiation damage affects involve chemical modifications of the protein, i.e. disulphide bond breakage or decarboxylation of glutamates and aspartates, which manifest as absent electron density in the refined structure. At higher doses the crystal eventually stops diffracting. Proteins that use metals or redox cofactors are highly susceptible to X-ray induced reduction of these sites and this can severely hamper the structure determination of reaction intermediates. Reduction of all metal centres and redox cofactors can occur within a few kGy of absorbed dose, equivalent to one or two images on a modern synchrotron source.

Single crystal spectroscopy can again be used during the diffraction experiment to monitor the progression of any changes to trapped intermediates. In many cases single crystal spectrometers have been integrated directly into crystallography beamlines, allowing for simultaneous collection of diffraction and spectroscopic data. It is now also possible, indeed encouraged, to deposit spectroscopic information in addition to the atomic coordinates and structure factors in the Protein Data Bank [15, 16].

It should be noted that radiation damage is not always regarded as a curse, but in some cases as a blessing. The electrons generated during the exposure can be used as a reductant to drive an enzymatic reaction. This approach has been most successfully applied to the study of heme containing enzymes, where composite datasets with increasing dose have revealed details of oxygen activation [17, 18].

18.4 Complementary Approaches for Larger Scale Changes

Electronic changes in redox centres or side chain reorientation during catalysis often require only small movements in the protein that can easily occur within the conformational restrictions imposed by the crystal lattice. However, larger conformational changes linked to function may be severely restricted by the crystal packing, resulting in either no or only partial reaction, or in the crystal shattering upon addition of substrate. There are also many proteins, for example membrane proteins or large protein complexes, for which crystallization is extremely challenging. For these systems, alternative methods are required in order to gain structural insights into their function.

Small-angle X-ray scattering (SAXS) data provides information on the solution structure of the macromolecule of interest. A dilute solution of randomly oriented macromolecules produces a radially isotropic scattering pattern. The raw data provides basic information, such as radius of gyration, maximum particle dimension. Fourier transformation of the curve results in a distance distribution function, which can give information on the overall shape of the macromolecule. SAXS data collection and processing is treated in more detail elsewhere in this volume. Time-resolved SAXS can be used to study protein folding and oligomerisation [19–21].

Hydroxyl-radical footprinting (OHFP) is another solution-based method which reports on the structure of macromolecules by mapping solvent-exposed regions on their surface. The probes used are hydroxyl radicals. Initial studies used chemical methods to generate these radicals [22]. Hydroxyl radicals can also be formed in solution by ionising radiation, such as γ- and X-rays [23]. Using ionising radiation, hydroxyl radicals are formed throughout the sample, and the use of synchrotron radiation has made it possible to resolve changes in macromolecular structure down to millisecond time resolution [24]. In proteins, hydroxyl radicals are able to oxidise 14 of the 20 naturally occurring amino acids, providing high sequence coverage for the mapping [25]. The resulting covalent modifications can be accurately detected by mass spectrometry, following an enzymatic digest [26].

Mass spectrometry methods also provide a route to the time-resolved study of conformational changes in non-crystalline samples. Probably the best known of the mass spectrometry techniques is hydrogen-deuterium exchange (HDX), where changes in conformation alter the rate at which backbone amide protons exchange with solvent. It should be noted that HDX experiments can also be analysed using NMR [27]. More recently, a new mass spectrometry technique for analysing conformational change has been developed. Ion-mobility mass spectrometry is able to resolve different conformers of the same species, as well as provide information on their cross-section. This has proved especially powerful for the study of large-scale conformational change and protein complex assembly [28].

A number of spectroscopic methods, including NMR, EPR and FRET are able to provide increasingly accurate distance information using both intrinsic metals and fluorophores as well as labelling techniques [29–31]. Finally, electron microscopy is an invaluable tool for the structural analysis of very large macromolecular complexes and is able to provide time-resolved structural information [32].

18.5 Summary & Outlook

There is increasing interest in studying the dynamic behaviour of macromolecules. This is accompanied by an increasing awareness of the various experimental tools, and the time and length-scales accessible to each. Successful synthesis of all this information into a coherent whole is reliant on the availability of software able to accept experimental data of all kinds, convert this to restraints and use these to restrain a model of how the structure of the macromolecule of choice changes during its reaction. The development of this more detailed understanding of functionally relevant protein motions will help us to understand their role in activity and regulation.

References

1. Mozzarelli A, Rossi G-L (1996) Annu Rev Biophys Biomol Struct 25:343–365
2. Makinen MW, Fink AL (1977) Annu Rev Biophys Bioeng 6:301–343
3. Wilmot CM, Sjögren T, Carlsson GH, Berglund GI, Hajdu J (2002) Methods Enzymol 353:301–318
4. Pearson AR, Owen RL (2009) Biochem Soc Trans 37:378–381
5. Sjögren T, Carlsson G, Larsson G, Hajdu A, Andersson C, Pettersson H, Hajdu J (2002) J Appl Cryst 35:113–116
6. Merli A, Brodersen DE, Morini B, Chen Z, Durley RC, Mathews FS, Davidson VL, Rossi GL (1996) J Biol Chem 271:9177–9180
7. Carey PR (2006) Ann Rev Phys Chem 57:527–554
8. Pearson AR, Wilmot CM (2003) Biochimica et biophysica acta 1647:381–389
9. Kovaleva EG, Lipscomb JD (2008) Biochemistry 47:11168–11170
10. Wilmot CM, Hajdu J, McPherson MJ, Knowles PF, Phillips SEV (1999) Science 286:1724–1728
11. Kovaleva EG, Lipscomb JD (2007) Science 316:453–457
12. Sjögren T, Hajdu J (2001) J Biol Chem 276:13072–13076
13. Douzou P (1977) Cryobiochemistry: an introduction. Academic, London/New York
14. Garman EF (2010) Acta Cryst D66:339–351
15. Orville AM, Buono R, Cowan M, Heroux A, Shea-McCarthy G, Schneider DK, Skinner JM, Skinner MJ, Stoner-Ma D, Sweet RM (2011) J Synch Rad 18:358–366
16. Pearson AR, Mozzarelli A (2011) Biochim Biophys Acta 1814:731–733
17. Schlichting I, Berendzen J, Chu K, Stock AM, Maves SA, Benson DE, Sweet RM, Ringe D, Petsko GA, Sligar SG (2000) Science 287:1615–1622
18. Berglund GI, Carlsson GH, Smith AT, Szöke H, Henriksen A, Hajdu J (2002) Nature 417:463–468
19. Pollack L, Tate MW, Darnton NC, Knight JB, Gruner SM, Eaton WA, Austin RH (1999) Proc Natl Acad Sci U S A 96:10115–10117
20. Pollack L, Tate MW, Finnefrock AC, Kalidas C, Trotter S, Darnton NC, Lurio L, Austin RH, Batt CA, Gruner SM, Mochrie SG (2001) Phys Rev Lett 86:4962–4965

21. Lamb JS, Zoltowski BD, Pabit SA, Crane BR, Pollack L (2008) J Am Chem Soc 130:12226–12227
22. Schmitz A, Galas DJ (1980) Nucleic Acids Res 8:487–506
23. Ward JF (1988) Prog Nucleic Acid Res Mol Biol 35:95–125
24. Chance MR, Sclavi B, Woodson SA, Brenowitz M (1997) Structure 5:865–869
25. Xu G, Takamoto K, Chance MR (2003) Anal Chem 75:6995–7007
26. Gupta S, Bavro VN, D'Mello R, Tucker SJ, Venien-Bryan C, Chance MR (2010) Structure 18:839–846
27. Raschke TM, Marqusee S (1998) Curr Opin Biotech 9:80–86
28. Knapman TW, Morton VL, Stonehouse NJ, Stockley PG, Ashcroft AE (2010) Rapid Commun Mass Spec 24:3033–3042
29. Schuler B, Eaton WA (2008) Curr Opin Struct Biol 18:16–26
30. Otting G (2010) Annu Rev Biophys 39:387–405
31. Van Wonderen J, Kostrz DN, Dennison C, MacMillan F (2013) Angew Chem Int Ed 52:1990–1993
32. White HD, Thirumurugan K, Walker ML, Trinick J (2003) J Struct Biol 144:246–252

Chapter 19
Molecular Dynamics Probed by Short X-ray Pulses from a Synchrotron

Michael Wulff, Qingyu Kong, Jae Hyuk Lee, Tae Kyu Kim, Marco Cammarata, Dmitry Khakhulin, Savo Bratos, Jean-Claude Leichnam, Friedrich Schotte, and P.A. Anfinrud

19.1 Introduction

The intensity of single X-ray pulses from third generation synchrotrons like the ESRF, APS and Spring8 is now so high that it is feasible to conduct time resolved experiment with diffraction, scattering and spectroscopic techniques on samples ranging from small molecules to proteins, with 100 ps time resolution. This limit is dictated by the X-ray pulse length from a synchrotron. In a time-resolved X-ray experiment one first needs to initiate the process of interest in the sample and then open the X-ray shutter and record the signal as a function of time. In the experiments shown in this review, the signals are recorded on a large CCD detector in order to record the signal efficiently in space. Unfortunately, the time-resolution of current CCDs is still too slow, 100 μs at best, to be useful for monitoring molecular processes on molecular time scales. The only way to explore faster phenomena is to

M. Wulff (✉) • D. Khakhulin
European Synchrotron Radiation Facility, Experiments Division, 38043 Grenoble, France
e-mail: wulff@esrf.fr

Q. Kong
Division Expériences, Synchrotron Soleil, Saint-Aubin, 91192 Gif-sur-Yvette, France

J.H. Lee
Department of Chemistry, Kaist, Daejeon, 305-701, Republic of Korea

T.K. Kim
Department of Chemistry, Pusan National University, Busan, Republic of Korea

M. Cammarata
Institut de Physique, Université de Rennes 1, 35042 Rennes, France

S. Bratos • J.-C. Leichnam
Laboratoire de Physique Théorique de la Matière Condensée,
Université Pierre & Marie Curie, 75252 Paris 05, France

F. Schotte • P.A. Anfinrud
NIDDK, Laboratory of Chemical Physics, NIH, Bethesda, MD 20892-0520, USA

J.A.K. Howard et al. (eds.), *The Future of Dynamic Structural Science*,
NATO Science for Peace and Security Series A: Chemistry and Biology,
DOI 10.1007/978-94-017-8550-1_19,

use the pump-probe method where the time resolution is obtained from varying the pump-probe delay. In the examples discussed here, the samples were excited by picosecond or nanosecond laser pulses and the excited samples were probed by 100 ps X-ray pulses from an undulator.

The pulse separation from synchrotrons is very short, nanoseconds to hundreds of nanoseconds, depending on the filling pattern of the storage ring. That is significantly shorter than the readout time of a CDD detector. To exploit the intrinsic time-resolution of a 100 ps X-ray pulse, one needs therefore to isolate pulses from their neighbours with a chopper and to accumulate the single-pulse shots on the CCD. Since a chopper greatly reduces the average intensity (ph/s) by factors of about 1,000, each pulse should be as intense as possible. In the examples shown here, the pink beam from the U17 undulator on beamline ID09 at the ESRF was used to maximize the pulse intensity. The spectrum of the U17 is shown in Fig. 19.1b. Note the semi Gaussian shape of the undulator fundamental. The trade-off is that the effective wavelength, which is needed to convert scattering angle to q, is less well defined due to this asymmetry. This problem can be solved by monochomatising the beam with multilayer crystals that are tailor made for the undulator spectrum. The best 1–3 % part of the undulator spectrum near the peak can then be used efficiently at the expense of loss of peak intensity from the reflectivity of the multilayers. In the following section we will briefly describe the pump-probe instruments on ID09. For more technical details about pump-probe beamlines at the ESRF, APS and KEK the reader is referred to Refs. [1–5].

19.2 Beamline for Pump-Probe Experiments

Figure 19.1a shows the laser pump and X-ray probe scheme implemented on ID09B. For small molecules in solution the setup comprises a closed-loop liquid jet that injects a new sample for every laser/X-ray pulse at 1 kHz, a short-pulse laser, a high-speed chopper and a CCD detector. In these small molecule experiments an ultra short (0.1–1 ps) laser pulse initiates a photo reaction in a molecule and a delayed 100 ps X-ray pulse takes a scattering snapshot of the structure. When a series of snapshots at different delays are stitched together, one gets a movie of the change in structure of the excited molecules. The instrumental time resolution is determined by the convolution of three terms: the pump and the probe pulse lengths plus their relative jitter, 100 ps, 1 ps and 1 ps respectively. After convolution, the time resolution is about 100 ps (FWHM values). Note that the shortest oscillatory signal that can be resolved with a 100 ps X-ray pulse is 200 ps according to Nyquist-Shannon's theorem.

The X-rays on ID09 are delivered by two in vacuum undulators, a U17 and a U20. The magnetic field from the undulator magnets forces the electrons into sinusoidal orbits. The X-rays are mainly emitted at the turning points where the acceleration is greatest. ESRF electrons are accelerated to 6 GeV in a booster synchrotron and then injected into a 844 m long storage ring which has 32 straight

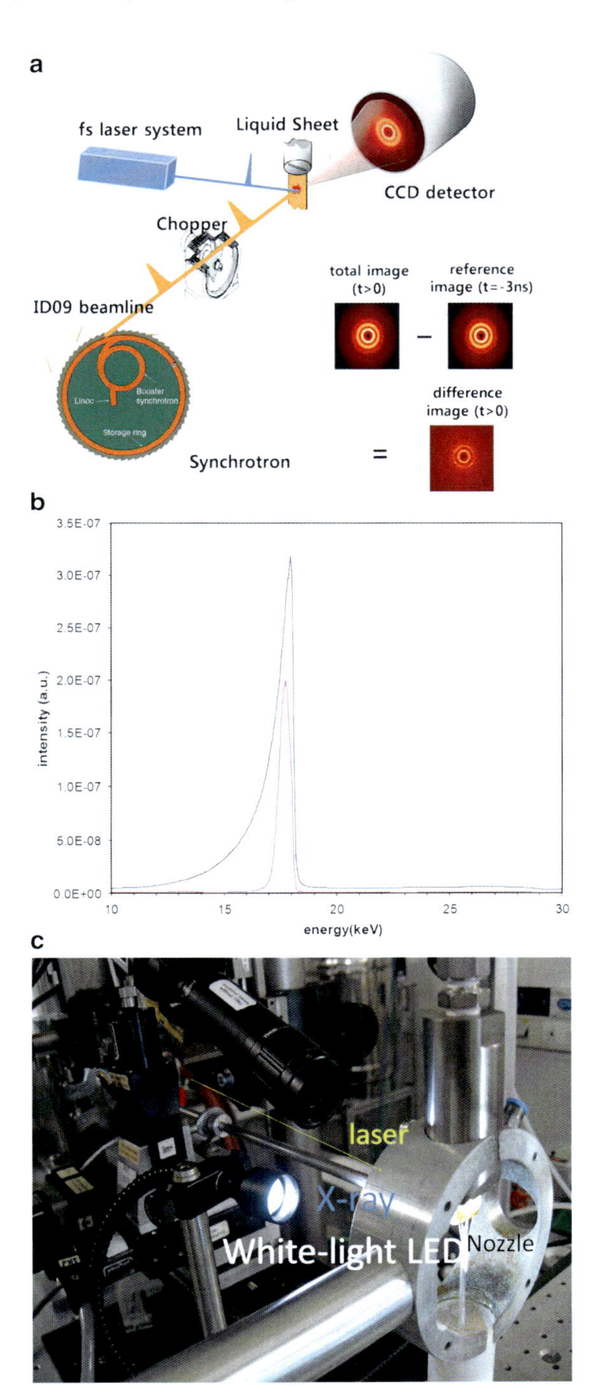

Fig. 19.1 (**a**) Experimental setup for a time resolved scattering experiment. X-rays from a synchrotron are guided through a beamline and single X-ray pulses are isolated by a high-speed chopper. The chopper consists of a flat triangular disk spinning at the same frequency as the laser. The laser pulses excite the sample and the time-delayed X-ray pulses probe the photo induced structural changes as diffraction images on the CCD detector. The difference diffraction image for a given positive time delay is generated by subtracting the reference image at −3 ns from the diffraction image. (**b**) Spectral shape of an X-ray pulse from the U17 undulator. To increase the signal-to-noise ratio the raw quasi-monochromatic (3 % bandwidth) X-ray beam is used. (**c**) Picture of the pump-probe setup on ID09B. The *blue* and *orange lines* represent the paths of the X-ray and laser beams, respectively. Both beams are overlapped in the most stable region of the jet (300 mm thickness). Diffraction patterns from the scattered X-rays are recorded by a MarCCD detector

sections for undulators. The electron bunches make one round trip in 2.82 μs which corresponds to a frequency of 354.6 kHz. In the uniform filling of the ring at the ESRF the electrons are evenly distributed over 992 bunches with a time separation of 2.84 ns. This mode is not suitable for the fastest pump-probe experiments as it is impossible to isolate single pulses with a chopper. That requires a time separation of at least 150 ns. At the ESRF the bunch modes for pump-probe experiments are: 4-bunch mode (705 ns, 40 mA), 16-bunch mode (176 ns, 90 mA) and the 7/8 mode (176 ns, 200 mA). In the 7/8 mode, 7/8th of the ring is filled uniformly with a single bunch placed in the centre of the gap. The 7/8 mode has the highest current, 200 mA, which is good for most experiments, while also providing a single bunch for timing experiments. The 7/8 mode is the default mode at the ESRF and it runs 75 % of the time. The single bunch current is low, 1.5–3.0 mA, but on the other hand the pulses are shorter, 60–80 ps (FWHM).

19.2.1 Undulator and Chopper

The X-ray pulses used in experiments discussed here were produced by the U17 undulator which has a magnetic period of 17 mm. The magnets are located inside the storage ring vacuum which makes it possible to move them as close as 3 mm to the electron beam and hereby increase the field without spoiling the lifetime and heating the magnets. The undulator spectrum is shifted in energy by varying the magnet gap. The U17 fundamental is at 15.2 keV at 6 mm gap and it is shifted to 20.3 keV when the gap is fully open at 30 mm. The best undulator gap for scattering experiments is ~9 mm. Here the spectrum is dominated by the first harmonics at 18 keV ($\lambda = 0.67$ Å) and higher harmonics are removed by the toroidal mirror which has a cut-off at 26 keV. For Laue diffraction the increased diffraction power at lower energies favours 15.2 keV at 6 mm gap. The bandwidth (bw) of the spectrum is 3.0 %. The shape of the spectrum is "half-Gaussian" with a rising edge from the red shift of the off-axis radiation. The pink beam is 250 times more intense than the beam from a conventional silicon monochromator. Many liquid experiments would not be possible with a normal monochromatic beam. The exposure time would be too long resulting in radiation damage from both the laser and X-rays. The X-rays are focused by a platinum-coated toroidal mirror into a 0.1×0.06 mm^2 spot which has up to 10^{10} X-ray photons per pulse in four bunch mode (10 mA per bunch).

The high energy of ESRF electrons, 6 GeV, makes it possible to produce harder X-rays with the undulator fundamental in the range 15–25 keV. That extends the q range to 12 Å^{-1} (and spatial resolution) and makes it easier to normalise the scattering to the atomic form-factors of the molecule of interest.

The frequency of the X-rays from the 16-bunch mode is 5.7 MHz which is much higher than the frequency from wavelength tuneable femtosecond lasers.

They normally operate in the 1–10 kHz range to have sufficient pulse energy for efficient wavelength conversion. Again since the laser and X-ray pulses have to arrive in tandem, the frequency of X-rays has to be lowered to match that of the laser. This is done by the high-speed chopper that normally isolates pulses at 1 kHz. The ID09 chopper is shown schematically in Fig. 19.1a. The rotor is a flat triangle spinning at 1 kHz. One of three edges on the rotor has two slits that form a semi-open tunnel. The tunnel is 0.126 mm high. When the chopper rotates at 1 kHz, the tunnel opens for 265 ns.

19.2.2 Laser System

Highly intense laser pulses are needed to excite, *via* photo absorption, the solutes in the sample. At solute concentrations around 10 mM, a (0.1) mm^3 sample contains 6×10^{12} solutes. To fully excite the sample, the laser pulse has to have at least 6×10^{12} photons per pulse. In addition the laser pulse has to be shorter than the X-ray pulse to keep the time resolution at 100 ps. In most liquid experiments we use a Ti-sapphire femtosecond laser with a picosecond amplifier (0.6 or 1.2 ps pulses). The pulse-energy at 800 nm is ~4.0 mJ per pulse at 1 kHz. The wavelength can be changed by an optical parametric amplifier (OPA), the TOPAS from Light Conversion, which works by mixing harmonics in a non linear crystal. The TOPAS covers 250–2000 nm. The laser is focused on the sample into a 0.2 mm^2 spot and the direction of the laser beam is usually 10° from the X-ray beam to facilitate the alignment on the sample microscope. Care has to be taken to avoid hot spots which could lead to multi-photon absorption in the sample.

In a pump-probe experiment the laser pulse has to be synchronized to the X-ray pulse to very high precision. As described in the previous section, the mechanical chopper isolates a train of pulses at 986.3 Hz, the 360th sub-harmonic of the RF clock. A frequency divider combined with a phase shifter produces a RF/4 signal at 88.05 MHz which is used to synchro-lock the femtosecond oscillator (MICRA, Coherent). A sub train of these are amplified by a chirped-pulse-amplifier, the Legend, which produces a beam at 800 nm at 986.3 Hz. The time-delay between the laser and X-ray pulse is controlled electronically to a resolution of 11 ps by an electronic board, the N354, developed by Christian Herve and Marco Camarata (ESRF). The short term jitter between the laser and the X-ray pulse is 1 (FWHM). On longer time scales, hours, the arrival time depends on the bunch charge due to the decay in the bunch current. Hence in 16 bunch mode, the X-rays arrive 10 ps earlier at the end of a fill due to the higher effective force on the bunch in the radio cavities. By measuring the delay on-line, this "bunch slippage" can be corrected in the analysis.

19.2.3 Spatial and Temporal Overlaps

To increase the signal-to-noise ratio and to define the accurate time-delay, the laser and X-ray beams have to be aligned in space *and* time. To check the temporal overlap we use a fast GaAs photo detector that is sensitive to both X-rays and visible light [6]. With the13 GHz oscilloscope on ID09B, the pump-probe delay can be measured to ~25 ps accuracy. In sub nanosecond experiments, the laser/X-ray delay is sometimes measured on-line with an X-ray diamond detector in transmission and with the GaAs detector in the scattered laser beam.

The spatial overlap between the X-ray and laser pulses is measured by replacing the sample with a 25 μm pinhole. The laser position is then adjusted by moving the focusing lens which is motorised in x, y and z. For fine adjusting the spatial overlap we use the CCD signal from the excited sample. Specifically we measure the *change in scattering* from the thermal expansion caused by the energy released from the excited molecules to the solvent. For the systems studied here that happens typically after 1 μs. We then calculate the ratio of the intensity inside and outside the liquid peak. When the sample expands, the liquid peak shifts to lower angles. The low-angle intensity increases and the high-angle decreases. The ratio of the two, the FOM (figure of merit), is a measure of the temperature rise in the solvent. The X-ray beam is typically 60 μmV and 100 μmH. The laser spot is circular with a 200 μm diameter.

19.2.4 CCD Detector and the Data Acquisition

In the examples below the scattered (or diffracted) X-rays were recorded on a 133 mm diameter Mar CCD which was replaced in 2009 by a faster Frelon camera from the ESRF. The detector is typically 45 mm from the sample for small molecules and 250 mm for proteins which gives q ranges of 0.3–8.6 $Å^{-1}$ and 0.02–2.4 $Å^{-1}$ respectively at 18 keV. The detector is centred on the incoming beam behind a Ø1.6 mm beam stop. The scattered X-rays are converted to light by a phosphor screen and the optical image is relayed to the CCD by a fibre-taper-optic. Scattering images are recorded on a 2048 × 2048 CCD with an effective pixel size of 64.276 μm (MAR CCD). The quantum efficiency at 18.0 keV is ~50 %. The CCD is cooled to −70 °C to reduce the dark current to 0.02 ADU/s. The readout noise is 0.1 ADU/pixel. The typical exposure time in a small molecule experiment is 5 s. The incoming X-ray flux is $\sim 5 \times 10^{11}$ ph/s of which $\sim 1 \times 10^{-3}$ are scattered on the detector.

Scattering data are typically collected for ten time-delays evenly spaced on a logarithmic scale from 100 ps to 3 μs. There images are interleaved by none-excited images at −3 ns, i.e. for laser pulses arriving after the X-rays which probes the non-excited signal. In this way the effect of slow drifts in beam position, drifts in the detector and radiation damage cancel out in the difference images. The time series is usually repeated 10 times and averaged. If a radial distribution curve deviates by more than 4–5 σ from the average, it is rejected.

With the high-speed Frelon camera the read-out time is reduced to 0.5 s per image. Together with improvements in the beamline network, the Frelon can collect up to 1,200 images per hour. In one experiment it is now possible to study geminate recombination inside a cage (0.01–1 ns), non geminate recombination from diffusive molecules (10 ns–10 μs) and the hydrodynamics of the liquid (0.1 ns–1 μs) in one time series. A special application is laser slicing where the X-ray pulse is cut in two by placing the laser inside the X-ray pulse in space. The excited sample is thus probed by a truncated X-ray pulse which is shorter than 100 ps. Note that a 100 ps X-ray pulse is the equivalent of a 30 mm long bullet in space. In certain cases, laser sliced data can be deconvoluated to a time resolution limited only by the step of the time series. The minimum step size at the ESRF is 11 ps. We should stress that deconvolution does not replace a shorter X-ray pulse from an XFEL for example, but it can enhance features that are present without slicing.

19.2.5 Liquid Jet

Two different flow systems have been used so far. The solution is either circulated through a capillary or through a sapphire nozzle. In the latter case the nozzle makes a 0.3 mm thick sheet of liquid which is 6–8 mm wide. The inner diameter of the capillary is 0.3–1.0 mm. The open jet has the advantage that there is no scattering from the capillary which reduces the background. That is important for weakly scattering samples (lowZ). In practice the capillary is often pierced by the laser or the flow is stopped due to deposits inside the capillary. For both systems the flow speed is typically 1 m/s which is sufficient for having a fresh sample for every new laser/X-ray pair at 1 kHz. The output from the jet is collected in a funnel. For reversible photo molecules, the solution can often be cycled through the beam many times, typically for 1–2 h.

19.3 Probing Solution Phase Dynamics with X-rays

Time resolved X-ray scattering has been used to capture the structures of photo molecules and to track complex reaction pathways as a function of time. In the following we will show some examples to illustrate the rich variety of phenomena that can be studied. Five type of systems have been studied, simple molecules [7–14], organometallic complexes [15], nanoparticles [16] and proteins [17, 18] covering a variety of structural problems in photochemistry. We will now show what can be learned from these studies.

19.3.1 Photo Dissociation of HgI_2 in Methanol

The photo dissociation dynamics of HgI_2 in the gas and solution phase has been widely studied by time-resolved spectroscopic techniques [19–21]. Zewail and co-workers studied two-body dissociation by laser-induced fluorescence [20] and two- and three-body dissociation by time-resolved mass spectrometry [21]. In solution the solvation dynamics of HgI_2 in ethanol was investigated by femtosecond absorption spectroscopy [22–24]. The vibrational relaxation and rotational dynamics of HgI_2 after impulsive dissociation were extensively studied in these works. Despite ample information about the dynamics on short time-scales, a comprehensive mechanism spanning from picoseconds to microseconds is still lacking in these early works. We applied time resolved X-ray scattering to HgI_2 and demonstrated that this information can be obtained directly. Figure 19.2a shows the experimental and theoretical intensities $q\Delta S(q)$. Figure 19.2b shows the corresponding experimental and theoretical radial intensity $r\,\Delta S(r)$. In our global-fitting analysis we used two-body dissociation $HgI_2 \rightarrow HgI + I$ as the primary pathway and optimized the rate constants for geminate and non-geminate recombination to HgI_2, non-geminate formation of molecular iodine and included the laser beam size for the thermal response of the liquid. The best fits are shown in Fig. 19.2a, b. The fitted theoretical curves successfully reproduce the experimental curves in q- and r-space. The global-fitting gives the population changes of all the chemical species (Fig. 19.2c) and the temperature and density change of the solvent (Fig. 19.2d). Figure 19.2e summarizes the results. With photo excitation at 267 nm, 34($\pm$3.8) % of the excited HgI_2 dissociate into HgI + I. Then 71 %($\pm$1.2) of the HgI radicals recombine non-geminately with atomic iodine (that escaped from other cages) to reform HgI_2. In parallel 29($\pm$0.5)% of the iodine atoms recombine to form I_2. The rate constants for the non-geminate recombination of HgI_2 and I_2 are 5.0 ($\pm$0.5) $\times$ 10^{10} M^{-1} s^{-1} and 1.65($\pm$0.25) $\times$ 10^{10} M^{-1} s^{-1}, respectively [25].

Before the thermal expansion starts after 100 ns, the solute-alone dynamics dominates over the cage and solvent terms thanks to the heavy atoms Hg and I. In Fig 19.3a, b, the decomposition into solute, cage, and solvent after 100 ps is shown. In the global-fitting the total difference signal was decomposed into these contributions. As expected, the signal from the solute dominates over the entire q range; however, the contributions from the cage and the solvent are not negligible at low q (Fig. 19.3a). In Fig. 19.3b, the negative peak at 2.6 Å is mainly from the depletion of the Hg–I bond in HgI_2 and the formation of a new Hg–I bond in transient HgI.

The negative peak at 5.3 Å corresponds to the I···I distance in the parent molecule. The positive peak at 3.5 Å is assigned to a reorganization of the solvent around the truncated solute (new cage) and solvent heating which leads to the broadening of methanol-methanol distance. These interpretations can also help in determining the primary reaction pathway in the dissociation of HgI_2 as follows. Four pathways were considered: $HgI_2 \rightarrow HgI + I$, $HgI_2 \rightarrow Hg + I + I$, $HgI_2 \rightarrow Hg + I_2$ and the linear isomer reaction $HgI_2 \rightarrow$ HgI–I The data at early times

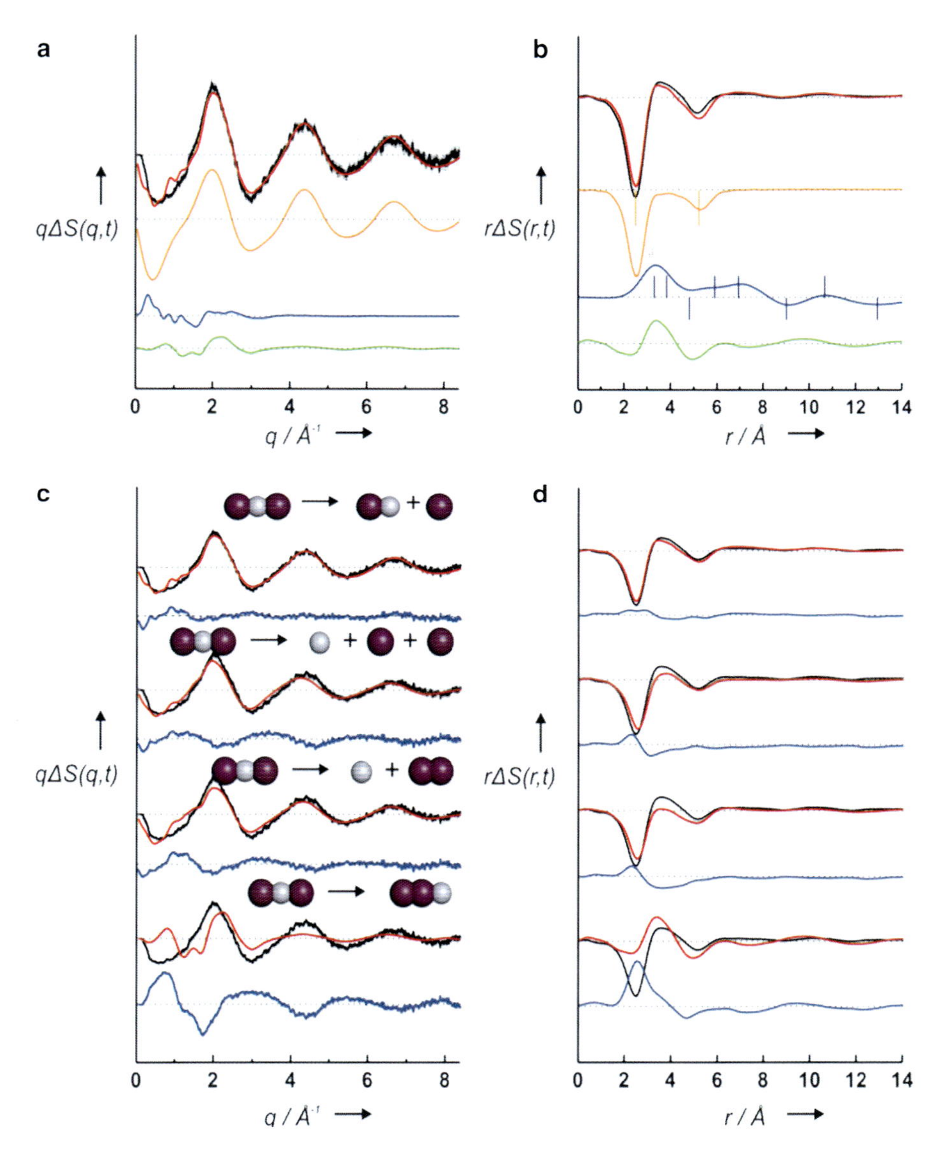

Fig. 19.2 Contributions from the solute-only, solute–solvent (cage), and solvent–solvent correlations to the difference intensities for HgI_2 in methanol at 100 ps. The curve fitting is based on MD and experimental solvent differentials. They include intramolecular and intermolecular contributions from the solutes and the solvent. The experimental (*black*) and theoretical (*red*) $q\Delta S(q)$ (**a**) and corresponding $r\Delta S(r)$ (**b**) at 100 ps are shown. Also given are the decomposed components: solute without cage (*orange*), cage effects (*blue*), and solvent contribution (*green*). Decomposed components of the cage (*blue*) and the solvent term (*green*) in (**b**) are multiplied by 3 to magnify the main peaks and valleys. (**c**) Determination of the photodissociation pathway of HgI_2 in methanol at 100 ps. Theoretical (*red*) and experimental (*black*) difference intensities for candidate channels are shown as $q\Delta S(q)$ curves (**c**) and $r\Delta S(r)$ curves (**d**). The difference residuals between theoretical and experimental curves are shown in *blue*. The two body dissociation pathway ($HgI_2 \rightarrow HgI + I$) gives the best fit

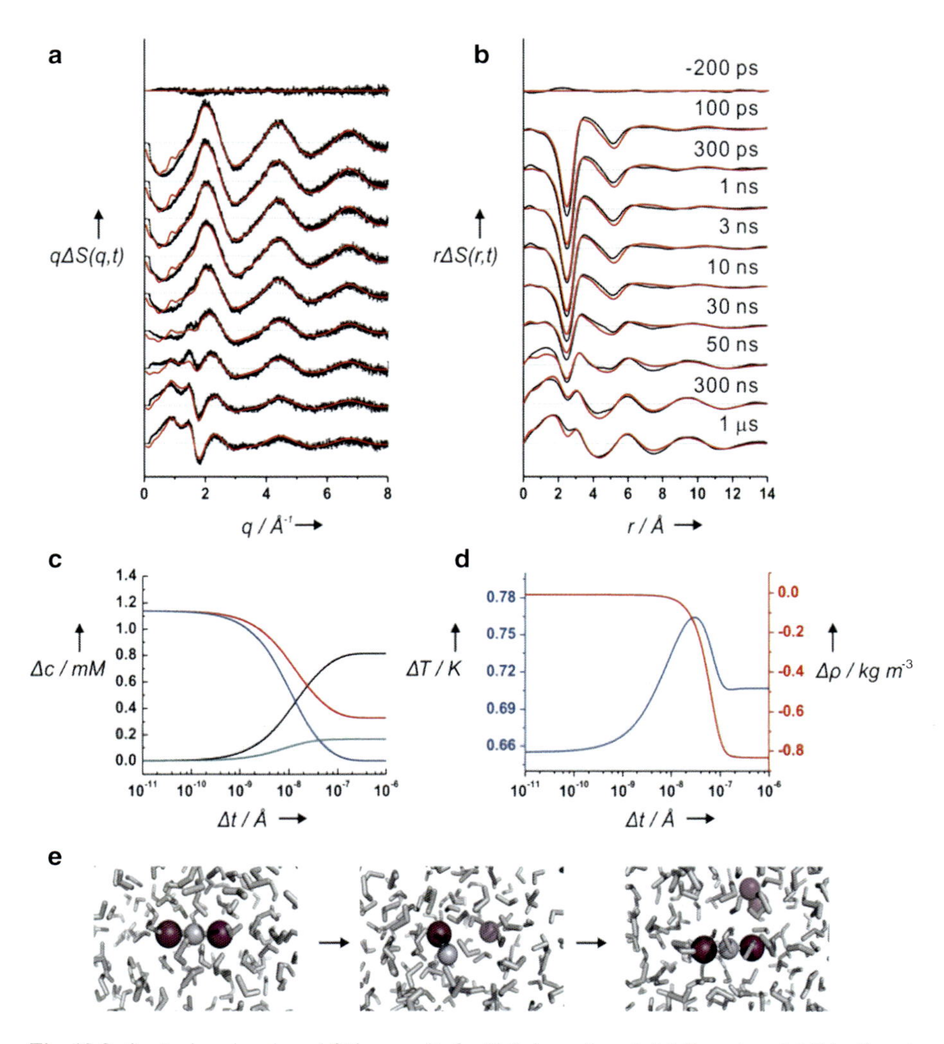

Fig. 19.3 Scattering signals and fitting results for HgI_2 in methanol. (**a**) Experimental (*black*) and theoretical (*red*) difference intensities qΔS(q) and their error bars in q space. The time delays from *top* to *bottom* are −100 ps, 100 ps, 300 ps, 1 ns, 3 ns, 10 ns, 30 ns, 50 ns, 300 ns, and 1 ms. (**b**) Difference RDFs, rΔS(r). Shown are the sine-Fourier transforms of the intensities in (**a**). (**c**) Concentration change of the photoreaction of HgI2 versus time. HgI, I, I_2 and HgI_2 are displayed in *red, blue, green*, and *black lines*, respectively. (**d**) Change in the solvent density (*red*) and temperature (*blue*) as a function of time. Note the duality between the decay in the excited-state population in (**c**) and the rise in temperature in (**d**). (**e**) Schematic reaction mechanism by TRXL for the photoreaction of HgI_2. The pictures are snapshots from the MD simulations

(100 ps, 300 ps, and 1 ns) are fitted globally using these pathways. Figure 19.3c, d show the fitted results for $q\Delta S(q)$ and $r\Delta S(r)$ for these trial pathways. We can rule out isomer formation because of the poor agreement between experiment and theory. The fits for two-body dissociation match perfectly well the experimental

data over entire q range and give the smallest χ^2 value. The fits for three-body dissociation and I_2 formation have a rather good figure of merit, but the fits are worse than the two-body case. In fact, in the fits for these channels, the disagreements are significant for the positive peak at 2 $Å^{-1}$ and the negative peak at 3 $Å^{-1}$. As shown in Fig. 19.3a, the contribution from the solvent is non negligible. Thus the disagreements in this region for these two channels indicate that the solvent contribution for these channels does not agree with the actual energy transfer to the solvent. This tells us that when the energy flow is included in the analysis, the solvent term can help to discriminate between channels with very small differences at high q. These results clearly demonstrate that TRWAXS determines concurrently the structures of excited solutes in their cages and the structure and hydrodynamics of the solvent. It is also shown that TRWAXS can be used as an ultrafast calorimeter since the determination of transient structure (the primary dissociation channel) is strongly correlated with the energy transfer from solute and solvent molecules.

19.3.2 The Transient States of $C_2H_4I_2$

Halogen dissociation from haloalkanes are of great interest since the reaction products is under stereo chemically controlled with respect to the C=C double bond [26–29]. Haloethyl radicals such as CH_2ICH_2* and CF_2ICF_2* are reaction intermediates from halogen dissociation and their structures are thought to account for the observed stereo selectivity. In the bridged structure, the halogen is shared equally between the two carbon atoms, whereas in the *classical mixture* structure (a mixture of anti and gauche conformers), the primary halide resides predominantly on one of carbon atoms [7]. A *bridged* structure prevents the rotation about the C–C bond, thereby maintaining the functional group positions in the product and providing stereo chemical control. However, despite numerous theoretical and experimental investigations, the structure of the intermediate in solution remained unknown. To unravel the structure of the intermediate and the related kinetics, we applied time resolved X-ray scattering (TRWAXS) to the iodine elimination reactions $C_2H_4I_2$ and $C_2F_4I_2$ both dissolved in methanol [7, 11].

To explain the experimental data theoretically and fit the signals globally, we included all putative structures in the fitting for both molecules. Our TRWAXS data show that the iodine elimination in $C_2H_4I_2$ and $C_2F_4I_2$ are completely different (Fig. 19.3c, e for $C_2H_4I_2$ and Fig. 19.3d, f for $C_2F_4I_2$). The global-fitting results for $C_2H_4I_2$ show that C_2H_4I and I are the dominant species after 100 ps and that the formation of C_2H_4 or I_2 is not observed. Subsequently C_2H_4I does not decay into C_2H_4 + I; rather it reacts with an iodine atom to form a linear isomer, $C_2H_4I{-}I$, with a bimolecular rate constant of $7.94(\pm3.48) \times 10^{11}\ M^{-1}\ s^{-1}$[25] which is larger by two orders of magnitude than the rate constant for non-geminate recombination of molecular iodine in CCl_4 solvent.

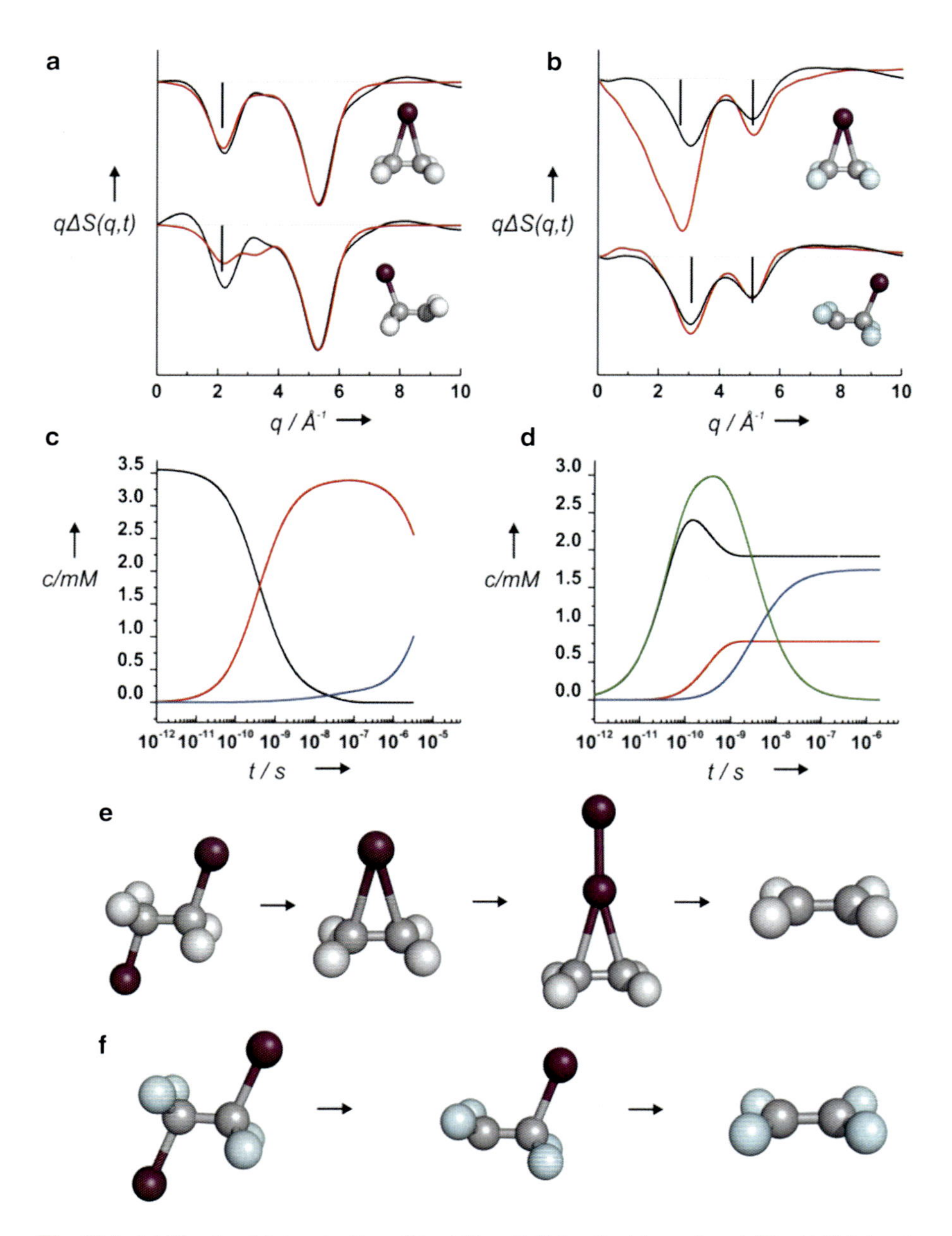

Fig. 19.4 (**a**) Structural determination of the 100 ps C_2H_4I radical in methanol. The C_2H_4I signal is obtained by subtracting other contributions from the raw data. This approach allows comparing the solution structure with gas-phase structures of the anti and bridged conformers. Experimental (*black*) and theoretical (*red*) RDFs for the two channels are shown together with their molecular structure (I: *purple* and C: *gray*). The *upper* and *lower curves* represent the bridged and the classical anti structure, respectively. (**b**) Structure determination of the 100 ps C_2F_4I radical in methanol. The solute-only term for the two candidate models (classical and bridged) are shown together with their molecular structure (I: *pink*, C: *white* and F: *yellow*). The data show that the C_2H_4I and C_2F_4I radical have the bridged and classical mixture structure, respectively. (**c**) and (**d**) Reaction kinetics and population changes of the relevant chemical species during the photo

This finding suggests that the isomer is mostly formed *via* in-cage recombination. The isomer, which is the major species in the nanoseconds regime, eventually decays into $C_2H_4 + I_2$ in microseconds with a rate constant of $1.99(\pm1.38) \times 10^5$ M^{-1} s^{-1} [25]. By contrast, C_2F_4I and I are major species at 100 ps as they are essentially formed within a few picoseconds. Complete dissociation into C_2F_4 and 2I is not observed in the investigated time regime. 20 ± 1.3 % of the C_2F_4I radicals decay to C_2F_4 + I with a time constant of 306 ± 48 ps. We can compare these values with gas-phase electron diffraction: 55 ± 5 % and 26 ± 7 ps [30, 31]. This illustrates how solvent molecules greatly reduce the rate and yield of secondary dissociation. That is not surprising as the available internal energy of the C_2F_4I radical in solution is lower than that in the gas phase due to the energy transfer to the solvent which typically occurs in tens of picoseconds. Molecular iodine I_2 is formed concurrently via recombination of two iodine atoms in about 100 ns with a time constant of $4.4(\pm1.25) \times 10^{10}$ M^{-1} s^{-1} [25], which is comparable with the rate constant for non-geminate recombination of molecular iodine in solution.

Of particular interest in these molecules is the structure of the transient species. As mentioned, TRWAXS is the only method that could answer whether the radical is *bridged* or *classically open*.

The results from this approach for $C_2H_4I_2$ and $C_2F_4I_2$ are shown in Fig. 19.4a, b, respectively. The transient species alone are found by subtracting the contributions from the solvent, cage and other nascent solutes and then compared with the calculated candidate structures. This finger print approach is identical to the one used in gas-phase time-resolved electron diffraction. The comparison between the 100 ps data and the putative intermediates demonstrates that the C_2H_4I radical has a *bridged* structure while C_2F_4I is a *mixture* of a *classical gauche* and an *anti* conformer. In C_2H_4I, the negative peak at 5 Å, the depletion of the I···I distance correlation in the parent molecule, is common for both models. The peaks between 1 and 3 Å are indeed dependent on the position of the I atom relative to the two carbons in the transient intermediate. Only the *bridged* structure matches the experimental peaks. Note that the *anti* model has two peaks from the two different C..I distances. For the C_2F_4I structure the broad peak between 2 and 4 Å is the fingerprint region. Only a *classical mixture* structure of C_2F_4I reproduces the broad negative peak in the fingerprint region. The position and line shape of the 3 Å peak are very sensitive to the position of the iodine atom relative to the two carbon atoms and the four fluorine atoms in the transient. Another argument for this structure assignment comes from the χ^2 values from the global-fits for the two models. In the elimination reaction of $C_2F_4I_2$, the χ^2 value for the *classical mixture* is 1.7, which is

Fig. 19.4 (continued) elimination reaction for $C_2H_4I_2$ (**c**) and $C_2F_4I_2$ (**d**) in methanol as a function of time. In (**c**), *black* is bridged C_2H_4I, *red* is the isomer, *blue* is the final productC_2H_4. In (**d**), *green* is the I atom, *black* is the classical form of C_2F_4I, *red* is C_2F_4 and *blue* is the I_2 molecule. (**e**) The reaction model of $C_2H_4I_2$. After photo-excitation, one I atom is detached from the parent molecule, and the C_2H_4I radical is bridged. Then C_2H_4I recombines with another I atom to form a linear isomer $C_2H_4I{-}I$. Then, the isomer breaks into C_2H_4 and I_2. (**f**) The reaction model of $C_2F_4I_2$. After photo-excitation, one I atom is detached from the parent molecule, and C_2F_4I takes the classical anti form. Some C_2F_4I decays into C_2F_4 + I. The I atoms finally recombine non-geminately into I_2

smaller than the χ^2 value of 3.2 for the *bridged* structure at all time-delays. When both models (*bridged and classical mixture*) are included in the global-fits, the *bridged* fraction is almost zero. These findings strongly suggest that C_2H_4I is *bridged* whereas C_2F_4I is a *classical mixture* and supports the fluorination effect on the radical which was predicted by quantum mechanical calculations.

19.3.3 Filming the Birth of a I_2 Molecule Inside a Cage

As already mentioned one can exploit the fact that the laser pulse (1 ps) and the synchronization (1 ps) are much shorter than the 100 ps X-ray pulse to explore phenomena on the sub 100 ps time scale. This was done in a recent study of the geminate recombination of I_2 in CCl_4 [12, 32]. The I_2 molecules were dissociated by a 1 ps laser pulse at 530 nm and the outcome probed by delayed 100 ps pink X-ray pulses from the U17 undulator ($E_f = 18$ keV, BW $= 3.5$ %). A large fraction of the molecules, 60 %, recombine with their original partner inside the cage and form vibrationally excited I_2 in the X state. In solution newly formed molecules cool down by transferring energy to the surrounding solvent, a process that takes tens of picoseconds. This cooling process is fundamental in chemistry and has so far never been visualised with a spatial probe like X-rays. A substantial gain in resolution was obtained by slicing the X-ray pulses in two with the 1 ps laser pulses and scanning the delay in steps of 11 ps. The laser sliced data was deconvoluted with a new algorithm that requires knowing the intensity profile of the X-ray pulse [32]. The final result of a lengthy analysis is shown in Fig. 19.5b. Figure 19.5b (a) shows the Morse Potential for I_2. For energies near the dissociation limit, the I atoms vibrate between the boundaries of the potential while gradually lowering their energy by heating the solvent. Note the bimodal character of the electron density between 0 and 40 ps. That stems from the fact that the vibrating molecules spend a large amount of time at the turning points of the Morse Potential. Note how the distribution sharpens up with time at the I_2 molecule reaches the ground state in about 100 ps. The ground state bond length of I_2 is 2.67 Å (Fig. 19.5).

19.3.4 Protein Dynamics in Crystals

Proteins are engaged in a multitude of tasks essential to life. To understand in detail how proteins function it is crucial to know the time ordering of events that give rise to their designed function. Time resolved Laue crystallography can, for certain radiation hard small proteins, be used to film small and large scale changes in structure in three dimensions, with a tremendous gain in information over other techniques. The proteins have so far been studied successfully, myoglobin (Mb),

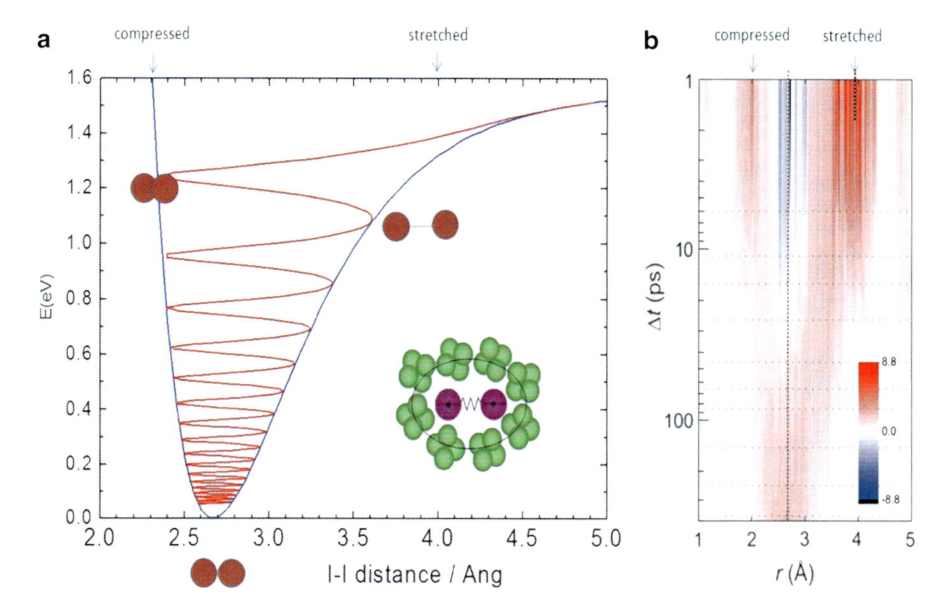

Fig. 19.5 (**a**) The I_2 (X) ground state potential constrains the vibrations of the newly formed I_2 molecule inside the liquid cage of CCl_4. A simplify trajectory to the ground state is shown in *red*. (**b**) The electron density distribution for the contracting I_2 molecule as it moves back to the ground state

[33, 34], hemoglobin (Hb), [35] the photo active yellow protein PYP [36–38] and the photosynthetic reaction centre RC [39].

For historical reasons we will only discuss some of the first results from myoglobin that examined the motion of photo dissociated CO inside the protein after the breaking of the Fe–CO bond by the laser. In fact MbCO was the first protein to be studied by time resolved Laue diffraction in 1994 at the ESRF. Note that in Laue diffraction the crystal is not rotated during the X-ray exposure and the polychromatic undulator spectrum ensures that the diffracted intensity is fully recorded. Myoglobin is a ligand-binding heme protein and it has long served as a model system for investigating ligand transport and binding in proteins. Using femtosecond time-resolved polarized IR spectroscopy, the dynamics of CO motion after photo detachment from MbCO were measured by Philip Anfinrud and co workers [40, 41]. Those studies revealed the time-dependent orientation of CO, the presence of a docking site that mediates the transport of ligands to and from the active binding site, as well as the dynamics of ligand binding and escape. In addition, site-specific mutations among the highly conserved residues circumscribing the ligand docking site were found to have a marked influence on the dynamics of ligand binding and escape. While much has been learned about ligand dynamics in Mb, much less is known about the structural evolution that accompanies ligand translocation.

Time-resolved Laue diffraction images were acquired using pump-probe: 0.5 ps laser pulse triggers ligand dissociation in a ~250 micron P6 MbCO crystal and a

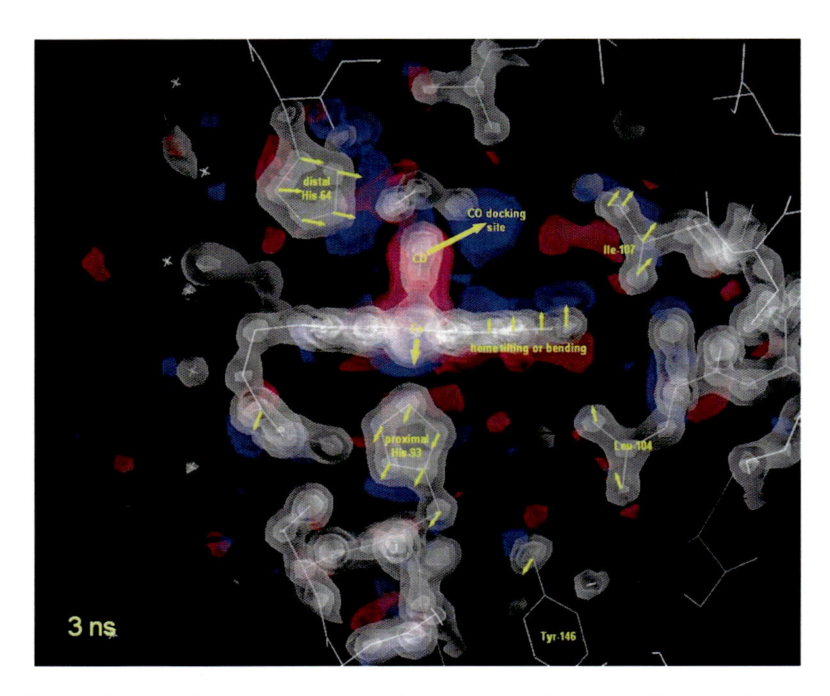

Fig. 19.6 Difference electron density map of MbCO determined 3 ns after photolysis of a P_6 MbCO crystal at 10 °C. The measured difference is imposed on the initial state with loss in density marked in *red*, gain in *blue*. The *arrows* indicate the direction of motion of the CO and the surrounding side chains. Note the heme dooming, i.e. the displacement of Fe and the tilt of the heme plane. The CO docking site prevents in effect CO recombination with Fe. This effect is in stark contrast to the (simple) recombination of I_2 in CCl_4 where 90 % of the dissociated I atoms recombine with their initial partner

delayed X-ray pulse probes the structure at that delay. The diffracted X-ray photons were detected on the Ø133 mm MAR CCD. Reconstruction of the protein structure with atomic resolution requires a series of diffraction images with crystal orientations spanning 60 degrees owing to the P_6 symmetry of the crystal. With the pink radiation from the U17, the 3.5 % bandwidth was sufficient to obtain redundant data with images collected every 2 degrees. To obtain high dynamic range diffraction images with the available x-ray flux, approximately 64 X-ray shots were integrated on the MAR CCD between readouts. Because the protein crystal requires sufficient time to recover between photolysis pulses, which are intense enough to excite a significant fraction of the protein molecules, the maximum repetition rate used was 3 Hz. Diffraction images were accumulated with and without photolysis to generate accurate difference-diffraction data. The synchrotron was operated in single bunch mode for the first experiments between 1994 and 2000, with a 15 mA single bunch of 150 ps (FWHM).

A high-resolution electron difference density map of MbCO is shown in Fig. 19.6. This image is the first of a time series that reveal, with atomic resolution, the order of events that accompany ligand translocation, from the Fe binding site to the nearby "docking site". Numerous features are observed including the displacement of the heme iron toward the proximal histidine, tilting of the heme, the docking of CO in a site near the heme iron, and the correlated motion of protein side chains in the vicinity of the active binding site. For example, the distal histidine (residue 64) shifts toward the site once occupied by CO, possibly raising the barrier to geminate recombination. The photo detached CO, found in a heme pocket docking site only about 2 Å away from the binding site, causes a displacement of the nearby isoleucine (residue 107). The CO translocation and heme displacement are also evident in a time-resolved structure determined at 150 ps, suggesting that the protein response to the ligand translocation event is ultrafast. On the time scale of a few hundred ns (not shown), the "docked" CO slips around to the proximal side of the heme and is found in the so-called Xe1 docking site, as observed in low temperature trapping studies [42, 43] as well as in earlier time-resolved studies [44]. Within a few microseconds, the CO escapes from this site into the surrounding solvent. The structural changes that accompany ligand translocation help explain how the protein is able to excrete toxic CO with high efficiency, even though the CO is temporarily located so close to the active binding site.

19.3.5 Protein Dynamics in Solution

One can examine many more proteins in solution than in crystals by time resolved wide-angle scattering (TRWAXS). The low protein concentration (few mM or less) is a challenge and their large size (more than a thousand times larger than small molecules) complicates the analysis. However recent TRWAXS data from model proteins have demonstrated that medium to large-scale dynamics of proteins is rich in information from ns to seconds [17]. Due to the inverse relationship between the interatomic distance and the scattering angle, the scattering from macromolecules appears at smaller scattering in the SAXS/WAXS range and we will term such time resolved measurements TR-SAXS/WAXS. The TR-SAXS/WAXS method has been applied to human haemoglobin (Hb), a tetrameric protein made of two identical αβ dimers that is known to have at least two different quaternary structures (a ligated stable "relaxed" (R) state and an unligated stable "tense" (T) structures) in solution. The tertiary and quaternary conformational changes of human haemoglobin triggered by laser induced ligand dissociation have been identified using this method. A preliminary analysis by the allosteric kinetic model gives a time scale for the R-T transition of ~1–3 μs, which is shorter than the time scale derived with time-resolved optical spectroscopy. In Fig. 19.7 the gas-phase scattering from the crystal structures of HbCO and Hb (deoxyHb) are shown together with myoglobin and a water molecule. In Fig. 19.7b the relative change from the transition HbCO → Hb is simulated for a 1 mM concentration.

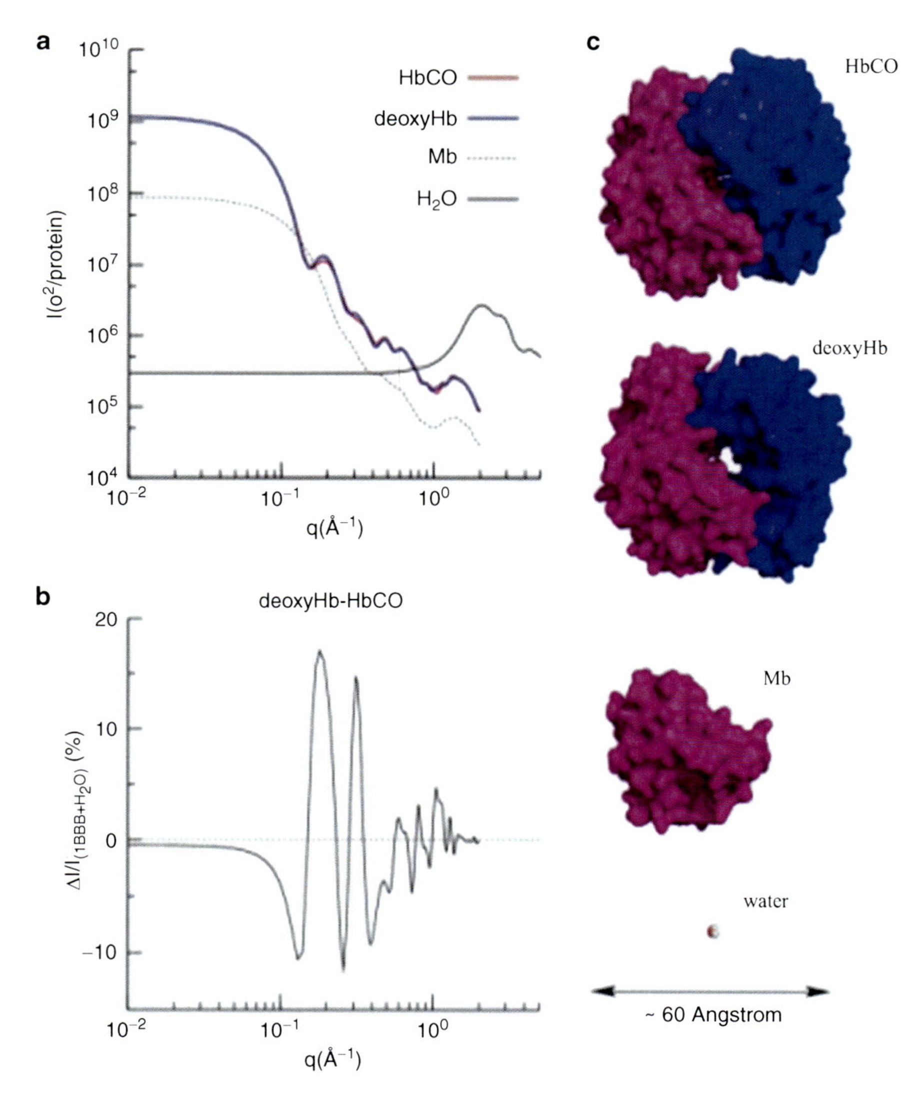

Fig. 19.7 (**a**) Calculated Debye scattering for haemoglobin (HbCO and Hb), myoglobin (Mb) and water. The calculations were performed with CRYOSOL using the relevant crystal structures adapted to the solution phase. (**b**) Relative change of the protein signal to the water background for the R-to-T transition (HbCO- > Hb) for an excited-state concentration of 1 mm. (**c**) Snapshots of the molecular structures used in the simulations

Note the good signal to background ratio between 0.1 and 1 $Å^{-1}$ due to the weak water background at low q (water plateau). Finally the structure of the proteins and water are shown in Fig. 19.7c. The optically induced tertiary relaxation of myoglobin and refolding of cytochrome c have also been studied with TR-SAXS/WAXS [17]. The advantage of TR-SAXS/WAXS over time-resolved X-ray protein crystallography is that it can probe irreversible reactions as illustrated with the folding of cytochrome c as well as reversible reactions such as ligand reactions in heme proteins. Although the diffraction patterns from proteins in solution contain structural information, the information content is probably insufficient to reconstruct protein structures in atomic scales. In this respect the use of structures from X-ray crystallography and NMR spectroscopy as a starting point is promising and the development of accurate theoretical analysis is in progress.

19.4 Comparison with Time-Resolved Optical Spectroscopy

To track time-dependent processes, various time-resolved spectroscopic tools have been developed and reaction dynamics with femtosecond time resolution can now be routinely investigated by such methods. Time-resolved optical spectroscopy has been applied to study ultrafast events in various areas of chemistry, physics and biology. A laser pulse triggers a photochemical reaction or a perturbation in the system and another laser pulse (with wavelength ranging from the ultraviolet to infrared (IR) and far-infrared) probes the optical responses of the system as a function of the delay between the pump and probe pulses as well as the wavelength of the probe pulse. This method permits the detection of specific short-lived species or energy states with high sensitivity. Unfortunately, the optical probe is not able to interfere with all atoms and thus experimental observables in optical spectroscopy such as transition energy and transition intensity cannot provide direct structural information such as atomic coordinates or at least bond lengths and bond angles. In contrast, since the X-ray scattering probes all atom-atom pairs in the molecular system, TRWAXS gives real-space unique access to the structural dynamics in the sample. In spite of the diffraction signal in TRWAXS being less sensitive to a specific transient than optical spectroscopy, its signal allows for the direct observation of major transient species and provides a global picture of the reactions with accurate branching ratios between multiple reaction pathways as shown in the examples in the previous sections. Since X-rays scatter from all atoms in the solution, solutes as well as solvent molecules, the analysis of TRWAXS data can provide the temporal behavior of the solvent as well as the structural progression of the solute molecules in *all* their reaction pathways, thus providing information of the rearrangement of the solvent around the transient solutes.

Vibrational transitions in a molecule can often be correlated to specific sites within molecular systems and for this reason time-resolved vibrational (IR and Raman) spectroscopies [45, 46] have been used for probing transient structures in chemical reactions. However, the vibrational signals bias species with high absorption cross-section and these techniques may fail to detect optically silent species. In addition, it is often difficult to obtain insight into global structure from vibrational spectroscopy and new approaches such as multidimensional IR and Raman spectroscopies have emerged which are very promising in this respect [47, 48].

Two-dimensional spectroscopy measures the vibration coupling between specific vibrational modes which are correlated with the relative orientation of the molecule. The vibration-vibration coupling is very sensitive to molecular structure by analogy to the spin-spin coupling in two-dimensional NMR spectroscopy. Great progress has been made recently in resolving molecular dynamic in biological systems by multi-dimensional spectroscopy.

As shown in the examples in Sect. 19.3, time-resolved optical spectroscopy and TRWAXS complement each other well. The former has a higher sensitivity and thus is useful for extremely dilute concentrations whereas TRWAXS provides direct structural information and accurate branching ratios.

19.5 Comparison with Time-Resolved X-ray Absorption Spectroscopy

X-ray absorption spectroscopy such as X-ray absorption near-edge structure (XANES) and extended X-ray absorption fine structure (EXAFS). EXAFS and XANES provide local structural information and therefore the local response to laser excitation can be obtained. The EXAFS region delivers quantitative structural information with sub Å accuracy via the so-called EXAFS equation [49], whereas XANES provides a quantitative fingerprint of the local chemical bonding geometry such as orbital hybridization in the vicinity of the atom(s) of interest. This can be useful for understanding molecular dynamics in solution where a lot of important chemistry occurs and where the environment substantially influences the reaction dynamics. Like TRWAXS, X-ray absorption spectroscopy also has been used in time-resolved studies with the laser-pump and X-ray probe scheme and has been applied to study charge transfer and spin-crossover processes in coordination chemistry compounds, solvation dynamics of atomic radicals and structural dynamics of photo dissociated intermediates in solution phase [50–54]. The structural information obtained by X-ray absorption is limited to the local environment of a particular atom. This feature is in contrast with the diffraction method which provides structural information for all species. The locality of the X-ray absorption signal has the advantage of a much higher sensitivity. In summary TRXL and time-resolved X-ray absorption spectroscopy are highly complementary and thus in the future, a combination of the two techniques will prove useful in providing more accurate results and such efforts are in progress within the upgrade program for ID09 at the ESRF for 2015.

19.6 Summary and Outlook

In this review we have outlined the experimental and theoretical background for TRWAXS and given examples from photochemical research in the liquid phase. Although TRWAXS has limitations, it provides direct structural and kinetic information about reaction dynamics of chemical reactions in liquids. TRWAXS addresses detailed information about structural changes and spatiotemporal kinetics of reaction intermediates in *all* reaction pathways, temporal rearrangements of solvents around solutes, and the related solvent hydrodynamics. For example, TRWAXS can aid in the understanding of major reaction pathways where TRWAXS results are different from time-resolved optical spectroscopy. TRXL also can identify new intermediates as shown for the photolysis of $Ru_3(CO)_{12}$ [15] where TRXL complements the results from time-resolved spectroscopy. Moreover TRXL is capable of providing direct structural information which is difficult to obtain from time-resolved optical spectroscopy, such as the molecular structure of intermediates involved in iodine elimination reactions from $C_2H_4I_2$ and $C_2F_4I_2$. In addition, TRXL has been applied to investigate structural dynamics of more complicated systems such as nanoparticles and proteins in solution.

Increasing q_{max} from the current 8 to 12 $Å^{-1}$ by using harder X-rays will provide more data and in particular improve the scaling of RDF curves by giving access to the *atomic limit* where molecules scatter as the sum of independent atoms without structure. Multilayer monochromators are now available that can extract high-intensity beam from higher undulator harmonics down to 0.50 Å. Application of the TRWAXS method to molecules without heavy-atoms is a challenge. Such systems require an even higher signal to noise ratio which means having a very liquid jet and sample. For example to extract a 10 times weaker signal (signal is proportional to Z^2), the acquisition time needs to be 100 times longer. This is unrealistic with the current setup, but it might become feasible with the future X-ray sources with a much higher flux. Chemical substitution with heavy atoms may also provide a solution.

The time-resolution of TRWAXS performed on synchrotron sources is currently limited to 100 ps. This time-resolution has recently been improved to 0.1–1 ps at the LCLS at Stanford with XFEL radiation. It is now possible to probe molecules on the time scale of vibrations and image atomic motions along the potential energy surface, monitor bond breaking/formation and isomerization processes. True quantum effects should be within reach with XFEL sources and that will no doubt have great impacts on the study of reaction dynamics.

References

1. Schotte F, Techert S, Anfinrud PA, Srajer V, Moffat K, Wulff M (2002) Picosecond structural studies using pulsed synchrotron radiation. In: Dennis M (ed) Third-generation hard X-ray synchrotron radiation sources. Wiley, New York, pp 345–401. ISBN 0-471-31433-1
2. Wulff M, Plech A, Eybert L, Randler R, Schotte F, Anfinrud P (2002) Realisation of sub-nanosecond pump and probe experiments at the ESRF. Faraday Discuss 122:13–26
3. Guerin L, Kong Q, Khakhulin D, Cammarata M, Ihee H, Wulff M (2012) Tracking atomic positions in molecular reactions by picosecond X-ray scattering at the ESRF. Synchrotron Radiat News 25(2):25–31

4. Graber T, Anderson S, Brewer H, Chen Y-S, Cho HS, Dashdorj N, Henning RW, Kosheleva I, Macha G, Meron M, Pahl R, Ren Z, Ruan S, Schotte F, Srajer V, Viccaro PJ, Westferro F, Anfinrud P, Moffat K (2011) BioCARS: a synchrotron resource for time-resolved X-ray science. J Synchrotron Rad 18:658–670
5. Nozawa S, Adachi S-i, Takahashi J-i, Ryoko Tazaki L, Guerin MD, Tomita A, Sato T, Chollet M, Collet E, Cailleau H, Yamamoto S, Tsuchiya K, Shioya T, Sasaki H, Mori T, Ichiyanagi K, Sawa H, Kawata H, Koshihara S-y (2007) Developing 100 ps-resolved X-ray structural analysis capabilities on beamline NW14A at the photon factory advanced ring. J Synchrotron Rad 14:313–319
6. Wrobel R, Brullot B, Dainciart F, Doublier J, Eloy J-F, Marmoret R, Vilette B, Mathon O, Tucoulou R, Freund A (1998) Characterisation of ultrafast X-ray detectors at the European Synchrotron Radiation Facility. SPIE 3451:156–161
7. Ihee H, Lorenc M, Kim TK, Kong QY, Cammarata M, Lee JH, Bratos S, Wulff M (2005) Ultrafast X-ray diffraction of transient molecular structures in solution. Science 309:1223–1227
8. Kim TK, Lorenc M, Lee JH, Russo M, Kim J, Cammarata M, Kong QY, Noel S, Plech A, Wulff M, Ihee H (2006) Spatiotemporal reaction kinetics probed by picosecond X-ray diffraction. Proc Natl Acad Sci U S A 103:9410–9415
9. Kong Q, Wulff M, Lee JH, Bratos S, Ihee H (2007) Photochemical reaction pathways of carbon tetrabromide in solution probed by picosecond X-ray diffraction. J Am Chem Soc 129:13584–13591
10. Lee JH, Kim J, Cammarata M, Kong Q, Kim KH, Choi J, Kim TK, Wulff M, Ihee H (2008) Transient X-ray diffraction reveals global and major reaction pathways for the photolysis of iodoform in solution. Angew Chem Int Ed 47:1047–1050
11. Lee JH, Kim TK, Kim J, Kong Q, Cammarata M, Lorenc M, Wulff M, Ihee H (2008) Capturing transient structures in the elimination reaction of haloalkane in solution by transient X-ray diffraction. J Am Chem Soc 130:5834–5835
12. Plech A, Wulff M, Bratos S, Mirloup F, Vuilleumier R, Schotte F, Anfinrud PA (2004) Visualizing chemical reactions in solution by picosecond x-ray diffraction. Phys Rev Lett 92:125505
13. Wulff M, Bratos S, Plech A, Vuilleumier R, Mirloup F, Lorenc M, Kong Q, Ihee H (2006) Recombination of photodissociated iodine: a time-resolved x-ray diffraction study. J Chem Phys 124:034501
14. Cammarata M, Lorenc M, Kim TK, Lee JH, Kong QY, Pontecorvo E, Lo Russo M, Schiro G, Cupane A, Wulff M, Ihee H (2006) Impulsive solvent heating probed by picosecond X-ray diffraction. J Chem Phys 124:124504
15. Kong QY, Lee JH, Plech A, Wulff M, Ihee H, Koch MHJ (2008) Ultrafast X-ray solution scattering reveals an unknown reaction intermediate in the photolysis of Ru3(CO)12. Angew Chem Int Ed 47:5550–5553
16. Plech A, Kotaidis V, Lorenc M, Wulff M (2005) Thermal dynamics in laser excited metal nanoparticles. Chem Phys Lett 401:565–569
17. Cammarata M, Levantino M, Schotte F, Anfinrud PA, Bowman RM, Gruebele M, Zewail AH, Ewald F, Choi J, Cupane A, Wulff M, Ihee H (2008) Tracking the structural dynamics of proteins in solution using time-resolved wide-angle scattering. Nat Methods 5:881–887
18. Cho HS, Dashdorj N, Schotte F, Graber T, Henning R, Anfinrud P (2010) Protein structural dynamics in solution unveiled via 100-ps time-resolved x-ray scattering. Proc Natl Acad Sci U S A 107(16):7281–7286
19. Baumert T, Pederson S, Zewail AH (1993) Femtosecond real-time probing of reactions. Vectorial dynamics of transition states. J Phys Chem 97:12447–12459
20. Dantus M, Bowman RM, Gruebele M, Zewail AH (1989) Femtosecond real-time probing of reactions: the reaction of IHgI. J Chem Phys 91:7437–7450
21. Zhong D, Zewail AH (1998) Femtosecond real-time probing of dynamics and structure in charge-transfer reactions. J Phys Chem A 102:4031–4058

22. Pugliano N, Szarka AZ, Gnanakaran S, Triechel M, Hochstrasser RM (1995) Vibrational population-dynamics of the HgI photofragment in ethanol solution. J Chem Phys 103:6498–6511
23. Pugliano N, Szarka AZ, Hochstrasser RM (1996) Relaxation of the product state coherence generated through the photolysis of HgI2 in solution. J Chem Phys 104:5062–5079
24. Volk M, Gnanakaran S, Gooding E, Kholodenko Y, Pugliano N, Hochstrasser RM (1997) Anisotropy measurements of solvated HgI2 dissociation: transition state and fragment rotational dynamics. J Phys Chem A 101:638–643
25. The values for rate coefficients and time constants are corrected by a factor of 2 with respect to the published values due to a simple error of doubling the rate constant in the global-fitting program for TRXL data
26. Fossey J, Lefort D, Sorba J (1995) Free radicals in organic chemistry. Wiley, New York
27. Ihee H, Kua J, Goddard WA, Zewail AH (2001) CF2XCF2X and CF2XCF2 center dot radicals (X = Cl, Br, I): Ab initio and DFT studies and comparison with experiments. J Phys Chem A 105:3623–3632
28. Ihee H, Zewail AH, Goddard WA (1999) Conformations and barriers of haloethyl radicals (CH2XCH2, X = F, Cl, Br, I): Ab initio studies. J Phys Chem A 103:6638–6649
29. Skell PS, Tuleen DL, Readio PD (1963) Stereochemical evidence of bridged radicals. J Am Chem Soc 85:2849–2850
30. Ihee H, Lobastov VA, Gomez UM, Goodson BM, Srinivasan R, Ruan CY, Zewail AH (2001) Direct imaging of transient molecular structures with ultrafast diffraction. Science 291:458–462
31. Ihee H, Goodson BM, Srinivasan R, Lobastov VA, Zewail AH (2002) Ultrafast electron diffraction and structural dynamics: transient intermediates in the elimination reaction of C2F4I2. J Phys Chem A 106:4087–4103
32. Lee JH, Wulff M, Bratos S, Petersen J, Guerin L, Leicknam J-C, Cammarata M, Kong Kim Q, Møller KB, Ihee H (2013) Filming the birth of molecules and accompanying solvent rearrangement. J Am Chem Soc 135:3255–3261
33. Srajer V, Teng T, Ursby T, Pradervand C, Ren Z, Adachi S, Schildkamp W, Bourgeois D, Wulff M, Moffat K (1996) Photolysis of the carbon monoxide complex of myoglobin: nanosecond time-resolved crystallography. Science 274:1726–1729
34. Schotte F, Lim M, Jackson TA, Smirnov A, Soman J, Olson JS, Phillips GN, Wulff M, Anfinrud PA (2003) Watching a protein as it functions with 150-ps time-resolved X-ray crystallography. Science 300:1944–1947
35. Schotte F, Cho HS, Soman J, Wulff M, Olson JS, Anfinrud P (2013) Real-time tracking of CO migration and binding in the alpha and beta subunits of human hemoglobin via 150-ps time-resolved Laue crystallography. Chem Phys D-12:00661R1
36. Perman B, Srajer V, Ren Z, Teng T-Y, Pradervand C, Ursby T, Schotte F, Wulff M, Kort R, Hellingwerf K, Moffat K (1998) Energy transduction on the nanosecond time scale: early structural events in a xanthopsin photocycle. Science 279:1946–1950
37. Schotte F, Cho HS, Kail Ville RI, Kamikubo H, Dashdorja N, Henry ER, Graber TJ, Henning R, Wulff M, Hummer G, Kataoka M, Anfinrud PA (2012) Watching a signaling protein function in real time via 100-ps time-resolved Laue crystallography. Proc Natl Acad Sci U S A 109(47):19256–19261
38. Jung YO, Lee JH, Kim J, Schidt M, Moffat K, Srajer V, Ihee H (2013) Volume-conserving trans-cis isomerization pathways in photoactive yellow protein visualized by picosecond X-ray crystallography. Nat Chem 5(3):212–220
39. Wöhri AB, Katona G, Johansson LC, Fritz E, Malmerberg E, Andersson M, Vincent J, Eklund M, Cammarata M, Wulff M, Davidsson J, Groenhof G, Neutze R (2010) Light-induced structural changes in a photosynthetic reaction center caught by Laue diffraction. Science 328(5978):630–633
40. Lim M, Jackson TA, Anfinrud PA (1995) Binding of CO to myoglobin from a hemepocket docking site to form nearly linear Fe-C-O. Science 269:962

41. Lim M, Jackson TA, Anfinrud PA (1997) Ultrafast rotation and trapping of carbon monoxide dissociated from myoglobin. Nat Struct Biol 4:209
42. Chu K, Vojtchovsky J, McMahon BH, Sweet RM, Berendzen J, Schlichting I (2000) Structure of a ligand-binding intermediate in wild-type carbonmonooxy myoglobin. Nature 403:921
43. Ostermann A, Waschipky R, Parak FG, Nienhaus GU (2000) Ligand binding and conformational motions in myoglobin. Nature 404(6774):205–208
44. Srajer V, Ren Z, Teng TY, Schmidt M, Ursby T, Bourgeois D, Pradervand C, Schildkamp W, Wulff M, Moffat K (2001) Protein conformational relaxation and ligand migration in myoglobin: a nanosecond to millisecond molecular movie from time-resolved Laue X-ray diffraction. Biochemistry 40:13802
45. Nibbering ETJ, Fidder H, Pines E (2005) Ultrafast chemistry: using time-resolved vibrational spectroscopy for interrogation of structural dynamics. Annu Rev Phys Chem 56:337–367
46. Kukura P, McCamant DW, Mathies RA (2007) Femtosecond stimulated Raman spectroscopy. Annu Rev Phys Chem 58:461–488
47. Hamm P, Lim MH, Hochstrasser RM (1998) Structure of the amide I band of peptides measured by femtosecond nonlinear-infrared spectroscopy. J Phys Chem B 102:6123–6138
48. Zheng J, Kwak K, Fayer MD (2007) Ultrafast 2D IR vibrational echo spectroscopy. Acc Chem Res 40:75–83
49. Bressler C, Chergui M (2004) Ultrafast X-ray absorption spectroscopy. Chem Rev 104:1781–1812
50. Khalil M, Marcus MA, Smeigh AL, McCusker JK, Chong HHW, Schoenlein RW (2006) Picosecond x-ray absorption spectroscopy of a photoinduced iron(II) spin crossover reaction in solution. J Phys Chem A 110:38–44
51. Gawelda W, Johnson M, de Groot FMF, Abela R, Bressler C, Chergui M (2006) Electronic and molecular structure of photoexcited [Ru-II(bpy)(3)](2+) probed by picosecond X-ray absorption spectroscopy. J Am Chem Soc 128:5001–5009
52. Pham VT, Gawelda W, Zaushitsyn Y, Kaiser M, Grolimund D, Johnson SL, Abela R, Bressler C, Chergui M (2007) Observation of the solvent shell reorganization around photoexcited atomic solutes by picosecond X-ray absorption spectroscopy. J Am Chem Soc 129:1530–1531
53. Gawelda W, Pham VT, Benfatto M, Zaushitsyn Y, Kaiser M, Grolimund D, Johnson SL, Abela R, Hauser A, Bressler C, Chergui M (2007) Structural determination of a short-lived excited iron(II) complex by picosecond x-ray absorption spectroscopy. Phys Rev Lett 98:057401
54. Chen LX, Jager WJH, Jennings G, Gosztola DJ, Munkholm A, Hessler JP (2001) Capturing a photoexcited molecular structure through time-domain X-ray absorption fine structure. Science 292:262–264

Printed by Publishers' Graphics LLC
DBT140226.15.19.9